GOING GREEN AND GETTING REGULATION RIGHT: A PRIMER FOR ENERGY EFFICIENCY

GOING GREEN AND GETTING REGULATION RIGHT: A PRIMER FOR ENERGY EFFICIENCY

Charles J. Cicchetti, Ph.D.

2009
Public Utilities Reports, Inc.
Vienna, Virginia

© Public Utilities Reports, Inc. 2009

All rights reserved. No part of this publication may be reproduced, stored in a retrieval system, or transmitted in any form or by any means, electronic, mechanical, photocopying, recording, or otherwise, without the prior written permission of the publisher.

This publication is designed to provide accurate and authoritative information in regard to the subject matter covered. It is sold with the understanding that the publisher is not engaged in rendering legal, accounting or other professional service. If legal advice or other expert assistance is required, the services of a competent professional person should be sought. (From a *Declaration of Principles* jointly adopted by a *Committee of the American Bar Association and a Committee of Publishers*.)

First Printing, March 2009

Library of Congress Cataloging-in-Publication Data
Cicchetti, Charles J.
  Going green and getting regulation right: a primer for energy efficiency/Charles J. Cicchetti
  p. cm
Includes bibliographical references and index.
ISBN 978-0-910325-21-9
  1. Energy policy – United States. 2. Energy conservation – United States. I. Title.

HD9502.U52C55 2009

333.790973 – dc22

2009005114

Printed in the United States of America

Table of Contents

Acknowledgements

Section I: What and Why 1
    Chapter 1: Introduction: Why This Primer 3
    Chapter 2: Utility-Sponsored Conservation and Other Social Mandates in the United States 7
    Chapter 3: The Role of Utilities in Conservation: Regulatory Details That Matter 19
    Chapter 4: Regulating the Path to Achieve Successful Utility-Sponsored Energy Efficiency 27

Section II: What Has Been Happening Around the Nation 37
    Chapter 5: A Review of Utility-Sponsored Energy Efficiency and Load Management Activities 39
    Chapter 6: Relative State-by-State Comparisons Over Time 67
    Chapter 7: Benchmarking Effort and Performance 87

Section III: Regulating Energy Efficiency 109
    Chapter 8: An Overview of Energy Efficiency Regulations 111
    Chapter 9: The Economics of Utility-Sponsored Energy Efficiency When There are Externalities 121
    Chapter 10: External Benefits and Energy Efficiency 131
    Chapter 11: Some Quantitative Evidence to Analyze External Benefits 135
    Chapter 12: Combining Data and Theory 145

Section IV: The Cost of Energy Efficiency 151
    Chapter 13: The Cost of Energy Efficiency and Load Management 153
    Chapter 14: The Per-Unit Costs of Energy Efficiency and Load Management 167
    Chapter 15: The Per-Unit Cost of Load Management 181
    Chapter 16: How Regulation Affects the Costs That Customers Pay for Demand-Side Programs 187
    Chapter 17: Cost-of-Service Regulation and Demand-Side Management 191
    Chapter 18: Why Cost-of-Service Regulation May be Given Too Much Current Weight 195

Section V: Testing the Theory That Incentives Matter 207
    Chapter 19: California Shows Mandates and Rate Riders Work:
    A Brief History of California's Energy Efficiency Efforts 209
    Chapter 20: Mixing Mandates and Incentives 217
    Chapter 21: How States Are Decoupling 225
    Chapter 22: Direct Financial Incentives for Demand Response Programs 235
    Chapter 23: Testing the Statistical Significance of Regulatory Incentives for Energy Efficiency 249
    Chapter 24: The Statistical Significance of Incentives for Regulatory Policy Targets 271

Section VI: The Path Ahead 275
    Chapter 25: Save-a-Watt: A New Paradigm 277
    Chapter 26: Renewables and Energy Efficiency 283
    Chapter 27: Conclusions 287

Bibliography 297
Index 305

ACKNOWLEDGEMENTS

I owe a great deal of thanks to many people who have increased my knowledge and provided me with much insight into the economics and regulation of energy efficiency. This began during my post-doctoral work at Resources For the Future (RFF) under the stimulation from and guidance of John Krutilla, Allen Kneese, and Joseph Fisher. I also benefited from my collaborations with Cliff Russell, Bob Haveman, Rick Freeman, Kerry Smith, and Tony Fisher. The lessons learned made me understand how pricing, capping, trading, and quantifying externalities could change public policy and alter behavior.

My true insight into regulated energy utilities came from my five-year association as the first economist at the Environmental Defense Fund (EDF). During those years, I worked with Rod Cameron, Ed Berlin, William Gillen, Hasty Habricht, and Phil Mause. We believe that this collaboration at EDF brought marginal cost and time-of-use pricing to the forefront of electricity regulation and ushered in the first wave of utility-sponsored conservation.

During the same period, David Freeman was directing the Ford Foundation's Energy Policy Project, which provided financial and intellectual guidance to my colleagues and I. This enabled us to collaborate with Ralph Turvey, William Vickrey, James Bonbright, and Amory Lovins. These four giants of regulation have left their mark on many topics. This work owes them much.

My colleagues at the University of Wisconsin also encouraged both my research and direct involvement in public policy debates and regulatory proceedings during the 1970s. I cannot imagine any other major university that could have been more supportive of our efforts to reduce energy demand through utility tariff reform, mandating conservation, and internalizing environmental externalities. During this period, Wesley Foell introduced me to the experience of other nations. This added cogeneration and central heating to the utility reform agenda that we promoted.

Most important, Governor Patrick J. Lucey appointed me to head his Energy Office and to Chair the Public Service Commission of Wisconsin. This brought me into direct contact with Congress, National Association of Regulatory Utility Commissioners (NARUC), and the Executive Branch during the second world oil crisis. These lessons learned helped to achieve the passage of the Public Utility Regulatory Policies Act (PURPA). This has led to restructuring, utility tariff reforms, as well as more energy efficiency and renewables.

The next thanks go to Alfred Kahn, Irwin Stelzer, Paul Joskow, Larry Ruff, Joe Pace, Howard Pifer, Bill Hogan, and William Dickenson. These individuals were far more than consulting colleagues in the 1980s. We argued and I learned much. This also helped me to discover how to explain regulatory policy so as to change both regulation and consumer behavior.

During this current energy crisis, I have benefited from the work and insight of my colleagues Mark Lowry and Larry Kaufmann. They helped me to understand the details of decoupling. Very special thanks go to James Lin and Colin Long. Much of this work reflects their hard work as well as the assistance of Danielle Neveu.

My motivation to write this book is James Rogers, CEO of Duke Energy. We have been friends for a long time. He has a boatload of responsibility running Duke Energy. Nevertheless, he added a great deal more work to these responsibilities when he proposed to dramatically alter the culture, regulation, and business model for regulated entities, in order to help the nation become more energy efficient. This example has caused me to alter my work and responsibilities in order to add my forty years of regulatory experience to help expand energy efficiency and renewable energy.

My work is independent. I take full responsibility for any flaws or errors. I owe much to many. The most important source of support and inspiration is my family, which is centered around my soul mate and lovely wife, Sally. I write this book in the hope of helping my children and their children benefit from what I believe is the third round of energy efficiency. I think that this time we will get it done. I am now convinced that energy efficiency and renewables will no longer need to be reintroduced. These programs will become part of the utility industry's regulatory and business fabric, and with sensible regulation, the current efforts will increase and, most importantly, they will be sustained in the future.

SECTION I: WHAT AND WHY

This section sets the table for a detailed discussion of energy efficiency. History is important. The nation has, over the past forty years, had at least two previous periods when conservation seemed necessary and needed support to become a sustained new policy. This mostly did not happen to the degree needed or predicted. The past tells us "what" has worked, "what hasn't" and "why." In all of this, regulated energy utilities emerge as both an opportunity and solution.

This discussion begins with a brief history of "what" regulators and utilities have been doing to increase energy efficiency. Some jurisdictions remain on track and have achieved a fairly consistent, sustained, and enviable record-making energy efficiency and load management part of their states' electricity industry. Other jurisdictions once had significant utility-sponsored energy efficiency programs that were allowed to disappear. Differences in regulatory incentives and public opinion emerge as important drivers to help explain these very different results across the nation.

CHAPTER 1
INTRODUCTION: WHY THIS PRIMER

Support for increased energy efficiency is growing exponentially. That is mostly a good thing considering the current, real threats related to environmental, economic, and national security. These various efforts in the United States are primarily concentrated in the electric utility sector. These businesses are either comprehensively or partially regulated. This makes it rather easy to mandate specific utility programs or to offer incentives for them to expand energy efficiency.

Spending more for energy efficiency is likely, regardless of the "stick" or "carrot" nature of the regulatory and political inducements applied. Suppose the reason for energy efficiency is simply to help customers and voters "feel" good. Utility-sponsored green investments would be viewed positively even though most consumers only spend a miniscule amount more in terms of the prices paid per unit of energy they consume.

If spending more on energy efficiency is the goal, utility regulation is very well suited to accomplish such a "feel good" goal. Utilities have significant cash flow and access to capital. Regulators can readily use "pipes and wires" to attach new spending requirements for energy efficiency. There is also much experience with designing cost recovery tariffs based upon widely known and much used regulatory cost-of-service principles. Such a purpose would not do as much as energy efficiency reforms that add incentives and encourage innovative packaging and marketing and add financial incentives that depend upon utility performance.

A primer is a fundamental text that answers basic questions related to a particular subject. The subject here is energy efficiency, which is at the nexus of energy, environment, and economics. These matters are individually complex. Their intersection brings both synergy and additional complexity. The subtext is going green with sensible regulatory support. This analysis uses the most complete up-to-date data to gauge the performance and experiences of the nation's electric utilities as well as the various states.

The first set of questions involves basic definitions. What is energy efficiency? How does it relate to demand-side or load management? Next, there are matters of performance. Does energy efficiency work? Is it reliable? These questions often lead to comparisons such as: where does it work better and why?

Performance questions point to policy matters. What is the optimal amount of energy efficiency? Why don't individuals purchase the optimal or cost-effective amount of energy efficiency? How can public policy help close the gap between what individuals do and the socially optimal amount of energy efficiency?

In the United States, there have been past periods when these basic questions were answered in two ways. First, new appliance standards and building codes were changed to increase energy efficiency and labeling was expanded to help consumers make wise new and replacement choices. These informational programs work particularly well when the consumer seeks solutions to energy and/or environmental problems. In fact, these external reasons often seem more important than information that simply helps consumers to save money.

Second, energy regulators have mandated active roles for investor-owned utilities (IOUs) to install or cause more energy efficiency in the retail electricity and natural gas markets. This primer focuses on this regulatory approach to determine how some jurisdictions and utilities have succeeded, while others have simply lost interest. Nevertheless, these matters are not independent. The economy and public opinion matter a great deal and likely cause regulators to seek approaches that work and last.

The additional regulatory questions begin with these basic matters. What can IOUs do to stimulate energy efficiency? How can they be compensated for doing so? Are financial incentives necessary? What works best?

There are also very detailed and specific policy matters. Are the energy efficiency costs significant? How can costs be recovered in tariffs? What role should nonparticipating customers play and how much should specific customers pay? How significant are external costs in answering such tariff questions and in determining the public policy targets for the amount of energy saved relative to the amount sold, or the amount the utility spends per unit of energy saved?

This primer addresses all these questions individually and attempts to fuse them together in a broader collective framework to establish specific performance targets for regulators and utilities. The second goal is to determine the relevance of sensible incentives that will achieve the public policy goals and sustain the various efforts to expand significantly the latest commitments to make energy efficiency a new paradigm for electricity and natural gas suppliers.[1]

---

[1] The theory and analyses of this book apply to both these regulated industries. However, the data collected focuses on electricity. In addition, greenhouse gas curtailment also focuses more attention on electricity. Therefore, this analysis reflects these facts.

This book is divided into six distinct sections that can be read alone, because these specific topics are mostly written in a self-contained manner. Section I addresses the meaning and rationale for energy efficiency. This section sets the stage for the discussion that follows.

Section II reviews what the states have been dong to achieve energy efficiency and demand-side objectives. This section introduces the use of incentives and the elimination of disincentives. A major theme throughout is how to sustain energy efficiency efforts when, as seems inevitable given the history of energy efficiency, public and political support wanes.

Section III reviews the past debates that the nation has had concerning the role utilities can and should play in providing energy efficiency. There are three core questions: (1) Should nonparticipating customers subsidize energy efficiency? (2) Do external benefits matter? and (3) What do regulatory tests of benefits and costs mean and show?

Section IV assumes that regulators and utilities determine to pursue energy efficiency. The purpose of this discussion is to quantify the costs and to review the regulatory approaches that states adopt to encourage energy efficiency. Additionally, this discussion addresses the manner in which some states blunt the inherent disincentives of cost-of-service regulation in order to stimulate energy efficiency.

Section V describes various statistical analyses of the factors, including positive incentives and decoupling to remove disincentives, which are believed to affect and sustain energy efficiency efforts. This analysis identifies how and why some states spend more and save more. This discussion brings a modicum of restraint to the discussion. The benefits of energy efficiency are easy to comprehend. However, energy efficiency has costs and virtually all the successful energy efficiency programs find a way to provide incentives. The statistical analyses in this section put such matters in perspective.

Section VI shows the path ahead, if the nation is to get it right this time. Throughout this analysis, there is a setting that includes other related policies, such as renewable energy. These solutions affect the case for energy efficiency and often share a common set of societal objectives. These concluding chapters discuss both the complementary and conflicting aspects of these various "green" utility efforts. This discussion points to the path ahead and the need to consider new business and regulatory paradigms to improve the energy and environmental policies in the nation.

CHAPTER 2
UTILITY-SPONSORED CONSERVATION AND OTHER SOCIAL MANDATES IN THE UNITED STATES

This chapter reviews the intellectual and political roots of the current heightened interest in utility-sponsored energy efficiency. There are two reasons for this discussion. First, there are lessons to learn and current regulators would benefit from knowing what their predecessors did and why. Second, some states have previously expanded energy efficiency, mostly adding some form of profits to this new utility service. Other states treated conservation as a political "soup of the day" that at some point became unappetizing. With a return of the need for this sustenance, they are trying to discover how to prepare this soup. The lessons learned begin with a brief history that begins about forty years ago.

In the early 1970s, environmental groups sometimes referred to the United States as the "cowboy" economy. They averred that the United States had not accepted the fact that resources were limited and the nation could no longer pretend that it could move on to the next valley, water source, or whatever. The United States was lectured to and made to accept blame, because with just five percent of the world's population, we consumed a third of its resources.

Rachel Carson[1] and Garret Hardin[2] explained that the world was resource and environmentally constrained. The innate resilience of the environment had limits and resources were finite. There were also the popularized mathematical forecasting models of the Club of Rome that articulated that too many people and not enough land, water, energy, and other resources would cause food shortages and future environmental catastrophes. The seventies were mostly a decade of blame and pessimism that reflected a world that needed to understand and accept the full consequences of these limits.[3]

There were many things wrong. One that emerged in the utility sector was volume discount and other promotional policies that were adopted to reduce the per unit prices to recover the

---

[1] Carson, Rachel. (1962). *The Silent Spring*. New York: Houghton Mifflin.
[2] Hardin, Garret. (1968). The Tragedy of the Commons. *Science*, Vol. 162.
[3] Meadows, Donella H., Dennis L. Meadows, Jorgen Randers, and William H. Behrens III. (1972). *The Limits to Growth*. New York: Universe Books.

fixed costs of a regulated energy utility's services. Some early critics saw two-part and declining block utility tariffs as striking examples of a nation that had not come to grips with any semblance of limits to growth. Reducing average prices when use expanded was simply condemned as being out of touch with the world's limited natural resources and mounting social cost of pollution.

About the same time these growth-related environmental issues were being raised, environmentalists began to rely increasingly on courts to provide a forum for their pro-environmental and anti-growth agenda. Coincidentally, public interest legal entities, such as the Sierra Club, Environmental Defense Fund (EDF), Natural Resources Defense Council (NRDC), Center for Law and Social Policy, and others, began to challenge federal government actions, particularly the Army Corps of Engineers and Bureau of Reclamation's water projects as causing extraordinary environmental damages. One of the primary environmental claims was that federal water resource projects were legally required to compare benefits to costs.

Environmentalists and their attorneys asserted that the federal agencies that built such projects were innately biased in favor of inflated benefits and understated costs, which were combined with very low discount rates to make it seem as if every development project passed the necessary requirement that benefits exceed costs. The failure to access or collect reasonable user fees removed some of the potential restraints towards development. Regardless, the failure to collect reasonable user fees added another anti-big government and taxpayer dimension to these various battles.

Even when environmentalists were learning to make basic economic arguments, they did not forget their core values and argued strenuously that damages to the natural environment, what economists call "negative externalities," should be included in a "just and reasonable" benefit-to-cost comparison. The Sierra Club made this particular argument before the Federal Power Commission (FPC) (now the Federal Energy Regulatory Commission [FERC]) in a proceeding that decided whether a hydroelectric dam should be licensed on the Snake River at Hell's Canyon.

The FPC denied the Sierra Club's arguments to assign economic value to preservation. This was, in part, because there had been no quantification of damages or external costs. Subsequently, the U.S. Supreme Court reversed the FPC's decision and directed the FPC to commence a new licensing proceeding to compare the alleged preservation values to net hydroelectric benefits.

After turning down requests from both regional electricity interests and the Sierra Club to serve as an expert, Dr. John V. Krutilla, his post-doctoral associate, Dr. Charles J. Cicchetti, and another researcher, Dr. Jack Knetsch, who were all working for Resources for the Future

(RFF), agreed to provide testimony on the relative values of environmental preservation and development on behalf of the staff of the FPC for fees equal to $1.00. Their quantitative analysis[4] resulted in the FPC denying the very license for hydro development that the agency had previously approved, because the present value of the benefits of preservation on this wild river exceeded the net benefits of hydroelectricity at this location.

At about the same time as the FPC was hearing the Hell's Canyon case to determine the benefit of preservation, Congress enacted the National Environmental Policy Act (NEPA) in 1969. This act required all federal decisions, not just water resource development, to compare in a comprehensive Environmental Impact Statement (EIS) the direct benefits, costs, alternative actions (including not developing), and long-term irreversible consequences. The first policy review under NEPA was whether a license should be granted to build and operate a 48 inch in diameter, approximately 800-mile long pipeline from the Alaskan North Slope to Valdez, Alaska. This pipeline would carry up to two million barrels of crude oil per day to Valdez across pristine wilderness with high levels of seismic activity, as well as serious permafrost, and other ecological challenges. There, shippers would transport the crude oil using oceangoing supertankers traversing the pristine Prince William Sound, with its frequent weather-related challenges. This first NEPA test was granting a federal permit to oil companies to construct the Trans-Alaska Pipeline System (TAPS). This was no small federal decision. At the time, this was the largest single privately-financed project in the United States.

For some, it might have been surprising that environmentalist and their lawyers did not oppose Alaskan North Slope oil development, despite the fact that the United States had a Mandatory Oil Import Quota program in place to reduce the nation's use of what were then relatively cheap foreign oil imports. Instead, the environmentalists concluded and argued that a fair and objective NEPA review would show that it was far more advantageous for the environment, energy security, and consumers to build an all-land pipeline across Alaska and Canada. Such an all-land transportation choice would serve the already energy-short Midwest and eastern U.S. markets, not the relatively well-supplied western portion of the United States, which was then adequately supplied with California crude.[5]

After years of legal wrangling, the United States Department of the Interior (DOI) prepared a nine volume Environmental Impact Statement (EIS) that agreed with virtually every specific claim the environmentalists made. There were two matters that became the exceptions. The first was that a Trans-Canadian Pipeline (TCP) was environmentally preferable under

---

[4] Krutilla, John V. and Charles J. Cicchetti. (1972). Evaluating Benefits of Environmental Resources With Special Application to the Hells Canyon. *Natural Resources Journal*, Vol. 12, No.1.

[5] Cicchetti, Charles J. (1972). *Alaskan Oil: Alternative Routes and Markets*. Baltimore and London: Johns Hopkins University Press.

every comparison (there were hundreds), except one. The exception was the TCP was longer and would, therefore, come in contact with more abiotic resources. This is a technical phrase to say a 3,000 mile TCP would touch more "dirt" than would an 800-mile TAPS.

This led the author of the Executive Summary to conclude that there was no alternative considered that was superior for *every* environmental comparison. Additional DOI analyses also agreed with the environmental conclusion that the relative economics of the alternatives favored the TCP in every manner of comparison to TAPS. Despite the incredibly overwhelming environmental advantages (except dirt) and all the economic advantages that the TCP choice had over TAPs, the Executive Summary of the EIS did not favor TCP. Instead, the EIS process added national security as a third factor in the ultimate comparison.

The national security analysis concluded, in effect, that an all-American line was, by definition, more secure than a pipeline across a foreign nation, Canada. Environmentalists were shocked. They responded in three ways: (1) They pointed out the United Kingdom owned BP oil, which was a principal owner and developer on the North Slope of Alaska; (2) Canada's primary population, which centers in Ontario and Quebec, were entirely supplied with oil and much of their natural gas through pipelines that flowed across the United States; and (3) there was strong evidence that the major oil companies would attempt to export Alaskan oil in exchange for greater imports of cheaper foreign crude in excess of the then current import quotas in the eastern portion of the United States.

Subsequently, Congress sided with the Executive Branch after President Nixon ended the Mandatory Oil Import Quota Program in 1973. This effectively ended the environmentalists' legal challenges to TAPS. After the initial anger and disappointment, NEPA became known as, at best, a tool to make the facts known and federal actions transparent. It was *not*, the environmentalists decided, a tool that would cause federal actions to be based upon objective economic and environmental analyses. The aftermath of the TAPS decision was made even worse tasting, when a tie vote in the United States Senate to stop TAPS was broken, when Vice President Spiro Agnew, weeks before he resigned in disgrace, voted to approve TAPS and to shut down any further legal challenges.

This collective anger caused many of the major environmental lawyers and their financial supporters to seek to establish a new strategy and a different legal forum. The *new* environmental litigation strategy that arose from this major environmental defeat had two core principles. First, it emphasized demand reduction and conservation, not just opposition to new energy sources. This principle proved prescient, since it pre-dated the 1973 Arab Oil Embargo that caused the first worldwide oil crisis in the fall of 1973. Second, many environmental lawyers, often with EDF and NRDC in the lead, also decided to concentrate on state regulatory agencies, not federal courts, because these quasi-judicial entities, as the FPC demonstrated with its Hell's Canyon decision, could entertain, understand, and were more likely to evaluate objective scientific and economic evidence.

An environmental legal action plan emerged to provide financial support for economists and lawyers to bring their environmental opposition to utility programs that environmentalists alleged promoted growth. The environmentalists initially targeted four states that were considered to have a large number of experienced commission staff. These four states were Wisconsin, Michigan, New York, and California.

The environmentalists' specific policy target was the declining block or volume discount pricing of electricity. The basic economic arguments against volume discounts were introduced in Wisconsin in 1972 through Dr. Charles Cicchetti. He was joined in Michigan and New York by Dr. Ralph Turvey and Dr. William Vickrey. Financial support came from the Ford Foundation, National Science Foundation, the California Planning and Conservation Foundation, and most importantly the EDF.

The environmentalists' legal strategy was to start slowly in Wisconsin and Michigan to perfect their evidence before introducing these various data and theories into the more populous venues in California and New York. As it turned out, all four states were very receptive to the arguments in various hearings. However, Wisconsin took center stage. As the economists developed their theories, two concepts emerged to replace declining block prices. These were marginal cost, including environmental externalities, and time-of-use tariffs. Over the next two decades, these concepts became core concepts in the energy efficiency, load management, and integrated resource planning policy formulation. With David Freeman's leadership at the Ford Foundation's Energy Policy Project and ultimately in the White House, these became the core ideas found in the Public Utility Regulatory Policies Act of 1978.

The first major effort to apply economic efficiency arguments to an electricity rate design case began in 1972, when the EDF, a national group, joined forces with two local groups, Wisconsin Environmental Decade and Capitol Community Citizens, and intervened in a Madison Gas & Electric Company (MG&E) proceeding, in an attempt to force MG&E to replace declining block pricing with marginal cost pricing. The MG&E case turned out to be a landmark state regulatory decision. One reason for this is that all major electric utilities in Wisconsin decided to join forces to oppose the intervenors in this case, which were generally associated with environmental advocacy groups. Additionally, Wisconsin has generally been recognized as a policy trendsetter among state utility regulatory commissions.

The utilities' first set of arguments had three components. First, tradition required utility commissions to measure actual (*i.e.*, what had already happened) historic costs. Second, marginal cost was an interesting, perhaps irrelevant, academic concept (later, the argument against marginal cost was based on "second best" problems). Third, customers' demands were not sensitive to price changes. Indeed, the utilities' initial argument maintained that demand was completely price inelastic. In such circumstances, the utilities averred that the regulators should simply ensure that fair and accurate costs were allocated to each customer's bill and not be bothered by tariff structure.

The environmental intervention involved a detailed set of issues. The arguments were complex, and involved both the concepts of marginal cost and its measurement. In the electric utility case, there were also environmental costs associated with producing electricity. The intervenors attempted to communicate these concepts and procedures in simple and straightforward terms. In considering the question of estimating the marginal cost of services, the utilities argued that marginal cost was a theoretical concept, difficult to define in terms that allowed "practical" measurement. The intervenors, in contrast, contended that efforts to move the process closer to achieving economic efficiency (*i.e.*, maximizing net social welfare) were more important than the ease of measuring average accounting costs.[6]

The second component of the environmental intervenors' case concerned the fact that prices mattered. Demand *would* respond to price changes. Furthermore, external costs and time-of-use were also important tariff design concepts. Consequently, investments in capacity and the magnitude of environmental costs, both of which are associated with generating electricity, can be expected to be influenced by pricing decisions. The second point directly follows from the first. Economic efficiency requires that prices signal full marginal cost (*i.e.*, the real resource cost) of electricity. This was not the practice of regulation in the past. Precedent was not required, in their view, to correct past mistakes. They asserted that the position of the utilities represented a "nothing should ever be done for the first time" perspective.

With these two straightforward arguments, the intervenors attempted to demonstrate the technical feasibility of their underlying arguments—marginal costs could be measured and price responsiveness did exist. Empirical studies were introduced into evidence, simple procedures for estimating marginal costs were developed and expanded, and computer simulations were presented to indicate the substantial difference that the cost concept and measurement would make. As the staff and the commission began to respond to these principles and evidence, the Wisconsin utilities replaced their original consultants and organized their own team of economic experts. Their charge seemed to be that if economics and econometrics

---

[6] Ironically, the proceedings revealed that accounting costs were also not precisely measured because there were nearly 30 conflicting recognized approaches for analyzing embedded average costs.

are to replace traditional accounting approaches for defining prices (*i.e.*, rate design), then the utilities and their experts would show how it could be done properly.

Economists on both sides of this regulatory hearing were able to readily agree on three issues:[7] (1) Due to the significant time-of-use and voltage differences associated with delivering electricity to the meter, inverted block pricing made no more sense (in terms of economic efficiency) than declining block pricing; (2) the price elasticity of demand was not zero and may well differ during peak and off-peak demand periods and across different customer classes. Thus, it would be desirable to examine both the effects of time-of-use prices (especially since costs were expected to vary rather dramatically with time of day, week and season of use), as well as the price responsiveness of different customers on tariff design; and finally (3) external social costs should be included in estimates of utilities' marginal costs.

Two tasks remained. First, it was necessary to implement a practical marginal cost pricing system, which meant the need to estimate marginal costs. Both the United Kingdom and France had undertaken efforts to develop and to implement marginal cost pricing arrangements in their electricity tariff and marketing arrangements. This is where the esteemed British economist Dr. Ralph Turvey was able to communicate two ideas. First, tariffs and marginal costs based on proper theory had to be only "roughly right;" and regardless, the electric utility's system engineers, not the utility accountants, already measured marginal costs in their system lambda determinations. Second, as he so eloquently put it: "'Tis better to be roughly right than precisely wrong." This often-repeated phrase helped to refocus regulators away from growth-promoting tariffs to regulations favoring conservation.

With the support of the National Science Foundation, economists and engineers from Europe and the United States organized to facilitate direct communication among energy experts and economists. As a result of these discussions, much analytical and institutional information relevant to the initiation of marginal cost-based electricity tariffs was obtained. Most fundamentally, these discussions revealed that a close surrogate to true marginal cost was already being calculated and used by Wisconsin utilities in their system dispatching and planning functions. Professor Turvey was deemed to have been correct. Utility engineers had been estimating short- and long-run marginal costs, without calling them that, in order to respond to their charge to minimize short-run operating (dispatch) costs, as well as to minimize long-run system expansion costs.

---

[7] The observation that agreement was relatively easy should be considered in relative terms. The actual decisions involved many months, pages of testimony, and extensive hearings. Nonetheless, *ex post* examination of the records indicates clear agreement among the economists on the principles involved.

Since the utility ratemaking function was largely the domain of utility accountants charged with collecting authorized revenue requirements, the existing marginal cost calculations of the engineers remained unknown to the rate makers. As a result, the merging of the two disparate information sources known to the Wisconsin utilities made it possible to implement new time-of-use (TOU) tariffs and to establish new transparent integrated resource planning.

The Wisconsin commission, in its Final Order, Docket 2-U-7423, found marginal cost pricing to be the correct conceptual basis for pricing utility services in Wisconsin.[8] It required all Wisconsin utilities to estimate marginal costs, prepare tariffs and develop implementation plans, before initiating any requests for a rate increase. At the time, time-of-use (TOU) demand studies were not available for any class of customers. These were necessary in order to estimate the consequences of full marginal cost pricing. For example, the practical issues associated with deciding the number of different prices to be charged by time of day, week, and year, and the degree of responsiveness in them to customer responses had to be resolved. Equally important, metering costs were uncertain. The final order deferred these matters to subsequent future rate case filings. This order drew heavily from the theory, as well as the empirical and implementation knowledge introduced into the case. The initial economist who put these arguments before these four state commissions became the Chair of the Public Service Commission of Wisconsin in 1977 and implemented these earlier decisions in future rate cases and integrated planning in Wisconsin in the late 1970s and 1980.

The MG&E rate case is an example of a situation in which the application of economic analysis and data was able to directly influence the direction of policy. Several reasons account for this. First, the issue of increasing relative scarcity of energy fuels and of the environmental consequences of using them was high on the research agenda of a large number of economists and engineers at the time. A wealth of talent and expertise with similar backgrounds and training could be brought to bear on this issue in a relatively short time period. Second, the basic questions at issue were relatively straightforward and had been worked on for decades in the economics fraternity—Does the consumption of electricity depend on its price? How should electricity tariffs be designed in the face of decreasing direct costs and increasing environmental costs? How can the utility revenue constraint be reconciled with economically efficient pricing rules? Third, unlike the case of the court system, the state regulatory commissions are quasi-legislative in nature, often supported by a highly trained professional organization. State regulators are often individuals with substantive knowledge in the area, and they have access to economists and engineers who serve as their staff. In such an environment, scientific and analytic information perhaps has its best chance of influencing decisions and shaping public policy. Finally, the issue at stake in the electrical pricing cases was between two parties (utilities and environmentalists) with divergent interests and

---

[8] See Madison Gas and Electric Co., 5 PUR4th 28 (Public Service Commission of Wisconsin, 1974).

objectives. The government was not a party. Therefore, the court was not placed in the position of interpreting and/or needing to second guess and override the executive or legislative branches of government. The high prices of the first oil crisis, starting in the fall of 1973, added political support to these environmental arguments.

These regulatory battles had several other important side effects as well. In the process of the hearings and the testimony, computer programs for calculating marginal cost that were capable of being brought into a hearing room were developed. National legislation requiring the collection of information that could be used to calculate marginal cost and time-of-use tariffs was made mandatory by the *Public Utility Regulatory Policies Act of 1978 (PURPA)*. A series of federally financed time-of-use electricity tariff experiments were undertaken throughout the nation. Perhaps most importantly, the National Association of Regulatory Utility Commissioners required the nation's electric utilities to embark on a multi-year study to resolve the various economic and statistical questions related to tariff reform. In short, the EDF set in motion a large number of very extensive electricity tariff reform research agendas, when it decided in 1972 to make utility rate reform a national priority.

Accomplishments in the policy arena, based upon the issues raised by this case, have been no less impressive. They began with various reforms in electricity tariffs, beginning with the virtual elimination of volume discount pricing. Eventually, these reforms encompassed time-of-use pricing and costing, marginal cost pricing for cogeneration, and long-range or integrated least-cost planning. This came to include demand-side programs, such as utility-sponsored conservation. There is also a direct link between industry restructuring through the introduction of competition in generation and the economic principles that these early cases raised.

Tariff reform ushered in renewed and increased interest in interruptible tariffs and load management. Real-time pricing emerged as a theory to send price signals during actual periods of peak demand and higher energy costs, not just potential peak periods. The federal government financed various electricity metering and time-of-use pricing experiments. PURPA also caused each state to design purchase power procedures and to consider marginal cost theory and quantification using a closely related concept of avoided costs. These were needed to encourage cogeneration and third party generated renewable energy sales to investor-owned utility companies. Some states also adopted time-sensitive rates for large customers and seasonal rates for all customers. Consumer advocacy groups joined the debate, particularly in the aftermath of even higher energy prices, when the second oil crisis hit in 1979 with the Iranian Revolution.

This led states to reconsider life line tariffs and no-disconnect rules. Some adopted utility consumer rights policies. Most significant, the common concerns with high prices caused environmental and consumer groups to come together to support utility-sponsored conservation and demand management programs to reduce utility bills and to protect the environment.

Nascent conservation efforts also started, because Americans seemed more interested in conservation and energy efficiency, because these could help consumers save money and would not require them to pay as much attention as time-of-use or real-time pricing would. PURPA also helped some states, such as California, expand renewable wind, geothermal power generation, and to promote new technologies to increase solar and energy efficiency. A combination of high inflation and economic recession in the early 1980s derailed some states' efforts.

Later, the falling crude oil prices in the 1990s and the economy caused some states to restructure their vertically integrated electric utility industry. Many decided that wholesale and retail competition should be introduced and concluded that this change would replace the need for regulatory mandates to redesign tariffs to reflect time-of-use marginal cost and to mandate load management. This also caused interest in regulated utility programs to wane. Indeed, one of the sometimes stated reasons to restructure the electric industry was for competition to take the "social engineering" ball out of utility regulators' hands.

Others followed California's example and introduced a so-called "public goods charge." This may have actually increased spending on conservation. California regulators decided that one of the prices for restructuring was a "wires charge," called a public goods (or benefits) charge. These dollars would be spent to expand conservation, energy efficiency, load management, renewable energy, and to provide low-income assistance. The national approach for energy efficiency became a "mixed bag." Perhaps an optimist might claim this divergence of state action as a fine example of the various distinct state "laboratories of democracy" at work. Some states expanded mandatory energy efficiency and others dropped their programs.

With economic expansion, those states still committed to the notion of utility sponsored energy efficiency also generally supported the notion that financial incentives would encourage their utilities to support energy efficiency. Further, some states accepted Maine commissioner David Moskovitz's reasoning that cost-of-service regulation was the problem. They agreed to remove the financial disincentives of rate based regulation through some form of sales, revenue or earnings decoupling.

In the late 1990s, the nation's wealth increased. The United States dollar grew stronger. Oil prices declined to about $10 per barrel, which was less than the U.S. price adjusted for inflation at the start of the first world oil crisis in 1973. Many states and customers effectively ignored energy efficiency. In the case of automobiles, many drivers and car companies returned to horsepower levels not seen since the 1950s and 1960s in the United States.

Post "9/11" national security and climate change, strongly reinforced with world crude prices that reached nearly $150 per barrel in the summer of 2008, have caused yet another regulatory run at energy efficiency. Both candidates for president in 2008 could not speak very long without praising energy efficiency.

In the chapters that follow, "how, why, and for whom" questions are addressed and analyzed. It is important to learn from the past and to pose the questions that might lead the nation's energy utilities and their regulators to set a course that will sustain the current efforts, and to avoid the missteps that caused the two previous national surges of interest in conservation to dissipate, if current political and economic reasons for supporting current conservation efforts change.

CHAPTER 3
THE ROLE OF UTILITIES IN CONSERVATION: REGULATORY DETAILS THAT MATTER

Introduction

Both "necessary" and "sufficient" conditions are required to reach a firm and logical conclusion. This discussion starts with the "necessary" aspect of utility-sponsored energy efficiency. The questions focus on: (1) why energy efficiency; (2) why utilities; and (3) why not energy consumers? Later chapters address the "sufficient" aspects of energy efficiency such as: (1) is it cost effective; (2) is it reliable; and (3) how can regulators help to make it succeed and be sustained?

Utility-sponsored energy efficiency is not a universal focus of regulation. Other concerns and specific utility conditions often combine to determine the extent of the utility customers' and the extent of political support for utility-sponsored energy efficiency.

Some of the factors that favor increased utility efforts to promote energy efficiency include: (1) increased environmental concerns most recently related to climate change; (2) applications to build conventional power stations often cause conservation to emerge as an offset; (3) states that restructure and seek stranded cost recovery, along with other charges utilities seek, often establish "public benefits charges" that can finance energy efficiency; (4) economic and national security concerns have recently combined to bring additional public support for reduced energy consumption; and (5) general rate increases can often provide the opportunity for various parties to seek energy efficiency.

The term energy efficiency is primarily used as a broad concept. It includes both conservation and enhanced demand response/load management. The same factors and conditions that favor more utility-sponsored energy efficiency would also likely lead to expanded efforts to require utilities to invest in renewable energy projects. There are some differences.

Renewable energy projects are often viewed as complementary supply-side investments that will achieve many of the same environmental, economic, and national security objectives as energy efficiency. However, renewable energy projects produce electricity, which is a core utility product that can be sold and directly metered. As such, there are fewer or no disputes with respect to who pays for renewable energy or who benefits. There are no lost revenue concerns. Similarly, there is less or no concern due to any inherent inconsistency with traditional rate base regulation. A utility that builds new renewable energy can expect to earn regulated returns for generation that uses wind power, solar, bio-fuels, waste products, and some particularly small hydroelectricity projects.

The Public Utility Regulatory Policies Act of 1978 (PURPA) provided the legal means for states to direct utilities to purchase "renewables" from qualifying facilities (QFs), owned by third party generators, at prices state regulators would determine using the avoided cost concept. This concept is a very close cousin to the marginal cost concept many environmentalists advocated in the early 1970s to promote more economically efficient utility tariffs, less utility load growth promotion, and improved environmental quality.

More recently, state regulators and some legislative bodies have been approving programs to establish Renewable Portfolio Standards (RPS). These would complement the third party or QF-supplied renewable energy with utility-owned, operated, and/or financed renewable energy. Restructured states have also found that renewables can be encouraged to enter wholesale capacity markets; used to provide "green" retail products; and are amenable to the same public benefits charge financing as energy efficiency.

Both energy efficiency and renewables can be encouraged through state or utility targets. These targets are typically stated as a percentage of a utility's or state's electricity consumption. The more aggressive states like California are setting the bar fairly high. California seeks at least 20 percent renewables in about 10 years. Others often set the bar for renewables at 15 percent. These are somewhat more aggressive than the approximate 10 percent savings that the top ten utilities, the high achievers, have been able to achieve for energy efficiency.

The differences may be coincidental. More likely, the supply-side nature of renewables gain utility support, because these avoid the inherent difficulty that both regulators and utilities often confront with energy conservation. The main difficulties are that energy efficiency causes less utility sales and underrecovery of fixed costs. Some states bring renewables to customers' premises in the form of solar roof units, distributed energy (batteries), etc. These hybrid renewable products often share some of the same challenges and characteristics that regulators attach to energy efficiency. Central station renewables and third party renewable sales to utilities are more straightforward and not any different from a regulatory perspective than electricity provided using other fuels.

The central focus of this book is energy efficiency. Nevertheless, where appropriate, renewables will be addressed given both the complementary and potential competitive aspects of utility-sponsored renewable energy policies and energy efficiency. Another purpose of this analysis is to determine how utilities, regulators, and consumers can design utility-sponsored energy efficiency programs that work, are sustainable, and yield net benefits to the various stakeholders. The same PURPA-styled QF principles have mostly not been applied to energy efficiency. These will be examined, along with other regulatory, political, and marketing principles, to determine any lessons that renewables can provide for regulating energy efficiency. Both are necessary, if the nation is to achieve what political leaders are currently targeting.

Why is Energy Efficiency the Focus of Public Policy

There are three broad reasons for society to invest in energy efficiency. Put simply, the energy not used (ENU) does not pollute, drain dollars from the economy, or raise climate change issues. ENU does not require fossil fuels. ENU avoids generation and frees up existing environmentally friendly supply-side resources (like wind power) to displace fossil fuel-fired generation that cause more greenhouse gas (GHG) emissions.

Conserving energy also reduces the need for imported energy. In the case of electricity, the energy independence goal is much more relevant for natural gas (including liquefied natural gas [LNG]) than it is for coal, because the U.S. has vast coal reserves. Using less natural gas to generate electricity would also mean there is more natural gas available for transportation. T. Boone Pickens is spending a chunk of his personal fortune to tell Americans that wind power and energy efficiency could release natural gas used to make electricity to fuel compressed natural gas (CNG) vehicles. This would help to ease crude oil imports for the nation and improve the environment. If the United States expands energy efficiency and helps to reduce crude oil imports, the benefits would also include national security; some relief for the U.S. dollar; and improved macroeconomic conditions as the nation imports less and this reduces the oil imports' drag on the economy.

Energy efficiency can also help consumers, because it would reduce the volume of utility energy consumed each month. There would be additional consumer benefits, if the avoided or marginal cost of electricity exceeds regulated retail prices. This would mean that growth in energy consumption would raise regulated revenue requirements, while cost-effective conservation would not.

The term cost-effective conservation means energy efficiency for which the marginal cost of energy efficiency is less than either the regulated price or the marginal cost of energy. If retail prices exceed the marginal cost of efficiency, well-informed and rationally behaving utility customers should purchase more efficiency and less electricity. If the marginal cost of electricity exceeds the marginal cost of energy efficiency, all utility customers could gain in terms of lower annual regulated revenue requirements for more expensive new generation, under utility-sponsored programs that substitute efficiency.

These environmental, economic and national security, and economic efficiency (cost effectiveness) criteria combine to create the political rationale for utility-sponsored energy efficiency. This is because regulated prices would increase, when utilities are required to internalize and reduce the external costs for things such as GHG or climate change. Accordingly, new policies like carbon taxes or cap-and-trade would cause electric utilities to act to curtail GHG or pay higher prices when they emit carbon or GHG. The final reason to encourage energy efficiency is the likely reduction in higher future compliance costs related to climate change policies such as carbon taxes or "cap-and-trade."

Why Do People Focus on Energy Utilities

There are numerous reasons to explain why energy utility customers do not purchase cost-effective energy efficient light bulbs, appliances, heating and cooling systems, etc. Some think utility bills are too confusing and the industry should have followed Thomas Edison's advice and charged people for the services they use, such as lighting, water heating, cooling, appliance use, etc., and not for something as mysterious as metered kilowatt-hours (kWhs) that are billed a month after the actual use.

Others claim consumers are not well informed about the performance, economic life, and other advantages related to new products, such as efficient light bulbs. Regardless, consumers seem far more likely to replace the old, with a similar product, particularly if it initially costs less. Consumers simply do not often like to experiment with new products even when labels and advertising tell them these new products are very cost effective over time and perform as well as or better than expected. The American consumer is exposed to so much advertising that many become rather jaded and tune out claims to "buy me because I am a better light bulb," etc.

Another explanation is the time value of money. Some cost-effective energy efficiency products require capital investments, such as new heating and cooling systems for a home or building. Replacing existing systems that "work" may strike some consumers as unreasonable despite the calculations and sales pitches that show the extra investments would have a relatively quick pay-back period. Therefore, waiting to spend money is not surprising.

A related reason for consumer apathy may be the fact that energy efficiency is mostly a purchase that is not an observable consumption or investment. Consumers see and experience a new dress or great meal. Energy efficiency does not register in the consumers' brains until the electricity bill arrives a month later. At that point, many other things change, such as weather. Untangling energy efficiency is difficult. The decision to purchase it is often masked. These can cause consumers to become indifferent or even apathetic towards energy efficiency. At the very least, while generally supporting energy efficiency and conservation, Americans are broadly unwilling to actually do so.

In a similar manner, some consumers delay acting on energy efficiency products and choices, because they believe a better light bulb or appliance is just around the corner. Apathy is easy, because it requires no additional time to gain necessary information, and consumers do not need to do anything differently than they have been doing.

Public policy and proactive regulators favor utility-sponsored mandates for energy efficiency, given the significant social and cost effectiveness of increased energy efficiency. Consumers would likely benefit considerably, if they became better informed and purchased more energy efficiency. Utility consumer indifference is why regulators often feel compelled to step in and make this happen.

Customer-owned and municipally-owned utilities can act, when collective action seems justified. For investor-owned utilities, the leverage for regulatory action is the franchise status of regulated utilities. This franchise status means that there is a distribution network with natural monopoly characteristics. There is an explicit social contract that in exchange for exclusive "wires" or "pipes" services, the utility would not face competition. The franchise would not need to pay duplicative costs and could expect scale economies to pay for its distribution systems. Neither wholesale competition for generation, open access for transmission, or even retail choice fundamentally alter the distribution franchises' natural regulated monopoly status, which its mostly non-bypassable wires establish.

Politically, regulators know that the utility franchise has value. They also know this can be used to require the franchise possessor to engage in cost effective, socially beneficial things, such as mandated energy efficiency. Regulators seeking more energy efficiency would view this as being similar to public safety, economic development, and other broader social matters. Mandating socially beneficial things is not new. There is more to consider, because some participating consumers that consume less would also pay less each month for electricity and others benefit because eternal costs are reduced.

A more reasonable public policy approach for energy efficiency is to recognize the "shared" responsibility nature of conservation. The regulated wires company could and most likely should play a major role in providing information and certifying legitimate vendors and acceptable products. There is also full justification for subsidies to encourage energy efficiency, when energy efficiency reduces utility costs or would do so prospectively by reducing avoidable or marginal costs including external costs.

Customers who accept or install utility-sponsored energy efficiency benefit both from lower consumption volumes and the likelihood of lower future utility prices. Accordingly, customers that participate directly in utility-sponsored energy efficiency should expect to pay some of the costs of energy efficiency. Investor-owned utilities would still need to package and market energy efficiency products and investments, arrange their proper installation, and help consumers understand how to use these new energy efficiency products.

In effect, this is a new business or service line for investor-owned utilities. It is unreasonable to expect that any new business could succeed unless it has a reasonable chance to earn sufficient income. For new energy efficiency services, the utility has direct costs and indirect costs. It also requires new management skills and marketing expertise. With reasonable cost markups or sales margins, energy efficiency would not reasonably become a sustainable and successful business.

There are two additional concerns. First, utilities have a franchise monopoly. This means the utility has direct retail customers access. This is favorable to expanding a new-energy efficiency enterprise. At the same time, this exclusive franchise condition raises questions related to the ability of others to compete against utility-sponsored energy efficiency. The fact that regulators could also add subsidies to encourage more energy efficiency is also a potential competitive challenge for other vendors.

Some states resolve the inherent competitive problems directly. Utilities hire third party firms to install energy efficiency services. Other states require utilities to work with retailers to reduce the purchase price of efficient light bulbs, and to provide rebates for efficient appliances and other home improvements that expand energy efficiency. Competitors often learn that the utility and its relatively secure cash flow can make their market even when the average consumer remains uninformed and apathetic.

Second, rate base or cost-of-service regulation historically has emphasized fixed cost recovery with a relatively controlled element of risk in regulated utility tariffs. This means that a regulated utility that promotes conservation or energy efficiency may be conflicted and experience significant fixed cost recovery risks. Energy efficiency requires regulators to address current tariffs and revenue requirements.

Some states have solved this lost margin problem by approving direct profit incentives for energy efficiency. Others adopt some form of what regulators call revenue or income decoupling. The latter means that losses in utility income or fixed cost recovery would be neutralized using various tariff adjustments.

In the 1980s and early 1990s, all states had traditional cost-of-service regulation. Various states experimented with different combinations of energy efficiency mandates, direct incentives, decoupling, and tariff riders or adjustment mechanisms. When California restructured in the late 1990s, the state required a "public goods charge" rider for regulated "wires and pipes" utilities to ensure that energy efficiency, low-income assistance, and renewable energy programs would continue. In other states, there was some concern that mandated energy efficiency subsidies were either inconsistent with competition or would morph into new product and marketing aspects that could harm other competitive retailers. The problem was addressed in various fashions and was not a detriment to using wires charges to promote energy efficiency and green electricity generation.

Nevertheless, retail choice often failed to deliver the same energy efficiency efforts as traditional regulation. Several states and FERC have begun to seek new approaches for attaching energy efficiency and demand response to wholesale market competition. Several states are once again revisiting the regulated distribution wires to attach rate riders to promote conservation and renewable energy.

The nation's energy utility companies are contemplating a third run at utility-sponsored energy efficiency. As with the two prior efforts, high world oil prices matter. Current support seems more broad-based politically than simply concerns about high prices. Indeed, most Americans also have climate change, macroeconomic, and national security concerns. The next chapter reviews some approaches that seem to work.

CHAPTER 4
REGULATING THE PATH TO ACHIEVE SUCCESSFUL UTILITY-SPONSORED ENERGY EFFICIENCY

In the early 1970s, America became aware that energy was scarce and expensive and the nation was far from what would prove to be the elusive goal of energy independence. As distant world markets combined into a global market, energy independence became a national and economic security concern. There was also a greater appreciation of the attendant external costs of energy supply. Nascent utility-sponsored conservation and demand-side management programs emerged. These efforts lost consumer and regulatory support as energy prices receded.

The marginal cost and time-of-use pricing debates in the early years, which the Public Utility Regulatory Policies Act (PURPA) encouraged, may also have softened regulatory interest in direct programs to encourage conservation and demand-side management. Regulators may have perceived that eliminating the promotional utility pricing policies, like declining block or volume discount pricing, was enough to cause consumers to conserve energy.

In the middle 1980s and early 1990s, there was a second wave of interest in energy efficiency. Utility-sponsored renewables and energy efficient customer premise devices, such as efficient light bulbs, heating/cooling, and other efficiency-enhancing purchases, were added to the utility-sponsored portfolio of choices. Utility-sponsored efforts were, however, primarily focused on conservation and demand response programs. The regulatory debates in this second wave were often related to how to protect non-participating utility consumers from paying undue subsidies for energy efficiency. As with the first wave, declining prices, along with new interest in retail competition and retail choice, may have caused some regulators to reduce their interest in utility-sponsored energy efficiency. In other words, the prospect of real-time wholesale pricing on the horizon, along with a reduced need for regulators to approve new expensive rate-based generating stations, meant there was less regulator interest in utility-sponsored energy efficiency.

Both prior waves of heightened regulatory interest in energy efficiency were marked with similar regulatory attitudes: (1) The utility obligation to serve, not the need for a new "for-profit" retail service, was the prevailing regulatory reason to support utility-sponsored energy efficiency; and (2) there was a strong, nearly ubiquitous regulatory predisposition to ensure there would be no undue subsidies ("no losers") associated with utility-sponsored energy efficiency programs. These can be deal killers, if the goal is to design a *sustained* energy efficiency market centered around regulated utilities.

Consumers and regulators are once again recognizing that energy is expensive. The associated external costs of energy consumption seem even more important today with the addition of new heightened public concern with climate change, national security, and the potential adverse macroeconomic consequences related to a weaker dollar and lost consumer confidence as energy prices soar. History suggests that utility regulators will not likely ignore this renewed customer and voter interest in energy efficiency. The question remains: Will they get it right this time?

There is more than political/customer support that regulators should contemplate, if they truly want energy efficiency to succeed and to be maintained this third time "around the regulatory block." The current political support for energy efficiency makes this possible. There are two essential components of any utility-sponsored energy efficiency program. These require regulators to avoid the past distractions caused by too much emphasis on "duty" and too much weight given to consumers that do not participate.

First, regulators must recognize that there is significant customer reluctance to do even the simple, no-hassle, and decidedly cost-effective energy efficient things they can do on their own. This is a threshold regulatory decision that, once reached, suggests the "need" for utility-sponsored energy efficiency programs.

Some may not be readily convinced that this is so and prefer to grasp at high energy prices and consumer education as likely to be sufficient. Past experience says "this ain't so." Just as relevant for the more active regulators, there is or should be greater acceptance that this will not be an easy utility task. This is true even with strong customer financial and hassle-reducing inducements to increase energy efficiency.

Second, John Rowe, CEO of Exelon, in another life and during a previous wave of regulatory interest in energy efficiency, got it absolutely right. He explained that "the rat must smell the cheese."[1] He quite elegantly observed that if regulators want utilities to embrace, expand, and make utility-sponsored energy efficiency work, it is rather simple. Show them the money (*e.g.*, the cheese) and let them seek it out.

---

[1] Kushler, Martin. (2007, March 21). "Addressing the Crucial Issue of Utility Disincentives Regarding Energy Efficiency: Basic Concepts and Current Status." Presentation to the American Council for an Energy-Efficient Economy (ACEEE) MT Symposium.

Chapter 4: Regulating the Path to Achieve Successful Utility-Sponsored Energy Efficiency

California's sustained success with energy efficiency relied on very significant shareholder financial incentives. California introduced revenue "decoupling" in 1982 in order for utilities not to sacrifice earnings when customers conserve. This removed one disincentive for utility-sponsored conservation. In 1991, California added shared savings for investors to further encourage energy efficiency.[2] Current regulators may now believe they overfed the utilities' bottom line. Regardless, there are no plans to abandon using financial inducements to sustain and expand California's utility companies' interest in the growing utility energy efficiency service.

California is not alone in using income-enhancing energy efficiency performance regulations. Some states are rather benign. They simply permit utilities to recover cost and/or they adjust for lost revenue. The timing of rate relief due to lost sales also matters. These mainly make utilities indifferent to any lost sales related to energy efficiency. Higher utility earnings would make energy efficiency more than a cost pass-through. This can take the form of cost-of-goods markups or margins, shared savings, or higher authorized earnings.

There are nineteen states that are adding "cheese" to their expanded energy efficiency efforts.[3] They have some form of direct financial incentive for regulated electric utilities. This is the key to making energy efficiency a viable, sustained new regulated service. There are also five states that have broad decoupling and six that have some form of lost margin adjustment mechanism. Decoupling proposals are also pending in nine states.

The term "decoupling" can be broadly applied to a rather broad range of regulatory programs to encourage energy efficiency. At one end, the term characterizes a sales or revenue neutrality in the form of a "true-up" type mechanism that adjusts for differences between predicted, often regulatory test year, sales of energy and actual annual sales. This type of adjustment would include a myriad of factors that would affect sales, such as weather and economic conditions, as well as energy efficiency. Other states simply adjust their test years to anticipate energy efficiency. At best, these regulatory approaches are neutral for energy efficiency. They express the concept that energy efficiency would not make the utility lose money. Furthermore, in many circumstances, the specific effects of energy efficiency are lost when a utility has a broad adjustment for any change in sales.

At the other extreme, the decoupling term applies to lost earnings or margins on energy sales that energy efficiency displaces. The purpose of such an adjustment is to estimate the differences between sales prices and avoided (mostly fuel and variable) costs. Either actual or

---

[2] Grueneich, Dian M., Commissioner. (2007). "California's Policy Framework to Advance Demand-Side Management." California Public Utilities Commission, International Energy Agency. Paris, France (DSM Workshop).

[3] These are discussed in detail in Table 21-3 of Chapter 21.

projected energy savings are credited with a lost margin recovery amount. This makes lost margin adjustments a form of financial inducement, based on shared savings, because the utility is able to recover the cost of energy efficiency, plus some amount that represents the margin it would have expected to recover on energy sales.

The amount of sharing between shareholders and customers would depend on the program's details, particular facts, and regulatory assumptions. These specific factors can cause lost margin incentives to approach levels consistent with more direct financial incentives or fall to the more neutral incentives associated with lost sales adjustments.

Regulators throughout North America are considering several means to encourage utility-sponsored energy efficiency. A second objective, although one less likely to be explicitly recognized, is the need to design such programs so that they are sustainable. Financial incentives are necessary for sustainability. The debate is not happening in isolation. Other changes affect what different jurisdictions might do to expand energy efficiency.

There is growing regulatory consideration for what to do when rate base does not represent a useful proxy for determining the value of an enterprise or the services a utility provides. The most obvious concerns are tied to U.S. states and Canadian provinces that have restructured, leaving retailers or wires companies that have virtually no (or a declining) rate base. In these jurisdictions, using a cost-of-goods-sold (COGS) markup of margin would be more akin to the delivery service or retailer business focus.

Businesses that no longer build, own, or operate (BOO) generation need a different regulatory paradigm than rate base cost-of-service regulation. Some jurisdictions, for example, Alberta and Maryland, have adopted retail margins that are quite similar to those in the competitive retail sector. In such jurisdictions, a similar approach can also be applied to energy efficiency products and services. If regulated income is simply based, in part, on marking up the cost of electricity that is either secured in wholesale markets and/or delivered to retail consumers, it is relatively straightforward to add similar markups to the cost of energy efficiency secured at wholesale or provided directly through the wires company or retail company to end users.

Retailers and wires companies that have a captive retail customer base can also be encouraged, or even required, to provide energy efficiency to customers at below the direct cost of goods sold or installed. The earnings shortfall could, in such circumstances, be made up in the costs assigned to all customers. A similar approach can be used to expand energy efficiency through direct incentives in more traditionally regulated jurisdictions. In these circumstances, the utility could be required to spend a specific amount to achieve a specific savings goal or to offer and provide a bundle of various energy services.

Cost-of-service regulation could pass through direct and indirect costs, a portion of common costs, and a reasonable margin or return to provide incentives for energy efficiency. True-ups, monitoring, measurement, and verification might also be added to such utility-sponsored energy efficiency programs.

These utility-sponsored programs can very often lead some to conclude that energy efficiency should be tightly regulated. If this sentiment goes too far, a form of a sometimes not very successful command and control approach emerges for energy efficiency. This does not typically work. The programs that seem to work best have a degree of utility flexibility combined with incentives that increase with success.

Some jurisdictions are considering symmetry for success-based incentives. This introduces the possibility of penalties, if a utility underachieves. The concept is reasonable. That said, flexibility, learning-by-doing, marketing, and packaging are not mere programmatic buzz words. Jurisdictions that consider energy efficiency penalties should think through when it would make sense to apply these penalties in the life cycle of mandated utility-sponsored energy efficiency programs.

Other regulated jurisdictions with vertically integrated utilities have been considering what is, in effect, a virtual rate base approach. Under such a regulated system, the utility "builds" demand-side replacements for power plants that it would, "but for" energy efficiency, need to build. Regulators would add a return to the virtual rate base.

Jurisdictions with integrated resource planning (IRP) have a ready proceeding to determine the relative values of building new generation or expanding energy efficiency to avoid or postpone new rate base construction. On a conceptual basis, the cost of the avoided rate base could be used to determine the implicit or virtual value of the rate base not built. Regulators could then review the arguments ranging from whether the avoided rate base should be discounted to help hard-pressed consumers or increased to encourage more energy efficiency, because it would save money and avoid negative externalities.

There are conceptually similar analyses that can be used in jurisdictions where the utility purchases electricity. This approach works for utilities that "buy" in organized competitive wholesale markets or as part of a more traditionally regulated vertically integrated utility. The purchase price metric for supply-side energy serves as a guide to determine how much energy efficiency would be economic. Again, a higher price might be assigned to energy efficiency, because it would likely reduce negative externalities.

The drawback of either the virtual rate base or avoided buy price approach arises from the fact that utilities sell energy for dollars. Energy efficiency could cause utilities to lose margin, because they would be selling less energy and not necessarily recovering their fixed costs and/or rate base returns "on" supply-side investments. In addition, energy efficiency does not necessarily secure sufficient payment from participating retail consumers to cover both its direct and assigned costs.

Despite these competing concepts, several jurisdictions consider avoided cost and reduced rate base to establish energy efficiency targets and/or mandates. The vexing questions are mostly focused on: (1) how to establish a just and reasonable income; (2) the extent and degree to which direct financial incentives are needed; and (3) on what to base the incentives.

Shared savings, as California demonstrates, has been a long-held approach to performance-based utility regulation. Prior to PURPA and the expanded wholesale markets for electricity, adjacent utility companies would often agree to exchange energy for reliability and economic reasons. In many regions of North America, utilities would agree to exchange electricity in real time, using the average marginal cost of the buyer and seller to set the exchange price. This price would split the benefits of shared savings equally between the seller and the buyer.

A similar approach can be conceptually applied to energy savings. The buyer saves the retail price and the seller would often spend less to provide some types of energy efficiency. Regulators are asked to consider splitting this difference between customers and shareholders. Such a regulatory approach is fair. If retail prices are reasonably aligned with the marginal cost of supplying electricity, then splitting the difference between retail electricity prices and the marginal cost of energy efficiency would also improve economic efficiency. This is due to the fact that selling more services with lower marginal costs (energy efficiency) would be substituted for a product with higher marginal costs (electricity).

There are two problems that regulators and utilities confront, under this conceptual approach. First, the bar for energy efficiency may be set fairly low, because economic efficiency would increase up to the point where the two respective marginal costs are equal. In addition, external costs would likely mean even more energy efficiency would be justified. However, if the incentive is to split the difference, this would tend to cap (conceptually) the amount utilities would be willing to spend on energy efficiency and, more important, this could constrain the types of energy efficiency a utility might willingly provide. Such concerns often invite aggressive or heavy-handed command and control approaches to energy efficiency regulation.

Second, rational consumer choice also comes into play. In theory, consumers confronted with the choice to conserve energy at a specific regulated price or purchase energy efficiency on their own at a lower price or marginal cost would be expected to behave rationally and purchase the least expensive choice that provides comparable light, cooling, heat, hot water, motor force, etc. However, in a great many instances and for various reasons, utility consumers do not make decisions that "experts" would define as rational. This could be due to consumers not understanding utility tariffs, the specific energy consumption of various applications, the performance or cost of energy efficient alternatives, inertia, lack of access to capital, apathy, etc. Regardless, regulators and utilities often agree not to charge customers that choose to participate in specific utility-sponsored energy efficiency programs (participants) either the full costs of energy efficiency or anything tied to the value of the new bundle of services the customers receive.

This means that other consumers (nonparticipants) are required to make up some of the costs of energy efficiency, as well as the lost margin due to the utility selling less regulated energy. Shared savings may be conceptually sound as justification for determining a direct financial incentive or program cost markup. This is about as far as this approach is likely to go, because, for the most part, utility-sponsored energy efficiency seems more focused on achieving the targeted levels and not on the more economically sensible objective of getting prices right and encouraging rational choices or decision making. Experience over the last few decades also supports the conclusion that this conceptual disconnect may be a primary reason why utility-sponsored energy efficiency programs often fail to be sustained when the initial energy, environmental, or economic crisis wanes and consumers/politicians move on to a new concern or interest.

Performance-based regulation (PBR) has already taken hold in various jurisdictions. One fairly common approach uses external and internal factors to determine how well a regulated company is performing relative to others and the utility's prior "self." Good performance means a higher authorized return on rate base, or even wires' charges. Failing to achieve would mean lower authorized earnings or at least no upside, and perhaps delays in seeking rate relief.

Jurisdictions with PBR, regardless of any degree of industry restructuring, are prime candidates for using return on equity (ROE) adjustments to encourage energy efficiency. Utilities could earn more, when they meet or beat targeted energy efficiency savings (incremental and/or cumulative) or spending levels with some *ex post* review. Falling short of regulatory expectations would mean earnings would be set the lower end of the just and reasonable range of ROE.

The potential regulatory problems associated with using PBR are relatively minor. It is likely that other factors would affect utility earnings and performance, including climate and the economy. Regulators would need to determine whether to blur the effect of all such factors or to focus on specific elements. The latter would help to determine the effectiveness and success level of energy efficiency programs. Like so many regulatory matters, this could cause more regulatory involvement in the details of energy efficiency. This has not always worked well in practice.

The conceptually preferred and likely more sustainable approach would be to design and implement a new energy efficiency business. This business would be expected to package, market, and sell energy efficiency for a reasonable (competitive) profit. In theory, there would be no need for subsidies and no role for bossy, heavy-handed command and control regulation. Consumers could make rational choices and the market would determine winners/losers, and more.

As a conceptual matter, there also would be no particular role or need for a regulated energy company to be in such a business and, in effect, be forced to compete with itself. This is where reality and public need smack down theory and concepts. There is a compromise available whereby regulators design new energy efficiency products and tariffs coupled with a new regulatory paradigm.

Somewhere in this mix is the need for financial incentives that would make regulated utilities at least indifferent to selling traditional energy or energy efficiency services. This requires a modicum of unbundling and carefully separating the joint and common utility and social costs of both energy and efficiency. A better outcome would be to create incentives and tariffs that would promote energy efficiency over energy, because the latter often has negative external costs associated with its supply and consumption.

Management fees or mark-ups tied to performance or expenditures are sometimes components of the incentive approaches just discussed. Simplicity sometimes works rather well. A rather straightforward regulatory approach would be to direct the utility to do specific things. If it does them, it would be permitted to raise retail prices through a retail rate rider or flow-through charge mechanism. If it fails to do things, there could be credits or reductions in such flow-through charges or rate riders.

This form of energy efficiency regulation is direct and simple. One downside is that the utility becomes a rather neutral player in the energy efficiency game. It functions like a bank that collects fees when it gets a customer. If there are no takers, it collects less. This approach can either lead regulators to take control or to invite third parties to do the heavy lifting. Utilities that accept such a passive role cannot expect their increased income to be significant or to last very long.

Utilities that do the hard work and design programs that work are more likely to be in the energy efficiency business for the long-term. This is especially likely if they have a reasonable opportunity to earn direct financial incentives and to avoid hidden or specific regulatory penalties for doing so.

Table 4-1 summarizes some of the ways regulators can attach profit incentives to energy efficiency. In some cases, monitoring and verification may also be necessary.

TABLE 4-1
SOME REGULATORY APPROACHES TO ADD PROFITS AND INCENTIVES FOR ENERGY EFFICIENCY

i. Replace rate base earnings with retail sales margins for energy efficiency services, particularly in retail choice states.

ii. Add virtual rate base that replaces a fraction of avoided supply-side costs.

iii. Share the cost savings between customers and shareholders equal to the difference between supply-side and energy efficiency costs (program benefits).

iv. Adjust ROE and/or net utility income based on the utility achieving targets for energy efficiency.

v. Unbundle supply-side energy and energy efficiency and either sell new energy services with a cost markup or permit customers to sell energy efficiency to the utility.

vi. Adopt performance-based "management fees" based on a percentage of total program costs and performance.

It is also important for regulators to pay considerable attention to the "just and reasonableness" of the financial incentives they place on energy efficiency. This means that regulators need to consider the difficulty that "the market" seemingly has, when it comes to inducing retail customers to become more efficient. High energy prices do not seem to affect all consumers, nor do they induce the socially optimal level of energy efficiency expenditures. This is no small matter. If utilities can overcome such stiff customer indifference, they should receive rather generous rewards.

Accordingly, the initial energy efficiency retail margins/incentives should be rather generous to prove the concept of energy efficiency works. This regulatory inducement is particularly necessary to gain support during the initial trial period of renewed or expanded utility-sponsored energy efficiency. Just as California did after some prolonged period of success, it may be possible for other states to reduce the level of these incentives over time. Perhaps an even better approach than reducing incentives, as California is now doing after a generous high-incentive trial period, is to introduce a monitoring and verification program to ensure that the programs that customers, particularly nonparticipants, are helping to finance are working and that the estimated benefits are realized. As long as benefits exceed costs, regulators should not deny the "cheese" of financial incentives for utilities that find the means of encouraging consumers to do more than high energy prices have done.

A serious national effort to do good should not be constructed on an "it's too good to make a profit" premise. Energy efficiency should reasonably be self-directed. The fact that high prices and public support do not work means more needs to be done. Accordingly, regulators should begin their renewed efforts to expand utility-sponsored energy efficiency with generous incentives and, over time, introduce sensible monitoring and verification. These are mostly high-level reasons for regulators to act. The next section puts some detail into the regulatory debate that would likely follow a decision to do "something." Subsequent chapters test the significance of different regulations and add some detail to these concepts.

SECTION II: WHAT HAS BEEN HAPPENING AROUND THE NATION

After discussing history and basic regulatory concepts, the next step is to assess what utilities and states have achieved. This section reviews the period from 1992 to 2006. This is when the federal government collected fairly consistent data for energy efficiency (conservation) and load management. This discussion reviews both individual utilities and specific state achievements.

CHAPTER 5
A REVIEW OF UTILITY-SPONSORED ENERGY EFFICIENCY AND LOAD MANAGEMENT ACTIVITIES

This chapter reviews what various utilities have accomplished in terms of energy efficiency and load management. There are significant differences across jurisdictions and over time. Today, many propose an expanded role for energy efficiency. Before this can be done, it is important to determine what has been done.

This chapter describes what has been accomplished. The differences over time and across jurisdictions are startling. The most complete and consistent picture of utility activities is provided by the Energy Information Administration (EIA) at the Department of Energy (DOE). The EIA collects annual data from the nation's electric utilities for demand-side management (DSM),[1] which EIA separates into two broad categories: energy efficiency (EE) and load management (LM). There are some differences in EIA's cost data over the fifteen years, 1992 to 2006, that this data has published. Regardless, the "savings" data were consistently collated and reported.

The EIA utilizes two different conceptual definitions related to the megawatt-hours (MWhs) of energy and megawatts (MW) of capacity saved. One is called "incremental" and represents the annual addition of new savings from either new customers added to existing EE or LM programs or new EE or LM programs introduced in a calendar year. The other is called "annual" and represents the cumulative savings from all current and past programs that save energy or capacity in the reported calendar year.

Recent State Activities

Energy and capacity savings need to be normalized before comparing differences across utilities. Obviously, larger utilities, other things being equal, would spend and/or save more. Therefore, rankings that do not adjust for such differences would be misleading. The percent a utility saves in terms of the total annual MWh volume sold is a relatively good measure to compare utilities to each other, as well as over time. The percent of peak load or MW demand a utility defers is also a reasonable basis for comparing load management savings across utilities and over time.

---

[1] Some use the DSM term as an umbrella for utility-sponsored conservation, energy efficiency, and load management. Others use energy efficiency as the umbrella.

Table 5-1 shows the incremental MWhs saved stated as a percent of retail MWh sales in 2004 through 2006 for the 100 largest utilities determined in terms of 2006 revenue privately, *i.e.*, investor-owned electric utilities (IOUs) in the U.S. These are shown for the three most recent years the data was available. After the 44th entry, the remaining IOUs had either minimal or no active programs in these years.

TABLE 5-1
INCREMENTAL MWhs SAVED FROM BOTH EE & LM / TOTAL MWhs
(RANKED BY 2006 %)

Rank	Utility Name	2004	2005	2006
1	United Illuminating Co	2.86%	1.33%	1.29%
2	Narragansett Electric Co	0.71%	0.83%	1.24%
3	Massachusetts Electric Co	1.12%	0.90%	1.18%
4	Connecticut Light & Power Co	0.91%	0.95%	1.12%
5	Western Massachusetts Electric Co	1.07%	0.97%	1.09%
6	Pacific Gas & Electric Co	0.71%	1.45%	0.92%
7	Southern California Edison Co	1.13%	1.44%	0.89%
8	Interstate Power & Light Co	0.66%	0.75%	0.84%
9	Puget Sound Energy Inc	0.87%	0.76%	0.72%
10	Sierra Pacific Power Co	0.15%	0.16%	0.71%
11	Northern States Power Co MN	0.76%	0.73%	0.71%
12	Nevada Power Co	0.20%	0.34%	0.68%
13	MidAmerican Energy Co	0.52%	0.51%	0.67%
14	Public Service Co of NH	0.57%	0.70%	0.64%
15	Wisconsin Power & Light Co	0.67%	0.57%	0.63%
16	Avista Corp	0.42%	0.66%	0.53%
17	Idaho Power Co	0.06%	0.31%	0.51%
18	Hawaiian Electric Co Inc	0.00%	0.00%	0.49%
19	NorthWestern Energy LLC	0.30%	0.29%	0.39%
20	PacifiCorp	0.30%	0.30%	0.37%
21	Arizona Public Service Co	0.00%	0.00%	0.29%
22	Wisconsin Electric Power Co	0.00%	0.00%	0.20%
23	Florida Power & Light Co	0.16%	0.18%	0.19%
24	Public Service Co of Colorado	0.24%	0.38%	0.17%
25	Minnesota Power Inc	1.01%	1.51%	0.17%
26	Public Service Electric & Gas Co	0.01%	0.18%	0.16%
27	Jersey Central Power & Light Co	0.31%	0.26%	0.14%
28	Gulf Power Co	0.14%	0.20%	0.01%
29	Progress Energy Florida Inc	0.07%	0.08%	0.09%
30	Tampa Electric Co	0.17%	0.01%	0.09%

TABLE 5-1 (CONT'D)
INCREMENTAL MWhs SAVED FROM BOTH EE & LM / TOTAL MWhs
(RANKED BY 2006 %)

Rank	Utility Name	2004	2005	2006
31	Consolidated Edison Co-NY Inc	0.00%	0.01%	0.07%
32	Southwestern Public Service Co	0.01%	0.10%	0.06%
33	El Paso Electric Co	0.06%	0.00%	0.06%
34	Tucson Electric Power Co	0.01%	0.00%	0.06%
35	Duke Energy Indiana Inc	0.03%	0.03%	0.04%
36	Louisville Gas & Electric Co	0.07%	0.14%	0.03%
37	Southwestern Electric Power Co	0.02%	0.12%	0.02%
38	Kentucky Power Co	0.03%	0.02%	0.02%
39	Kentucky Utilities Co	0.09%	0.05%	0.02%
40	Mississippi Power Co	0.01%	0.01%	0.01%
41	Indianapolis Power & Light Co	0.00%	0.02%	0.01%
42	Pennsylvania Electric Co	0.01%	0.01%	0.01%
43	Central Maine Power Co	0.00%	0.01%	0.01%
44	Metropolitan Edison Co	0.01%	0.01%	0.01%
45	Progress Energy Carolinas Inc	0.00%	0.00%	0.00%
46	Duke Energy Ohio Inc	0.00%	0.00%	0.00%
47	Georgia Power Co	0.00%	0.00%	0.00%
48	Alabama Power Co	0.00%	0.00%	0.00%
49	Appalachian Power Co	0.00%	0.00%	0.00%
50	Aquila Inc	0.00%	0.00%	0.00%
51	Atlantic City Electric Co	0.00%	0.00%	0.00%
52	Baltimore Gas & Electric Co	0.00%	0.00%	0.00%
53	Boston Edison Co	0.00%	1.01%	0.00%
54	Central Hudson Gas & Electric Corp	0.00%	0.00%	0.00%
55	Central Illinois Public Service Co	0.00%	0.00%	0.00%
56	Cleco Power LLC	0.00%	0.00%	0.00%
57	Cleveland Electric Illuminating Co	0.00%	0.00%	0.00%
58	Columbus Southern Power Co	0.00%	0.00%	0.00%
59	Commonwealth Edison Co	0.00%	0.00%	0.00%
60	Commonwealth Electric Co	0.68%	0.73%	0.00%
61	Consumers Energy Co	0.00%	0.00%	0.00%
62	Dayton Power & Light Co	0.00%	0.00%	0.00%
63	Delmarva Power & Light Co	0.00%	0.00%	0.00%
64	Detroit Edison Co	0.00%	0.00%	0.00%
65	Duke Energy Carolinas, LLC	0.00%	0.00%	0.00%

(Continued)

TABLE 5-1 (CONT'D)
INCREMENTAL MWhs SAVED FROM BOTH EE & LM / TOTAL MWhs
(RANKED BY 2006 %)

Rank	Utility Name	2004	2005	2006
66	Duquesne Light Co	0.00%	0.00%	0.00%
67	Entergy Arkansas Inc	0.00%	0.00%	0.00%
68	Entergy Gulf States Inc	0.00%	0.00%	0.00%
69	Entergy Louisiana Inc	0.00%	0.00%	0.00%
70	Entergy Mississippi Inc	0.00%	0.00%	0.00%
71	Entergy New Orleans Inc	0.00%	0.00%	0.00%
72	Illinois Power Co	0.00%	0.00%	0.00%
73	Indiana Michigan Power Co	0.00%	0.00%	0.00%
74	Kansas City Power & Light Co	0.00%	0.00%	0.00%
75	Kansas Gas & Electric Co	0.00%	0.00%	0.00%
76	Monongahela Power Co	0.00%	0.00%	0.00%
77	New York State Electric & Gas Corp	0.00%	0.00%	0.00%
78	Niagara Mohawk Power Corp	0.00%	0.00%	0.00%
79	Northern Indiana Public Service Co	0.00%	0.00%	0.00%
80	Northern States Power Co WI	0.00%	0.00%	0.00%
81	Ohio Edison Co	0.00%	0.00%	0.00%
82	Ohio Power Co	0.00%	0.00%	0.00%
83	Oklahoma Gas & Electric Co	0.00%	0.00%	0.00%
84	Orange & Rockland Utilities Inc	0.00%	0.00%	0.00%
85	PECO Energy Co	0.00%	0.00%	0.00%
86	Portland General Electric Co	0.00%	0.00%	0.00%
87	Potomac Electric Power Co	0.00%	0.00%	0.00%
88	PPL Electric Utilities Corp	0.00%	0.00%	0.00%
89	Public Service Co of NM	0.00%	0.00%	0.00%
90	Public Service Co of Oklahoma	0.00%	0.00%	0.00%
91	Rochester Gas & Electric Corp	0.00%	0.00%	0.00%
92	San Diego Gas & Electric Co	0.00%	0.00%	0.00%
93	South Carolina Electric & Gas Co	0.00%	0.00%	0.00%
94	The Potomac Edison Co	0.00%	0.00%	0.00%
95	Toledo Edison Co	0.00%	0.00%	0.00%
96	Union Electric Co	0.00%	0.00%	0.00%
97	Virginia Electric & Power Co	0.00%	0.00%	0.00%
98	West Penn Power Co	0.00%	0.00%	0.00%
99	Westar Energy Inc	0.00%	0.00%	0.00%
100	Wisconsin Public Service Corp	0.00%	0.00%	0.00%

Chapter 5: A Review of Utility-Sponsored Energy Efficiency and Load Management Activities

The top twenty IOUs incrementally saved at least .37 percent of their annual volume sold in 2006 based on either new programs or added participants to pre-existing ones. The seven highest achieving IOUs saved at least one percent incrementally in one or more of the most recent three years. Indeed, one percent incremental MWh savings each year seems to be a good candidate to become an aggressive, but still achievable, target for utility programs to strive to achieve each year after they reach a reasonable maturation point. The majority of the highest twenty performing utilities in terms of incremental energy savings are located in states that typically have high retail prices. These are also very often states that have restructured, both in California and New England. Often these utilities had a political process that required public goods or benefits charges as a *quid pro quo* for stranded cost recovery and/or perhaps other concessions. For example, two of the highest performing utilities are in California: Pacific Gas & Electric and Southern California Edison.

There are some notable exceptions to these initial observations. These are Northern States Power, primarily serving consumers in Minnesota, and Puget Sound Energy in the state of Washington. Both of these utilities have achieved incremental savings of a least .71 percent in each of the last three years. Neither is considered a high price jurisdiction. There are also other midwestern and west coast states that have performed in the top 15 without necessarily having either high retail prices or experiencing a restructuring process.

Table 5-2 shows the same type of data and rankings for cumulative "annual" energy savings expressed as percentages of retail sales in 2004 to 2006.

TABLE 5-2
CUMULATIVE MWhs SAVED FROM EE & LM / TOTAL MWhs
(RANKED BY 2006 %)

Rank	Utility Name	2004	2005	2006
1	United Illuminating Co	11.95%	11.35%	13.00%
2	Western Massachusetts Electric Co	10.86%	11.19%	12.31%
3	Northern States Power Co MN	10.33%	10.62%	11.25%
4	Avista Corp	9.93%	10.40%	10.64%
5	Minnesota Power Inc	8.26%	9.86%	10.25%
6	Southern California Edison Co	9.76%	10.32%	10.18%
7	Massachusetts Electric Co	8.30%	8.94%	9.95%
8	Connecticut Light & Power Co	8.03%	8.52%	9.42%
9	Puget Sound Energy Inc	10.08%	9.27%	8.77%
10	Northern States Power Co WI	9.07%	8.73%	8.69%
11	Wisconsin Power & Light Co	10.24%	9.15%	8.47%

(Continued)

TABLE 5-2 (CONT'D)
CUMULATIVE MWhs SAVED FROM EE & LM / TOTAL MWhs
(RANKED BY 2006 %)

Rank	Utility Name	2004	2005	2006
12	Pacific Gas & Electric Co	5.83%	7.61%	8.36%
13	Narragansett Electric Co	6.29%	8.51%	8.22%
14	Interstate Power & Light Co	6.60%	7.01%	7.93%
15	Potomac Electric Power Co	6.65%	6.49%	6.76%
16	Gulf Power Co	5.32%	5.43%	5.44%
17	NorthWestern Energy LLC	4.41%	4.54%	4.85%
18	PacifiCorp	3.58%	4.13%	4.23%
19	Public Service Co of NH	2.80%	2.81%	4.09%
20	Hawaiian Electric Co Inc	0.41%	0.45%	3.90%
21	Florida Power & Light Co	3.43%	3.59%	3.73%
22	MidAmerican Energy Co	2.52%	2.93%	3.39%
23	Tampa Electric Co	3.13%	3.15%	3.15%
24	Wisconsin Electric Power Co	3.08%	2.63%	2.99%
25	Public Service Co of Colorado	2.55%	2.86%	2.96%
26	Jersey Central Power & Light Co	2.51%	2.61%	2.85%
27	Progress Energy Florida Inc	2.41%	2.43%	2.67%
28	Idaho Power Co	1.79%	2.09%	2.47%
29	Duke Energy Indiana Inc	2.27%	2.20%	2.26%
30	Baltimore Gas & Electric Co	1.97%	1.92%	1.99%
31	Public Service Electric & Gas Co	1.93%	2.58%	1.98%
32	Nevada Power Co	0.20%	0.87%	1.50%
33	Tucson Electric Power Co	1.16%	1.14%	1.38%
34	Duke Energy Ohio Inc	1.21%	1.18%	1.22%
35	Sierra Pacific Power Co	0.26%	0.41%	1.01%
36	Southwestern Public Service Co	0.80%	0.88%	0.92%
37	Kentucky Power Co	0.74%	0.72%	0.76%
38	San Diego Gas & Electric Co	1.06%	1.88%	0.63%
39	Southwestern Electric Power Co	0.48%	0.59%	0.61%
40	Indianapolis Power & Light Co	0.54%	0.53%	0.56%
41	Georgia Power Co	0.33%	0.33%	0.32%
42	Columbus Southern Power Co	0.31%	0.30%	0.28%
43	El Paso Electric Co	0.22%	0.22%	0.27%
44	Virginia Electric & Power Co	0.22%	0.21%	0.22%
45	Ohio Power Co	0.20%	0.19%	0.21%
46	Mississippi Power Co	0.19%	0.22%	0.21%
47	Indiana Michigan Power Co	0.13%	0.13%	0.13%

TABLE 5-2 (CONT'D)
CUMULATIVE MWhs SAVED FROM EE & LM / TOTAL MWhs
(RANKED BY 2006 %)

Rank	Utility Name	2004	2005	2006
48	Louisville Gas & Electric Co	0.37%	0.45%	0.13%
49	Consolidated Edison Co-NY Inc	0.00%	0.01%	0.07%
50	Pennsylvania Electric Co	0.06%	0.06%	0.06%
51	Entergy Gulf States Inc	0.04%	0.04%	0.05%
52	Kentucky Utilities Co	0.01%	0.16%	0.05%
53	Metropolitan Edison Co	0.05%	0.04%	0.04%
54	Progress Energy Carolinas Inc	0.00%	0.00%	0.04%
55	Kansas City Power & Light Co	0.00%	0.00%	0.03%
56	South Carolina Electric & Gas Co	0.00%	0.00%	0.00%
57	Aquila Inc	0.00%	0.00%	0.00%
58	Oklahoma Gas & Electric Co	0.38%	0.01%	0.00%
59	Appalachian Power Co	0.00%	0.00%	0.00%
60	Alabama Power Co	0.00%	0.00%	0.00%
61	Arizona Public Service Co	0.00%	0.00%	0.00%
62	Atlantic City Electric Co	0.00%	0.00%	0.00%
63	Boston Edison Co	0.00%	8.45%	0.00%
64	Central Hudson Gas & Electric Corp	0.00%	0.00%	0.00%
65	Central Illinois Public Service Co	0.00%	0.00%	0.00%
66	Central Maine Power Co	0.09%	0.01%	0.00%
67	Cleco Power LLC	0.00%	0.00%	0.00%
68	Cleveland Electric Illuminating Co	0.00%	0.00%	0.00%
69	Commonwealth Edison Co	0.00%	0.00%	0.00%
70	Commonwealth Electric Co	4.94%	5.53%	0.00%
71	Consumers Energy Co	0.00%	0.00%	0.00%
72	Dayton Power & Light Co	0.00%	0.00%	0.00%
73	Delmarva Power & Light Co	0.00%	0.00%	0.00%
74	Detroit Edison Co	0.00%	0.00%	0.00%
75	Duke Energy Carolinas, LLC	0.00%	0.00%	0.00%
76	Duquesne Light Co	0.00%	0.00%	0.00%
77	Entergy Arkansas Inc	0.00%	0.00%	0.00%
78	Entergy Louisiana Inc	0.00%	0.00%	0.00%
79	Entergy Mississippi Inc	0.00%	0.00%	0.00%
80	Entergy New Orleans Inc	0.00%	0.00%	0.00%
81	Illinois Power Co	0.00%	0.00%	0.00%
82	Kansas Gas & Electric Co	0.00%	0.00%	0.00%

(Continued)

**TABLE 5-2 (CONT'D)
CUMULATIVE MWhs SAVED FROM EE & LM / TOTAL MWhs
(RANKED BY 2006 %)**

Rank	Utility Name	2004	2005	2006
83	Monongahela Power Co	0.00%	0.00%	0.00%
84	New York State Electric & Gas Corp	0.00%	0.00%	0.00%
85	Niagara Mohawk Power Corp	0.00%	0.00%	0.00%
86	Northern Indiana Public Service Co	0.00%	0.00%	0.00%
87	Ohio Edison Co	0.00%	0.00%	0.00%
88	Orange & Rockland Utilities Inc	0.00%	0.00%	0.00%
89	PECO Energy Co	0.00%	0.00%	0.00%
90	Portland General Electric Co	0.00%	0.00%	0.00%
91	PPL Electric Utilities Corp	0.00%	0.00%	0.00%
92	Public Service Co of NM	0.00%	0.00%	0.00%
93	Public Service Co of Oklahoma	0.00%	0.00%	0.00%
94	Rochester Gas & Electric Corp	0.00%	0.00%	0.00%
95	The Potomac Edison Co	0.00%	0.00%	0.00%
96	Toledo Edison Co	0.00%	0.00%	0.00%
97	Union Electric Co	0.00%	0.00%	0.00%
98	West Penn Power Co	0.00%	0.00%	0.00%
99	Westar Energy Inc	0.00%	0.00%	0.00%
100	Wisconsin Public Service Corp	0.00%	0.00%	0.00%

There are six IOUs that exceeded ten percent cumulative "annual" savings and 15 that exceed five percent annual savings in 2006.

A cumulative MWh percentage savings target of at least 10 percent of the annual volume of MWhs sold seems to be reasonably achievable. That said, since 2006, there has been increased political support for expanding energy efficiency programs. These new levels of support, coupled with the fact that seven of the top 100 IOUs in 2006 have virtually achieved or exceeded 10 percent, would suggest that targets as high as 15 percent might be achievable.

In addition, many states, provinces, and the U.S. Congress are debating the merits of aggressive renewable portfolio standards (RPS). California, for example, is moving towards a 20 percent RPS and voters are considering a proposition that would add two percent per year to the state's RPS target in order to hit 50 percent in 2025. This suggests some value in combining demand-side and renewables targets. Indeed, several jurisdictions are planning to replace all growth and to add some significant retirements of existing generation stations, particularly fossil-fuel fired generating stations. Both energy efficiency and renewables would seem necessary to do so.

The EIA data can also be used to compare and rank the top 100 IOUs in terms of the percent of capacity savings (MW) relative to their peak demand. There are two types of capacity savings and both are relevant. However, the capacity savings designated "actual" is gaining greater currency than the capacity savings designated "potential." The latter is mostly based on interruptible service programs, where the IOU can interrupt service under jointly agreed upon preset conditions. This type of LM tariff option has historically helped IOUs and their customers avoid capacity additions and long-term contracts, while still achieving reliable service. In effect, demand-side LM reduces the amount of supply-side reserves an IOU must hold. This saves consumers money.

There are some emerging conflicts between these worthy and cost-effective interruptible tariffs and the broader goals of energy efficiency. The primary potential conflict may occur when an IOU determines that it can purchase energy in competitive wholesale markets to sell to retail consumers under favorable economic terms. If these off-system purchases make economic sense, there would be little or no reason to interrupt any customer's service.

In effect, interruptible service often becomes a utility option or insurance approach. It likely does not reduce energy use. In fact, it may increase purchases for resale and these could cause externalities and less efficient operation. This does *not* mean that such tariff options serve no useful reliability purpose. To the contrary, they almost always do serve such a reliability purpose.

With technological progress, a new form of load management that is more consistent with the broader goals of energy efficiency has emerged. This is called "actual" LM. This category of capacity saving is more like a dispatchable supply-side resource. The newer technology permits the IOU to control customer premises equipment, often without the customer even being aware that the utility is shaving peak consumption, saving capacity, and displacing often very inefficient (*i.e.*, high heat rate) fossil fuel-fired generation.

Table 5-3 shows the "potential" savings percentages for incremental load management for the last three years for which the data was available. This table does not include estimated capacity savings related to energy efficiency or conservation, such as improved insulation or improved equipment efficiency.

TABLE 5-3
INCREMENTAL kW SAVED FROM LM (Potential) / PEAK kW
(RANKED BY 2006 %)

Rank	Utility Name	2004	2005	2006
1	PacifiCorp	0.00%	3.36%	3.83%
2	United Illuminating Co	1.42%	1.19%	3.71%
3	Connecticut Light & Power Co	0.62%	0.67%	3.28%
4	Alabama Power Co	0.79%	0.26%	2.46%
5	Duke Energy Indiana Inc	1.54%	3.37%	1.43%
6	Consolidated Edison Co-NY Inc	1.46%	1.25%	1.20%
7	Northern States Power Co WI	0.75%	1.46%	0.94%
8	Pacific Gas & Electric Co	3.87%	0.30%	0.87%
9	Hawaiian Electric Co Inc	0.00%	0.00%	0.71%
10	Indiana Michigan Power Co	0.84%	2.48%	0.69%
11	Appalachian Power Co	0.64%	0.00%	0.68%
12	Northern States Power Co MN	0.90%	0.72%	0.60%
13	Wisconsin Electric Power Co	0.62%	0.08%	0.39%
14	Louisville Gas & Electric Co	0.72%	0.36%	0.37%
15	Gulf Power Co	0.21%	0.12%	0.32%
16	Minnesota Power Inc	0.13%	0.06%	0.32%
17	Wisconsin Power & Light Co	0.00%	0.39%	0.31%
18	Public Service Co of Colorado	0.36%	0.25%	0.27%
19	Kentucky Utilities Co	0.21%	0.25%	0.26%
20	Florida Power & Light Co	0.06%	0.08%	0.24%
21	Idaho Power Co	0.21%	1.45%	0.23%
22	Progress Energy Florida Inc	0.08%	0.09%	0.18%
23	Nevada Power Co	0.50%	0.14%	0.14%
24	MidAmerican Energy Co	0.54%	0.27%	0.10%
25	Indianapolis Power & Light Co	-0.17%	0.67%	0.01%
26	Duke Energy Ohio Inc	0.00%	0.61%	0.05%
27	Georgia Power Co	0.11%	0.06%	0.03%
28	Tampa Electric Co	2.06%	2.39%	0.02%
29	Virginia Electric & Power Co	0.02%	0.01%	0.02%
30	Aquila Inc	0.00%	0.00%	0.00%
31	Arizona Public Service Co	0.00%	0.00%	0.00%
32	Atlantic City Electric Co	0.00%	0.00%	0.00%
33	Avista Corp	0.00%	0.00%	0.00%
34	Baltimore Gas & Electric Co	0.00%	0.00%	0.00%
35	Boston Edison Co	0.00%	0.00%	0.00%
36	Central Hudson Gas & Electric Corp	0.00%	0.00%	0.00%

TABLE 5-3 (CONT'D)
INCREMENTAL kW SAVED FROM LM (Potential) / PEAK kW
(RANKED BY 2006 %)

Rank	Utility Name	2004	2005	2006
37	Central Illinois Public Service Co		0.00%	0.00%
38	Central Maine Power Co	0.00%	0.00%	0.00%
39	Cleco Power LLC	0.00%	0.00%	0.00%
40	Cleveland Electric Illuminating Co	0.00%	0.00%	0.00%
41	Columbus Southern Power Co	0.22%	0.00%	0.00%
42	Commonwealth Edison Co	0.00%	0.00%	0.00%
43	Commonwealth Electric Co	0.00%	0.00%	0.00%
44	Consumers Electric Co	0.00%	0.00%	0.00%
45	Dayton Power & Light Co	0.00%	0.00%	0.00%
46	Delmarva Power & Light Co	0.00%	0.00%	0.00%
47	Detroit Edison Co	0.00%	0.00%	0.00%
48	Duke Energy Carolinas, LLC	0.00%	0.00%	0.00%
49	Duquesne Light Co	0.00%	0.00%	0.00%
50	El Paso Electric Co	0.00%	0.00%	0.00%
51	Entergy Arkansas Inc	0.00%	0.00%	0.00%
52	Entergy Gulf States Inc	0.00%	0.00%	0.00%
53	Entergy Louisiana Inc	0.00%	0.00%	0.00%
54	Entergy Mississippi Inc	0.00%	0.00%	0.00%
55	Entergy New Orleans Inc	0.00%	0.00%	0.00%
56	Illinois Power Co	0.00%	0.00%	0.00%
57	Interstate Power & Light Co	0.00%	0.00%	0.00%
58	Jersey Central Power & Light Co	0.00%	0.00%	0.00%
59	Kansas City Power & Light Co	0.00%	0.00%	0.00%
60	Kansas Gas & Electric Co	0.00%	0.00%	0.00%
61	Kentucky Power Co	0.00%	0.00%	0.00%
62	Massachusetts Electric Co			0.00%
63	Metropolitan Edison Co	0.00%	0.00%	0.00%
64	Mississippi Power Co	0.00%	0.00%	0.00%
65	Monongahela Power Co	0.00%	0.00%	0.00%
66	Narragansett Electric Co			0.00%
67	New York State Electric & Gas Corp	0.00%	0.00%	0.00%
68	Niagara Mohawk Power Corp	0.00%	0.00%	0.00%
69	Northern Indiana Public Service Co	0.00%	0.00%	0.00%
70	NorthWestern Energy LLC	0.00%	0.00%	0.00%
71	Ohio Edison Co	0.00%	0.00%	0.00%

(Continued)

TABLE 5-3 (CONT'D)
INCREMENTAL kW SAVED FROM LM (Potential) / PEAK kW
(RANKED BY 2006 %)

Rank	Utility Name	2004	2005	2006
72	Ohio Power Co	1.94%	0.00%	0.00%
73	Oklahoma Gas & Electric Co	0.00%	0.00%	0.00%
74	Orange & Rockland Utilities Inc	0.00%	0.00%	0.00%
75	PECO Energy Co	0.00%	0.00%	0.00%
76	Pennsylvania Electric Co	0.00%	0.00%	0.00%
77	Portland General Electric Co	0.00%	0.00%	0.00%
78	Potomac Electric Power Co	0.00%	0.00%	0.00%
79	PPL Electric Utilities Corp	0.00%	0.00%	0.00%
80	Progress Energy Carolinas Inc	0.00%	0.00%	0.00%
81	Public Service Co of NH	0.00%	0.00%	0.00%
82	Public Service Co of NM	0.00%	0.00%	0.00%
83	Public Service Co of Oklahoma	0.00%	0.00%	0.00%
84	Public Service Electric & Gas Co	0.00%	0.00%	0.00%
85	Puget Sound Energy Inc	0.00%	0.05%	0.00%
86	Rochester Gas & Electric Corp	0.00%	0.00%	0.00%
87	San Diego Gas & Electric Co	0.00%	0.00%	0.00%
88	Sierra Pacific Power Co	0.00%	0.00%	0.00%
89	South Carolina Electric & Gas Co	0.28%	0.58%	0.00%
90	Southern California Edison Co	0.00%	0.00%	0.00%
91	Southwestern Electric Power Co	0.00%	0.00%	0.00%
92	Southwestern Public Service Co	0.00%	0.00%	0.00%
93	The Potomac Edison Co	0.00%	0.00%	0.00%
94	Toledo Edison Co	0.00%	0.00%	0.00%
95	Tucson Electric Power Co	0.00%	0.00%	0.00%
96	Union Electric Co	0.00%	0.00%	0.00%
97	West Penn Power Co	0.00%	0.00%	0.00%
98	Westar Energy Inc	0.00%	0.00%	0.00%
99	Western Massachusetts Electric Co	0.00%	0.00%	0.00%
100	Wisconsin Public Service Corp	0.00%	0.00%	0.00%

Table 5-4 shows these same incremental capacity savings for "actual" load management over the last three years for which the data was available, without including any capacity savings that energy efficiency savings achieve.

TABLE 5-4
INCREMENTAL kW SAVED FROM LM (Actual) / PEAK kW
(RANKED BY 2006 %)

Rank	Utility Name	2004	2005	2006
1	United Illuminating Co	1.25%	1.19%	3.98%
2	Alabama Power Co	0.79%	0.00%	2.48%
3	Ohio Power Co	1.94%	0.00%	2.11%
4	PacifiCorp	0.00%	0.00%	1.41%
5	Consolidated Edison Co-NY Inc	0.00%	0.32%	1.12%
6	Northern States Power Co WI	0.75%	0.15%	0.94%
7	Appalachian Power Co	0.40%	0.00%	0.92%
8	Indiana Michigan Power Co	0.84%	2.48%	0.69%
9	Northern States Power Co MN	0.90%	0.72%	0.60%
10	Louisville Gas & Electric Co	0.72%	0.36%	0.37%
11	Pacific Gas & Electric Co	0.00%	0.05%	0.34%
12	Gulf Power Co	0.21%	0.12%	0.32%
13	Wisconsin Power & Light Co	0.00%	0.39%	0.31%
14	Public Service Co of Colorado	0.36%	0.25%	0.27%
15	Kentucky Utilities Co	0.21%	0.25%	0.26%
16	Progress Energy Florida Inc	0.00%	0.00%	0.18%
17	Nevada Power Co	0.18%	0.14%	0.14%
18	Wisconsin Electric Power Co	0.00%	0.02%	0.13%
19	Idaho Power Co	0.21%	1.35%	0.01%
20	Indianapolis Power & Light Co	-2.61%	0.67%	0.01%
21	MidAmerican Energy Co	0.54%	0.27%	0.08%
22	Georgia Power Co	0.11%	0.06%	0.03%
23	Virginia Electric & Power Co	0.01%	0.01%	0.02%
24	Florida Power & Light Co	0.06%	0.05%	0.02%
25	Aquila Inc	0.00%	0.00%	0.00%
26	Arizona Public Service Co	0.00%	0.00%	0.00%
27	Atlantic City Electric Co	0.00%	0.00%	0.00%
28	Avista Corp	0.00%	0.00%	0.00%
29	Baltimore Gas & Electric Co	0.00%	0.00%	0.00%
30	Boston Edison Co	0.00%	0.00%	0.00%
31	Central Hudson Gas & Electric Corp	0.00%	0.00%	0.00%
32	Central Illinois Public Service Co		0.00%	0.00%
33	Central Maine Power Co	0.00%	0.00%	0.00%
34	Cleco Power LLC	0.00%	0.00%	0.00%
35	Cleveland Electric Illuminating Co	0.00%	0.00%	0.00%

(Continued)

TABLE 5-4 (CONT'D)
INCREMENTAL kW SAVED FROM LM (Actual) / PEAK kW
(RANKED BY 2006 %)

Rank	Utility Name	2004	2005	2006
36	Columbus Southern Power Co	0.19%	0.00%	0.00%
37	Commonwealth Edison Co	0.00%	0.00%	0.00%
38	Commonwealth Electric Co	0.00%	0.00%	0.00%
39	Connecticut Light & Power Co	0.00%	0.00%	0.00%
40	Consumers Energy Co	0.00%	0.00%	0.00%
41	Dayton Power & Light Co	0.00%	0.00%	0.00%
42	Delmarva Power & Light Co	0.00%	0.00%	0.00%
43	Detroit Edison Co	0.00%	0.00%	0.00%
44	Duke Energy Carolinas, LLC	0.00%	0.00%	0.00%
45	Duke Energy Indiana Inc	0.87%	3.31%	0.00%
46	Duke Energy Ohio Inc	0.00%	0.61%	0.00%
47	Duquesne Light Co	0.00%	0.00%	0.00%
48	El Paso Electric Co	0.00%	0.00%	0.00%
49	Entergy Arkansas Inc	0.00%	0.00%	0.00%
50	Entergy Gulf States Inc	0.00%	0.00%	0.00%
51	Entergy Louisiana Inc	0.00%	0.00%	0.00%
52	Entergy Mississippi Inc	0.00%	0.00%	0.00%
53	Entergy New Orleans Inc	0.00%	0.00%	0.00%
54	Hawaiian Electric Co Inc	0.00%	0.00%	0.00%
55	Illinois Power Co	0.00%	0.00%	0.00%
56	Interstate Power & Light Co	0.00%	0.00%	0.00%
57	Jersey Central Power & Light Co	0.00%	0.00%	0.00%
58	Kansas City Power & Light Co	0.00%	0.00%	0.00%
59	Kansas Gas & Electric Co	0.00%	0.00%	0.00%
60	Kentucky Power Co	0.00%	0.00%	0.00%
61	Massachusetts Electric Co			0.00%
62	Metropolitan Edison Co	0.00%	0.00%	0.00%
63	Minnesota Power Inc	0.00%	0.00%	0.00%
64	Mississippi Power Co	0.00%	0.00%	0.00%
65	Monongahela Power Co	0.00%	0.00%	0.00%
66	Narragansett Electric Co			0.00%
67	New York State Electric & Gas Corp	0.00%	0.00%	0.00%
68	Niagara Mohawk Power Corp	0.00%	0.00%	0.00%
69	Northern Indiana Public Service Co	0.00%	0.00%	0.00%
70	NorthWestern Energy LLC	0.00%	0.00%	0.00%
71	Ohio Edison Co	0.00%	0.00%	0.00%

Chapter 5: A Review of Utility-Sponsored Energy Efficiency and Load Management Activities

TABLE 5-4 (CONT'D)
INCREMENTAL kW SAVED FROM LM (Actual) / PEAK kW
(RANKED BY 2006 %)

Rank	Utility Name	2004	2005	2006
72	Oklahoma Gas & Electric Co	0.00%	0.00%	0.00%
73	Orange & Rockland Utilities Inc	0.00%	0.00%	0.00%
74	PECO Energy Co	0.00%	0.00%	0.00%
75	Pennsylvania Electric Co	0.00%	0.00%	0.00%
76	Portland General Electric Co	0.00%	0.00%	0.00%
77	Potomac Electric Power Co	0.00%	0.00%	0.00%
78	PPL Electric Utilities Corp	0.00%	0.00%	0.00%
79	Progress Energy Carolinas Inc	0.00%	0.00%	0.00%
80	Public Service Co of NH	0.00%	0.00%	0.00%
81	Public Service Co of NM	0.00%	0.00%	0.00%
82	Public Service Co of Oklahoma	0.00%	0.00%	0.00%
83	Public Service Electric & Gas Co	0.00%	0.00%	0.00%
84	Puget Sound Energy Inc	0.00%	0.00%	0.00%
85	Rochester Gas & Electric Corp	0.00%	0.00%	0.00%
86	San Diego Gas & Electric Co	0.00%	0.00%	0.00%
87	Sierra Pacific Power Co	0.00%	0.00%	0.00%
88	South Carolina Electric & Gas Co	0.00%	0.00%	0.00%
89	Southern California Edison Co	0.00%	0.00%	0.00%
90	Southwestern Electric Power Co	0.00%	0.00%	0.00%
91	Southwestern Public Service Co	0.00%	0.00%	0.00%
92	Tampa Electric Co	0.00%	1.79%	0.00%
93	The Potomac Edison Co	0.00%	0.00%	0.00%
94	Toledo Edison Co	0.00%	0.00%	0.00%
95	Tucson Electric Power Co	0.00%	0.00%	0.00%
96	Union Electric Co	0.00%	0.00%	0.00%
97	West Penn Power Co	0.00%	0.00%	0.00%
98	Westar Energy Inc	0.00%	0.00%	0.00%
99	Western Massachusetts Electric Co	0.00%	0.00%	0.00%
100	Wisconsin Public Service Corp	0.00%	0.00%	0.00%

Both of these incremental capacity savings represent programs that IOUs and their regulators expect to use to reduce demand, as well as to improve reliability.

Table 5-5 adds the capacity savings when "actual" LM and energy efficiency-related capacity savings are combined for the 100 largest IOUs. These combined rankings show that IOUs performing in the top 10 can reasonably achieve at least an incremental savings of about one percent or more. This would represent a sizeable reduction in current IOU growth in much of North America. As with the one percent incremental target for incremental MWh savings, a goal of one percent in incremental MW capacity savings is well within reach for most of the nation's electric utility companies.

TABLE 5-5
INCREMENTAL kW SAVED FROM EE & LM (Actual) / PEAK kW
(RANKED BY 2006 %)

Rank	Utility Name	2004	2005	2006
1	United Illuminating Co	2.51%	2.30%	5.01%
2	Alabama Power Co	0.79%	0.00%	2.51%
3	Ohio Power Co	1.94%	0.00%	2.11%
4	PacifiCorp	0.00%	0.00%	1.41%
5	Consolidated Edison Co-NY Inc	0.00%	0.34%	1.22%
6	Northern States Power Co	1.58%	1.39%	1.18%
7	Pacific Gas & Electric Co	0.70%	1.30%	1.02%
8	Northern States Power Co	0.75%	0.15%	0.94%
9	Appalachian Power Co	0.40%	0.00%	0.92%
10	Connecticut Light & Power Co	0.87%	1.00%	0.89%
11	MidAmerican Energy Co	1.46%	1.02%	0.88%
12	Narragansett Electric Co			0.77%
13	Wisconsin Power & Light Co	0.49%	0.74%	0.71%
14	Western Massachusetts Electric Co	0.93%	0.73%	0.71%
15	Indiana Michigan Power Co	0.84%	2.48%	0.69%
16	Massachusetts Electric Co			0.65%
17	Progress Energy Florida Inc	0.31%	0.39%	0.60%
18	Nevada Power Co	0.46%	0.50%	0.59%
19	Southern California Edison Co	0.85%	1.09%	0.56%
20	Avista Corp	0.40%	0.60%	0.54%
21	Hawaiian Electric Co Inc	0.00%	0.00%	0.48%
22	Public Service Co of Colorado	0.65%	0.56%	0.46%
23	Gulf Power Co	0.41%	0.37%	0.40%
24	Louisville Gas & Electric Co	0.80%	0.54%	0.37%
25	Florida Power & Light Co	0.37%	0.38%	0.36%
26	Wisconsin Electric Power Co	0.00%	0.02%	0.35%
27	Kentucky Utilities Co	0.29%	0.32%	0.26%
28	Public Service Electric & Gas Co	0.05%	0.30%	0.26%
29	Minnesota Power Inc	0.40%	0.97%	0.25%

TABLE 5-5 (CONT'D)
INCREMENTAL kW SAVED FROM EE & LM (Actual) / PEAK kW
(RANKED BY 2006 %)

Rank	Utility Name	2004	2005	2006
30	NorthWestern Energy LLC	0.39%	0.56%	0.24%
31	Sierra Pacific Power Co	0.12%	0.29%	0.24%
32	Tucson Electric Power Co	0.01%	0.17%	0.21%
33	Jersey Central Power & Light Co	0.38%	0.40%	0.19%
34	Arizona Public Service Co	0.00%	0.00%	0.17%
35	Indianapolis Power & Light Co	-2.61%	0.71%	0.13%
36	Tampa Electric Co	10.49%	1.94%	0.12%
37	Southwestern Public Service Co	0.01%	0.13%	0.10%
38	Idaho Power Co	0.21%	1.35%	0.01%
39	Kentucky Power Co	0.12%	0.06%	0.06%
40	Duke Energy Indiana Inc	0.90%	3.33%	0.05%
41	Georgia Power Co	0.11%	0.06%	0.04%
42	Southwestern Electric Power Co	0.04%	0.08%	0.04%
43	Virginia Electric & Power Co	0.01%	0.01%	0.02%
44	Progress Energy Carolinas Inc	0.00%	0.00%	0.01%
45	Aquila Inc	0.00%	0.00%	0.00%
46	Atlantic City Electric Co	0.00%	0.00%	0.00%
47	Baltimore Gas & Electric Co	0.00%	0.00%	0.00%
48	Boston Edison Co	0.00%	0.64%	0.00%
49	Central Hudson Gas & Electric Corp	0.00%	0.00%	0.00%
50	Central Illinois Public Service Co		0.00%	0.00%
51	Central Maine Power Co	0.00%	0.00%	0.00%
52	Cleco Power LLC	0.00%	0.00%	0.00%
53	Cleveland Electric Illuminating Co	0.00%	0.00%	0.00%
54	Columbus Southern Power Co	0.19%	0.00%	0.00%
55	Commonwealth Edison Co	0.00%	0.00%	0.00%
56	Commonwealth Electric Co	0.57%	0.41%	0.00%
57	Consumers Energy Co	0.00%	0.00%	0.00%
58	Dayton Power & Light Co	0.00%	0.00%	0.00%
59	Delmarva Power & Light Co	0.00%	0.00%	0.00%
60	Detroit Edison Co	0.00%	0.00%	0.00%
61	Duke Energy Carolinas, LLC	0.00%	0.00%	0.00%
62	Duke Energy Ohio Inc	0.00%	0.61%	0.00%
63	Duquesne Light Co	0.00%	0.00%	0.00%
64	El Paso Electric Co	0.00%	0.00%	0.00%

(Continued)

TABLE 5-5 (CONT'D)
INCREMENTAL kW SAVED FROM EE & LM (Actual) / PEAK kW
(RANKED BY 2006 %)

Rank	Utility Name	2004	2005	2006
65	Entergy Arkansas Inc	0.00%	0.00%	0.00%
66	Entergy Gulf States Inc	0.00%	0.00%	0.00%
67	Entergy Louisiana Inc	0.00%	0.00%	0.00%
68	Entergy Mississippi Inc	0.00%	0.00%	0.00%
69	Entergy New Orleans Inc	0.00%	0.00%	0.00%
70	Illinois Power Co	0.00%	0.00%	0.00%
71	Kansas City Power & Light Co	0.00%	0.00%	0.00%
72	Kansas Gas & Electric Co	0.00%	0.00%	0.00%
73	Metropolitan Edison Co	0.00%	0.00%	0.00%
74	Mississippi Power Co	0.00%	0.00%	0.00%
75	Monongahela Power Co	0.00%	0.00%	0.00%
76	New York State Electric & Gas Corp	0.00%	0.00%	0.00%
77	Niagara Mohawk Power Corp	0.00%	0.00%	0.00%
78	Northern Indiana Public Service Co	0.00%	0.00%	0.00%
79	Ohio Edison Co	0.00%	0.00%	0.00%
80	Oklahoma Gas & Electric Co	0.00%	0.00%	0.00%
81	Orange & Rockland Utilities Inc	0.00%	0.00%	0.00%
82	PECO Energy Co	0.00%	0.00%	0.00%
83	Pennsylvania Electric Co	0.00%	0.00%	0.00%
84	Portland General Electric Co	0.00%	0.00%	0.00%
85	Potomac Electric Power Co	0.00%	0.00%	0.00%
86	PPL Electric Utilities Corp	0.00%	0.00%	0.00%
87	Public Service Co of NH	0.00%	0.00%	0.00%
88	Public Service Co of NM	0.00%	0.00%	0.00%
89	Public Service Co of Oklahoma	0.00%	0.00%	0.00%
90	Puget Sound Energy Inc	0.00%	0.00%	0.00%
91	Rochester Gas & Electric Corp	0.00%	0.00%	0.00%
92	San Diego Gas & Electric Co	0.00%	0.00%	0.00%
93	South Carolina Electric & Gas Co	0.00%	0.00%	0.00%
94	Southern Indiana Gas & Electric Co	0.00%	0.00%	0.00%
95	The Potomac Edison Co	0.00%	0.00%	0.00%
96	Toledo Edison Co	0.00%	0.00%	0.00%
97	Union Electric Co	0.00%	0.00%	0.00%
98	West Penn Power Co	0.00%	0.00%	0.00%
99	Westar Energy Inc	0.00%	0.00%	0.00%
100	Wisconsin Public Service Corp	0.00%	0.00%	0.00%

Chapter 5: A Review of Utility-Sponsored Energy Efficiency and Load Management Activities

Table 5-6 shows the cumulative potential capacity savings due to LM for the top 100 IOUs for the last three years for which the data was available. The top ten IOUs show that a potential target of 10 percent is achievable.

TABLE 5-6
CUMULATIVE kW SAVED FROM LM (Potential) / PEAK kW
(RANKED BY 2006 %)

Rank	Utility Name	2004	2005	2006
1	Northern States Power Co WI	14.94%	15.37%	14.54%
2	Progress Energy Florida Inc	7.74%	7.85%	11.99%
3	Alabama Power Co	12.14%	11.86%	11.56%
4	Northern States Power Co MN	9.31%	10.04%	10.05%
5	Interstate Power & Light Co	12.73%	11.37%	9.09%
6	Indiana Michigan Power Co	9.78%	9.44%	8.62%
7	Wisconsin Electric Power Co	7.24%	6.41%	8.05%
8	Duke Energy Indiana Inc	6.68%	5.47%	6.73%
9	Kansas Gas & Electric Co	6.62%	6.59%	6.26%
10	Southern California Edison Co	6.42%	7.11%	6.15%
11	Ohio Power Co	7.18%	6.37%	6.05%
12	Florida Power & Light Co	6.04%	5.63%	6.00%
13	Commonwealth Edison Co	6.73%	6.04%	5.48%
14	MidAmerican Energy Co	5.65%	5.13%	4.93%
15	South Carolina Electric & Gas Co	5.07%	5.39%	4.60%
16	Wisconsin Power & Light Co	4.99%	4.98%	4.52%
17	Pacific Gas & Electric Co	3.87%	4.00%	4.46%
18	Duke Energy Carolinas, LLC	4.85%	4.52%	4.16%
19	Public Service Co of Colorado	3.58%	3.58%	4.06%
20	Tampa Electric Co	4.12%	5.17%	4.01%
21	San Diego Gas & Electric Co	2.78%	12.00%	3.73%
22	Indianapolis Power & Light Co	3.33%	3.78%	3.69%
23	Oklahoma Gas & Electric Co	3.93%	3.69%	3.25%
24	Baltimore Gas & Electric Co	3.32%	3.27%	3.15%
25	Minnesota Power Inc	2.20%	2.27%	2.33%
26	Louisville Gas & Electric Co	1.61%	1.78%	2.27%
27	Central Illinois Public Service Co		12.37%	2.26%
28	Appalachian Power Co	2.09%	1.48%	2.13%
29	Georgia Power Co	2.85%	2.26%	2.11%
30	PECO Energy Co	2.31%	2.03%	1.96%
31	Westar Energy Inc	1.85%	0.00%	1.93%

(Continued)

TABLE 5-6 (CONT'D)
CUMULATIVE kW SAVED FROM LM (Potential) / PEAK kW
(RANKED BY 2006 %)

Rank	Utility Name	2004	2005	2006
32	Gulf Power Co	1.40%	1.52%	1.81%
33	Duke Energy Ohio Inc	2.53%	1.59%	1.54%
34	Idaho Power Co	0.25%	1.62%	1.49%
35	Metropolitan Edison Co	0.55%	1.65%	1.40%
36	Hawaiian Electric Co Inc	0.00%	0.49%	1.27%
37	Consolidated Edison Co-NY Inc	1.46%	1.25%	1.20%
38	Illinois Power Co	3.16%	1.74%	1.17%
39	Kentucky Utilities Co	0.69%	0.88%	1.16%
40	Public Service Electric & Gas Co	1.34%	1.16%	1.08%
41	Jersey Central Power & Light Co	0.97%	0.91%	0.84%
42	Kansas City Power & Light Co	0.38%	0.71%	0.81%
43	PPL Electric Utilities Corp	0.68%	0.68%	0.66%
44	Puget Sound Energy Inc	0.00%	0.57%	0.60%
45	Nevada Power Co	0.50%	0.40%	0.39%
46	Columbus Southern Power Co	0.55%	0.32%	0.29%
47	Pennsylvania Electric Co	0.39%	0.38%	0.26%
48	Virginia Electric & Power Co	0.25%	0.20%	0.17%
49	Duquesne Light Co	1.21%	1.14%	0.03%
50	Potomac Electric Power Co	0.54%	0.09%	0.01%
51	Aquila Inc	0.00%	0.00%	0.00%
52	Arizona Public Service Co	0.00%	0.00%	0.00%
53	Atlantic City Electric Co	0.00%	0.00%	0.00%
54	Avista Corp	0.00%	0.00%	0.00%
55	Boston Edison Co	0.00%	0.00%	0.00%
56	Central Hudson Gas & Electric Corp	0.00%	0.00%	0.00%
57	Central Maine Power Co	0.00%	0.00%	0.00%
58	Cleco Power LLC	0.00%	0.00%	0.00%
59	Cleveland Electric Illuminating Co	0.00%	0.00%	0.00%
60	Commonwealth Electric Co	0.00%	0.00%	0.00%
61	Connecticut Light & Power Co	0.62%	0.00%	0.00%
62	Consumers Energy Co	0.00%	0.00%	0.00%
63	Dayton Power & Light Co	0.00%	0.00%	0.00%
64	Delmarva Power & Light Co	0.00%	0.00%	0.00%
65	Detroit Edison Co	0.00%	0.00%	0.00%
66	El Paso Electric Co	0.00%	0.00%	0.00%
67	Entergy Arkansas Inc	0.00%	0.00%	0.00%

TABLE 5-6 (CONT'D)
CUMULATIVE kW SAVED FROM LM (Potential) / PEAK kW
(RANKED BY 2006 %)

Rank	Utility Name	2004	2005	2006
68	Entergy Gulf States Inc	0.00%	0.00%	0.00%
69	Entergy Louisiana Inc	0.00%	0.00%	0.00%
70	Entergy Mississippi Inc	0.00%	0.00%	0.00%
71	Entergy New Orleans Inc	0.00%	0.00%	0.00%
72	Kentucky Power Co	0.00%	0.00%	0.00%
73	Massachusetts Electric Co			0.00%
74	Mississippi Power Co	0.00%	0.00%	0.00%
75	Monongahela Power Co	0.00%	0.00%	0.00%
76	Narragansett Electric Co			0.00%
77	New York State Electric & Gas Corp	0.00%	0.00%	0.00%
78	Niagara Mohawk Power Corp	0.00%	0.00%	0.00%
79	Northern Indiana Public Service Co	0.00%	0.00%	0.00%
80	NorthWestern Energy LLC	0.00%	0.00%	0.00%
81	Ohio Edison Co	0.00%	0.00%	0.00%
82	Orange & Rockland Utilities Inc	0.00%	0.00%	0.00%
83	PacifiCorp	0.00%	0.00%	0.00%
84	Portland General Electric Co	0.00%	0.00%	0.00%
85	Progress Energy Carolinas Inc	0.00%	0.00%	0.00%
86	Public Service Co of NH	0.00%	0.00%	0.00%
87	Public Service Co of NM	0.00%	0.00%	0.00%
88	Public Service Co of Oklahoma	0.00%	0.00%	0.00%
89	Rochester Gas & Electric Corp	0.00%	0.00%	0.00%
90	Sierra Pacific Power Co	0.00%	0.00%	0.00%
91	Southwestern Electric Power Co	0.00%	0.00%	0.00%
92	Southwestern Public Service Co	0.08%	0.00%	0.00%
93	The Potomac Edison Co	0.00%	0.00%	0.00%
94	Toledo Edison Co	0.00%	0.00%	0.00%
95	Tucson Electric Power Co	0.00%	0.00%	0.00%
96	Union Electric Co	2.63%	1.71%	0.00%
97	United Illuminating Co	1.42%	2.45%	0.00%
98	West Penn Power Co	0.00%	0.00%	0.00%
99	Western Massachusetts Electric Co	0.00%	0.00%	0.00%
100	Wisconsin Public Service Corp	0.00%	0.00%	0.00%

Table 5-7 shows the actual cumulative capacity savings due to LM for the same top 100 IOUs for the same three years. These percentages are generally less, because newer technology just coming on-line is necessary. Sometimes, utilities report that actual capacity savings equal their potential capacity savings. There are two reasons for any such equivalence. First, actual would normally be a subset of potential. Therefore, IOUs with just load reductions controlled by the IOU would report the same savings in both categories. Second, while the definition is clear in the EIA survey data, some IOUs nevertheless might not always distinguish actual and potential LM.

TABLE 5-7
CUMULATIVE kW SAVED FROM LM (Actual) / PEAK kW
(RANKED BY 2006 %)

Rank	Utility Name	2004	2005	2006
1	Northern States Power Co WI	14.94%	14.07%	14.54%
2	Progress Energy Florida Inc	0.00%	0.00%	11.16%
3	Northern States Power Co MN	9.31%	10.04%	10.05%
4	Interstate Power & Light Co	12.73%	11.37%	9.09%
5	MidAmerican Energy Co	5.65%	5.13%	4.66%
6	Ohio Power Co	2.63%	2.23%	4.51%
7	Southern California Edison Co	5.45%	5.66%	4.50%
8	Wisconsin Power & Light Co	4.83%	4.84%	4.39%
9	Wisconsin Electric Power Co	3.80%	3.10%	4.14%
10	Public Service Co of Colorado	3.58%	3.58%	4.06%
11	Indiana Michigan Power Co	1.26%	3.70%	4.02%
12	Indianapolis Power & Light Co	0.89%	3.78%	3.69%
13	Pacific Gas & Electric Co	0.00%	3.08%	3.10%
14	Oklahoma Gas & Electric Co	0.33%	0.00%	2.59%
15	Alabama Power Co	0.88%	0.35%	2.46%
16	Louisville Gas & Electric Co	1.61%	1.78%	2.27%
17	Georgia Power Co	2.85%	2.26%	2.11%
18	PECO Energy Co	2.31%	2.03%	1.96%
19	Gulf Power Co	1.40%	1.52%	1.81%
20	Appalachian Power Co	0.58%	0.59%	1.50%
21	Metropolitan Edison Co	0.00%	1.58%	1.34%
22	Idaho Power Co	0.21%	1.55%	1.23%
23	Kentucky Utilities Co	0.69%	0.88%	1.16%
24	Consolidated Edison Co-NY Inc	0.00%	0.32%	1.12%
25	Public Service Electric & Gas Co	0.00%	1.16%	1.08%
26	San Diego Gas & Electric Co	0.96%	1.28%	0.96%
27	Illinois Power Co	0.00%	1.21%	0.89%
28	Jersey Central Power & Light Co	0.00%	0.91%	0.84%

TABLE 5-7 (CONT'D)
CUMULATIVE kW SAVED FROM LM (Actual) / PEAK kW
(RANKED BY 2006 %)

Rank	Utility Name	2004	2005	2006
29	Duke Energy Indiana Inc	3.11%	3.95%	0.73%
30	PPL Electric Utilities Corp	0.68%	0.68%	0.66%
31	Florida Power & Light Co	6.04%	3.84%	0.60%
32	Kansas City Power & Light Co	0.15%	0.71%	0.56%
33	Nevada Power Co	0.18%	0.40%	0.39%
34	Columbus Southern Power Co	0.30%	0.24%	0.18%
35	Virginia Electric & Power Co	0.07%	0.06%	0.17%
36	Pennsylvania Electric Co	0.00%	0.00%	0.13%
37	Aquila Inc	0.00%	0.00%	0.00%
38	Arizona Public Service Co	0.00%	0.00%	0.00%
39	Atlantic City Electric Co	0.00%	0.00%	0.00%
40	Avista Corp	0.00%	0.00%	0.00%
41	Baltimore Gas & Electric Co	0.00%	0.00%	0.00%
42	Boston Edison Co	0.00%	0.00%	0.00%
43	Central Hudson Gas & Electric Corp	0.00%	0.00%	0.00%
44	Central Illinois Public Service Co		2.07%	0.00%
45	Central Maine Power Co	0.00%	0.00%	0.00%
46	Cleco Power LLC	0.00%	0.00%	0.00%
47	Cleveland Electric Illuminating Co	0.00%	0.00%	0.00%
48	Commonwealth Edison Co	0.00%	0.00%	0.00%
49	Commonwealth Electric Co	0.00%	0.00%	0.00%
50	Connecticut Light & Power Co	0.00%	0.00%	0.00%
51	Consumers Energy Co	0.00%	0.00%	0.00%
52	Dayton Power & Light Co	0.00%	0.00%	0.00%
53	Delmarva Power & Light Co	0.00%	0.00%	0.00%
54	Detroit Edison Co	0.00%	0.00%	0.00%
55	Duke Energy Carolinas, LLC	0.00%	0.00%	0.00%
56	Duke Energy Ohio Inc	0.12%	0.27%	0.00%
57	Duquesne Light Co	0.00%	0.00%	0.00%
58	El Paso Electric Co	0.00%	0.00%	0.00%
59	Entergy Arkansas Inc	0.00%	0.00%	0.00%
60	Entergy Gulf States Inc	0.00%	0.00%	0.00%
61	Entergy Louisiana Inc	0.00%	0.00%	0.00%
62	Entergy Mississippi Inc	0.00%	0.00%	0.00%
63	Entergy New Orleans Inc	0.00%	0.00%	0.00%
64	Hawaiian Electric Co Inc	0.00%	0.00%	0.00%

(Continued)

TABLE 5-7 (CONT'D)
CUMULATIVE kW SAVED FROM LM (Actual) / PEAK kW
(RANKED BY 2006 %)

Rank	Utility Name	2004	2005	2006
65	Kansas Gas & Electric Co	0.00%	0.00%	0.00%
66	Kentucky Power Co	0.00%	0.00%	0.00%
67	Massachusetts Electric Co			0.00%
68	Minnesota Power Inc	0.00%	0.00%	0.00%
69	Mississippi Power Co	0.00%	0.00%	0.00%
70	Monongahela Power Co	0.00%	0.00%	0.00%
71	Narragansett Electric Co			0.00%
72	New York State Electric & Gas Corp	0.00%	0.00%	0.00%
73	Niagara Mohawk Power Corp	0.00%	0.00%	0.00%
74	Northern Indiana Public Service Co	0.00%	0.00%	0.00%
75	NorthWestern Energy LLC	0.00%	0.00%	0.00%
76	Ohio Edison Co	0.00%	0.00%	0.00%
77	Orange & Rockland Utilities Inc	0.00%	0.00%	0.00%
78	PacifiCorp	0.00%	0.00%	0.00%
79	Portland General Electric Co	0.00%	0.00%	0.00%
80	Potomac Electric Power Co	0.00%	0.00%	0.00%
81	Progress Energy Carolinas Inc	0.00%	0.00%	0.00%
82	Public Service Co of NH	0.00%	0.00%	0.00%
83	Public Service Co of NM	0.00%	0.00%	0.00%
84	Public Service Co of Oklahoma	0.00%	0.00%	0.00%
85	Puget Sound Energy Inc	0.00%	0.00%	0.00%
86	Rochester Gas & Electric Corp	0.00%	0.00%	0.00%
87	Sierra Pacific Power Co	0.00%	0.00%	0.00%
88	South Carolina Electric & Gas Co	0.00%	0.00%	0.00%
89	Southwestern Electric Power Co	0.00%	0.00%	0.00%
90	Southwestern Public Service Co	0.00%	0.00%	0.00%
91	Tampa Electric Co	0.00%	2.34%	0.00%
92	The Potomac Edison Co	0.00%	0.00%	0.00%
93	Toledo Edison Co	0.00%	0.00%	0.00%
94	Tucson Electric Power Co	0.00%	0.00%	0.00%
95	Union Electric Co	0.38%	0.00%	0.00%
96	United Illuminating Co	1.25%	2.30%	0.00%
97	West Penn Power Co	0.00%	0.00%	0.00%
98	Westar Energy Inc	0.00%	0.00%	0.00%
99	Western Massachusetts Electric Co	0.00%	0.00%	0.00%
100	Wisconsin Public Service Corp	0.00%	0.00%	0.00%

Chapter 5: A Review of Utility-Sponsored Energy Efficiency and Load Management Activities

Table 5-8 combines the peak capacity savings for both energy efficiency and actual LM on a cumulative basis for the last three years for which data are available. The top quartile of IOUs has achieved capacity savings equal to at least four percent of their peak load. The top ten have all beaten a 10 percent target at least once during these three years. Accordingly, regulators and utilities could reasonably target a 10 percent cumulative capacity savings just as these top performers have done. Furthermore, it is likely the top tier will continue to expand their own cumulative capacity saving accomplishments.

TABLE 5-8
CUMULATIVE kW SAVED FROM EE & LM (Actual) / PEAK kW
(RANKED BY 2006 %)

Rank	Utility Name	2004	2005	2006
1	Northern States Power Co WI	25.23%	23.60%	23.47%
2	Northern States Power Co MN	20.18%	21.62%	21.53%
3	Wisconsin Electric Power Co	21.66%	17.53%	19.09%
4	Progress Energy Florida Inc	5.41%	5.22%	17.21%
5	Southern California Edison Co	13.50%	14.36%	12.82%
6	Gulf Power Co	9.95%	10.37%	10.55%
7	United Illuminating Co	11.53%	12.41%	10.37%
8	Pacific Gas & Electric Co	6.43%	10.46%	10.28%
9	Western Massachusetts Electric Co	10.64%	10.51%	10.06%
10	Florida Power & Light Co	15.09%	12.48%	9.81%
11	Avista Corp	8.44%	9.16%	9.72%
12	Wisconsin Power & Light Co	11.88%	10.69%	9.66%
13	MidAmerican Energy Co	10.32%	9.90%	9.62%
14	Public Service Co of Colorado	7.69%	7.75%	8.68%
15	Connecticut Light & Power Co	7.66%	7.59%	7.93%
16	Minnesota Power Inc	5.94%	7.00%	7.31%
17	Massachusetts Electric Co			7.18%
18	Narragansett Electric Co			6.40%
19	Progress Energy Carolinas Inc	0.00%	0.00%	5.28%
20	Potomac Electric Power Co	5.75%	5.20%	5.04%
21	NorthWestern Energy LLC	4.46%	4.83%	4.99%
22	Public Service Electric & Gas Co	2.64%	4.03%	4.65%
23	Ohio Power Co	2.75%	2.34%	4.62%
24	Indianapolis Power & Light Co	1.58%	4.46%	4.40%
25	Indiana Michigan Power Co	1.33%	3.77%	4.09%
26	Hawaiian Electric Co Inc	0.32%	0.41%	3.88%
27	Oklahoma Gas & Electric Co	1.34%	0.00%	3.80%
28	Southern Indiana Gas & Electric Co	1.47%	4.00%	3.69%

(Continued)

TABLE 5-8 (CONT'D)
CUMULATIVE kW SAVED FROM EE & LM (Actual) / PEAK kW
(RANKED BY 2006 %)

Rank	Utility Name	2004	2005	2006
29	Tampa Electric Co	12.95%	5.37%	3.49%
30	Jersey Central Power & Light Co	2.36%	3.36%	3.33%
31	Duke Energy Indiana Inc	5.67%	6.27%	3.04%
32	Alabama Power Co	0.88%	0.35%	2.46%
33	Georgia Power Co	3.19%	2.59%	2.43%
34	Louisville Gas & Electric Co	2.05%	2.29%	2.38%
35	Tucson Electric Power Co	2.06%	1.99%	2.33%
36	Baltimore Gas & Electric Co	2.58%	2.26%	2.25%
37	Appalachian Power Co	1.20%	1.22%	2.12%
38	PECO Energy Co	2.31%	2.03%	1.96%
39	Nevada Power Co	0.46%	1.31%	1.74%
40	San Diego Gas & Electric Co	1.99%	2.93%	1.53%
41	Kentucky Power Co	1.36%	1.42%	1.53%
42	Metropolitan Edison Co	0.04%	1.61%	1.37%
43	Idaho Power Co	0.21%	1.55%	1.23%
44	Consolidated Edison Co-NY Inc	0.00%	0.34%	1.22%
45	Kentucky Utilities Co	0.82%	1.05%	1.19%
46	Duke Energy Ohio Inc	1.44%	1.40%	1.06%
47	Illinois Power Co	0.00%	1.21%	0.89%
48	Virginia Electric & Power Co	0.69%	0.61%	0.71%
49	Sierra Pacific Power Co	0.25%	0.46%	0.71%
50	PPL Electric Utilities Corp	0.68%	0.68%	0.66%
51	Kansas City Power & Light Co	0.15%	0.71%	0.62%
52	Southwestern Public Service Co	0.38%	0.54%	0.62%
53	Southwestern Electric Power Co	0.47%	0.51%	0.51%
54	Columbus Southern Power Co	0.55%	0.46%	0.38%
55	El Paso Electric Co	0.23%	0.22%	0.28%
56	Pennsylvania Electric Co	0.04%	0.03%	0.16%
57	Entergy Gulf States Inc	0.09%	0.10%	0.09%
58	Aquila Inc	0.11%	0.07%	0.00%
59	Arizona Public Service Co	0.00%	0.00%	0.00%
60	Atlantic City Electric Co	0.00%	0.00%	0.00%
61	Boston Edison Co	0.00%	7.85%	0.00%
62	Central Hudson Gas & Electric Corp	0.00%	0.00%	0.00%
63	Central Illinois Public Service Co		2.07%	0.00%
64	Central Maine Power Co	0.06%	0.00%	0.00%

TABLE 5-8 (CONT'D)
CUMULATIVE kW SAVED FROM EE & LM (Actual) / PEAK kW
(RANKED BY 2006 %)

Rank	Utility Name	2004	2005	2006
65	Cleco Power LLC	0.00%	0.00%	0.00%
66	Cleveland Electric Illuminating Co	0.00%	0.00%	0.00%
67	Commonwealth Edison Co	0.00%	0.00%	0.00%
68	Commonwealth Electric Co	4.58%	4.46%	0.00%
69	Consumers Energy Co	0.00%	0.00%	0.00%
70	Dayton Power & Light Co	0.00%	0.00%	0.00%
71	Delmarva Power & Light Co	0.00%	0.00%	0.00%
72	Detroit Edison Co	0.00%	0.00%	0.00%
73	Duke Energy Carolinas, LLC	0.00%	0.00%	0.00%
74	Duquesne Light Co	0.00%	0.00%	0.00%
75	Entergy Arkansas Inc	0.00%	0.00%	0.00%
76	Entergy Louisiana Inc	0.00%	0.00%	0.00%
77	Entergy Mississippi Inc	0.00%	0.00%	0.00%
78	Entergy New Orleans Inc	0.00%	0.00%	0.00%
79	Kansas Gas & Electric Co	0.00%	0.00%	0.00%
80	Mississippi Power Co	0.00%	0.00%	0.00%
81	Monongahela Power Co	0.00%	0.00%	0.00%
82	New York State Electric & Gas Corp	0.00%	0.00%	0.00%
83	Niagara Mohawk Power Corp	0.00%	0.00%	0.00%
84	Northern Indiana Public Service Co	0.00%	0.00%	0.00%
85	Ohio Edison Co	0.00%	0.00%	0.00%
86	Orange & Rockland Utilities Inc	0.00%	0.00%	0.00%
87	PacifiCorp	0.00%	0.00%	0.00%
88	Portland General Electric Co	0.00%	0.00%	0.00%
89	Public Service Co of NH	0.00%	0.00%	0.00%
90	Public Service Co of NM	0.00%	0.00%	0.00%
91	Public Service Co of Oklahoma	0.00%	0.00%	0.00%
92	Puget Sound Energy Inc	0.00%	0.00%	0.00%
93	Rochester Gas & Electric Corp	0.00%	0.00%	0.00%
94	South Carolina Electric & Gas Co	0.00%	0.00%	0.00%
95	The Potomac Edison Co	0.00%	0.00%	0.00%
96	Toledo Edison Co	0.00%	0.00%	0.00%
97	Union Electric Co	0.38%	0.00%	0.00%
98	West Penn Power Co	0.00%	0.00%	0.00%
99	Westar Energy Inc	0.00%	0.00%	0.00%
100	Wisconsin Public Service Corp	0.00%	0.00%	0.00%

In the past two years since the data was reported, the nation's IOUs have mostly increased these activities. New goals and targets are being established through both new regulation and legislation. As the nation continues to experience higher energy prices and has increasing environmental concerns, it is likely there will be more energy efficiency and load management. Some particularly high-priced jurisdictions, such as California and New England, have typically achieved greater savings. These are also states that often restructured and committed to spend money collected via "wires" charges to expand DSM. Other more conventionally regulated states, such as Minnesota and Wisconsin, have also done rather well. Wisconsin has also turned away from some of the investor-owned utilities using "independent," albeit utility financed, entities to operate the state's DSM programs. This distorts the more recent data that IOUs report and, shown in this form, understates the states' recent efforts.

One of the purposes of this primer is to determine what causes states to succeed and move to the top decile of DSM on both an "incremental" and sustained cumulative "annual" basis. Since regulation plays a role in the performance and operation of electric utilities, this primer will focus on tariff and cost allocation as conceptual matters and how regulators treat DSM empirically. These are addressed in subsequent chapters. Before doing so, the next chapter considers similar performance on a state-by-state basis over the past 15 years of data. The initial implication is that states that finance energy efficiency through utility tariffs have already done quite a lot. These look likely to continue as the top tier jurisdictions work to achieve sustained and very aggressive DSM and renewable energy programs. As other states seek to join the effort, there are lessons to learn and targets to establish.

CHAPTER 6
RELATIVE STATE-BY-STATE COMPARISONS OVER TIME

The previous analyses showed relative savings performance for the top 100 IOUs for the last three years for which EIA has reported data. In this discussion, the EIA data are aggregated to the state level for each of the 15 years that EIA has published this data. The approach used allocates the savings for multi-state utilities, based upon the respective proportion of system sales in each state served. This analysis also includes governmentally-owned and customer-owned utilities. There is also no cutoff for size. Virtually all electric utilities that report to EIA are included in this analysis.

Table 6-1 shows how the fifty states and the District of Columbia have performed in terms of their incremental savings in MWhs expressed as percentages of the volume of electricity sold statewide over the past 15 years for which the EIA data are available. These are ranked using their 2006 savings.

TABLE 6-1
ENERGY EFFICIENCY (EE) INCREMENTAL MWhs SAVED / TOTAL MWhs SOLD

Rank	State	1992	1993	1994	1995	1996	1997	1998	1999	2000	2001	2002	2003	2004	2005	2006
1	Rhode Island	1.2217%	1.8787%	0.9641%	0.7449%	0.5578%	0.5976%	0.6505%	0.8862%	0.0000%	0.7614%	0.6245%	0.6404%	0.6235%	0.7153%	1.0784%
2	Connecticut	0.6352%	0.6949%	0.5934%	0.5636%	0.3103%	0.5508%	0.4621%	0.4088%	0.8631%	0.9907%	0.7896%	0.4020%	1.1790%	0.9378%	1.0549%
3	California	0.3506%	0.8330%	0.7684%	0.6106%	0.4609%	0.4639%	0.2985%	0.3908%	0.4672%	0.5424%	0.3941%	0.2118%	0.6096%	0.9321%	0.6313%
4	New Hampshire	0.0878%	0.0618%	0.0635%	0.1666%	0.1383%	0.1787%	0.1311%	0.2561%	0.1155%	0.2639%	0.3856%	0.7650%	0.5297%	0.6381%	0.6215%
5	Nevada	0.2491%	0.7883%	0.1576%	0.2540%	0.0483%	0.0248%	0.0161%	0.0024%	0.0019%	0.1180%	0.1925%	0.1150%	0.1632%	0.2531%	0.6082%
6	Iowa	0.1708%	0.3587%	0.3830%	0.5095%	0.3976%	0.2850%	0.2361%	0.2571%	0.3030%	0.3410%	0.3549%	0.4584%	0.4561%	0.4865%	0.5939%
7	Hawaii	0.0247%	0.1495%	0.0219%	0.0154%	0.2015%	0.0778%	0.1049%	0.0522%	0.0493%	0.0912%	0.0635%	0.0680%	0.0573%	0.1144%	0.4569%
8	Idaho	0.2171%	0.5281%	0.3700%	0.3288%	0.2527%	0.1533%	0.1346%	0.0872%	0.0832%	0.3086%	0.1259%	0.1412%	0.1409%	0.3219%	0.4269%
9	Massachusetts	0.5978%	0.5782%	0.5373%	0.4563%	0.3785%	0.5365%	0.4971%	0.1242%	0.2613%	0.5689%	0.4503%	0.2873%	0.4802%	0.5953%	0.4243%
10	Minnesota	0.3608%	0.5447%	0.6796%	0.9194%	0.4844%	0.3474%	0.5872%	0.1559%	0.3998%	0.4317%	0.4716%	0.5432%	0.5595%	0.6106%	0.4130%
11	Washington	0.4738%	1.0509%	0.8187%	0.6338%	0.6987%	0.6120%	0.1231%	0.1515%	0.1596%	0.5058%	0.2685%	0.3456%	0.3398%	0.3373%	0.3832%
12	Utah	0.1708%	0.2938%	0.3217%	0.4996%	0.3779%	0.2586%	0.1792%	0.2245%	0.0957%	0.2831%	0.2328%	0.2840%	0.2451%	0.2438%	0.3010%
13	Montana	0.2828%	0.9181%	0.8890%	0.9104%	0.5787%	0.5784%	0.1060%	0.0497%	0.1027%	0.1849%	0.2129%	0.2324%	0.1821%	0.1812%	0.2505%
14	Wyoming	0.1285%	0.2204%	0.2426%	0.3787%	0.2754%	0.1930%	0.1339%	0.1665%	0.0684%	0.2058%	0.1649%	0.2027%	0.1740%	0.1689%	0.2059%
15	North Dakota	0.1899%	0.2849%	0.3417%	0.4652%	0.2697%	0.2564%	0.3256%	0.0986%	0.2235%	0.1985%	0.1883%	0.1900%	0.2096%	0.2102%	0.1956%
16	Wisconsin	1.0151%	1.0215%	0.9082%	0.6010%	0.7919%	0.3391%	0.5046%	0.4773%	0.4536%	0.3828%	0.2287%	0.0716%	0.1045%	0.0900%	0.1724%
17	Oregon	0.4007%	0.6286%	0.6684%	0.7277%	0.4817%	0.3907%	0.1699%	0.2151%	0.1797%	0.5507%	0.2683%	0.1778%	0.1520%	0.1482%	0.1709%
18	Arizona	0.1389%	0.1581%	0.1616%	0.1202%	0.0955%	0.0170%	0.0134%	0.0100%	0.0050%	0.0007%	0.0001%	0.0001%	0.0011%	0.0380%	0.1686%
19	South Dakota	0.1593%	0.2404%	0.2468%	0.4005%	0.1773%	0.1462%	0.2379%	0.0576%	0.1645%	0.1528%	0.1559%	0.1506%	0.1622%	0.1562%	0.1512%
20	Florida	0.1156%	0.1897%	0.2189%	0.2679%	0.3987%	0.2839%	0.1986%	0.1909%	0.1504%	0.1491%	0.1393%	0.1387%	0.1181%	0.1249%	0.1319%
21	Colorado	0.0053%	0.0021%	0.1723%	0.3452%	0.3882%	0.1488%	0.0342%	0.0249%	0.1358%	0.0813%	0.1150%	0.1391%	0.1510%	0.2430%	0.1211%
22	Vermont	0.5946%	0.9975%	0.7826%	0.7605%	0.5555%	0.5076%	0.4894%	0.5763%	0.1071%	0.0714%	0.0884%	0.0719%	0.0701%	0.0062%	0.1174%
23	New Jersey	0.1078%	0.1412%	0.2585%	0.9269%	0.9122%	0.2774%	0.1739%	0.1856%	0.1264%	0.0433%	0.1432%	0.0726%	0.0809%	0.1453%	0.1039%
24	New York	1.0697%	1.1309%	0.8531%	0.5630%	0.3348%	0.2739%	0.0277%	0.0504%	0.0350%	0.0130%	0.0082%	0.0118%	0.0436%	0.0479%	0.0728%
25	Texas	0.0528%	0.1335%	0.1356%	0.0996%	0.0631%	0.0316%	0.0529%	0.0423%	0.0602%	0.0607%	0.0874%	0.0459%	0.0431%	0.0546%	0.0664%
26	Kentucky	0.0177%	0.0267%	0.0538%	0.0246%	0.0335%	0.0360%	0.0392%	0.0337%	0.0306%	0.0475%	0.0666%	0.0462%	0.0606%	0.0606%	0.0502%

TABLE 6-1 (CONT'D)
ENERGY EFFICIENCY (EE) INCREMENTAL MWhs SAVED / TOTAL MWhs SOLD

Rank	State	1992	1993	1994	1995	1996	1997	1998	1999	2000	2001	2002	2003	2004	2005	2006
27	Mississippi	0.0149%	0.0245%	0.0333%	0.0261%	0.0263%	0.0209%	0.0131%	0.0201%	0.0223%	0.0285%	0.0348%	0.0288%	0.0279%	0.0278%	0.0257%
28	Alabama	0.0846%	0.0750%	0.0740%	0.0168%	0.0162%	0.0180%	0.0225%	0.0142%	0.0189%	0.0230%	0.0318%	0.0203%	0.0322%	0.0269%	0.0228%
29	New Mexico	0.0058%	0.0489%	0.0325%	0.0378%	0.0235%	0.0055%	0.0112%	0.0000%	0.0872%	0.0237%	0.0029%	0.0125%	0.0342%	0.0251%	0.0214%
30	Nebraska	0.0085%	0.0132%	0.0278%	0.0144%	0.0074%	0.0150%	0.0097%	0.0208%	0.0117%	0.0039%	0.0143%	0.0485%	0.0180%	0.0296%	0.0198%
31	Alaska	0.0158%	0.0269%	0.0251%	0.0251%	0.0181%	0.0149%	0.0521%	0.0159%	0.0141%	0.0100%	0.0118%	0.0081%	0.0106%	0.0229%	0.0185%
32	South Carolina	0.1133%	0.1448%	0.1374%	0.0700%	0.0616%	0.0280%	0.0155%	0.0127%	0.0106%	0.0108%	0.0019%	0.0032%	0.0026%	0.0462%	0.0185%
33	Tennessee	0.0045%	0.0118%	0.0239%	0.0079%	0.0065%	0.0098%	0.0102%	0.0125%	0.0116%	0.0135%	0.0291%	0.0201%	0.0169%	0.0161%	0.0179%
34	Illinois	0.0011%	0.0324%	0.0088%	0.0292%	0.0218%	0.0061%	0.0337%	0.0229%	0.0060%	0.0073%	0.0070%	0.0113%	0.0123%	0.0118%	0.0160%
35	Indiana	0.0418%	0.1920%	0.2530%	0.3147%	0.0958%	0.0461%	0.0164%	0.0228%	0.0110%	0.0134%	0.0085%	0.0107%	0.0081%	0.0010%	0.0120%
36	Michigan	0.1153%	0.4960%	0.1159%	0.1010%	0.1452%	0.0046%	0.0263%	0.0104%	0.0054%	0.0012%	0.0002%	0.0001%	0.0001%	0.0001%	0.0054%
37	Maine	0.6387%	0.4349%	0.3150%	0.4690%	0.3064%	0.4074%	0.0022%	0.2455%	0.0026%	0.0015%	0.0205%	0.0006%	0.0003%	0.0027%	0.0035%
38	Arkansas	0.0057%	0.0054%	0.0027%	0.0067%	0.0085%	0.0069%	0.0029%	0.0009%	0.0006%	0.0018%	0.0041%	0.0032%	0.0022%	0.0111%	0.0022%
39	North Carolina	0.0896%	0.0941%	0.1157%	0.0556%	0.0551%	0.0012%	0.0017%	0.0011%	0.0005%	0.0002%	0.0010%	0.0007%	0.0007%	0.0010%	0.0022%
40	Georgia	0.0780%	0.0709%	0.1389%	0.0633%	0.0095%	0.0078%	0.0076%	0.0020%	0.0018%	0.0030%	0.0014%	0.0015%	0.0005%	0.0009%	0.0019%
41	Louisiana	0.0194%	0.0026%	0.0068%	0.0040%	0.0050%	0.0041%	0.0024%	0.0003%	0.0001%	0.0013%	0.0057%	0.0101%	0.0014%	0.0089%	0.0017%
42	Pennsylvania	0.0532%	0.0414%	0.0152%	0.0471%	0.0351%	0.0119%	0.0026%	0.0025%	0.0020%	0.0009%	0.0020%	0.0011%	0.0015%	0.0015%	0.0014%
43	Missouri	0.0052%	0.0049%	0.0044%	0.0041%	0.0043%	0.0022%	0.0024%	0.0013%	0.0024%	0.0014%	0.0013%	0.0015%	0.0011%	0.0013%	0.0011%
44	Ohio	0.0333%	0.0822%	0.0966%	0.1737%	0.1286%	0.0734%	0.0735%	0.0678%	0.0094%	0.0019%	0.0005%	0.0006%	0.0007%	0.0005%	0.0007%
45	Oklahoma	0.0197%	0.0197%	0.0064%	0.0017%	0.0018%	0.0012%	0.0013%	0.0011%	0.0027%	0.0013%	0.0004%	0.0004%	0.0008%	0.0007%	0.0003%
46	Maryland	0.3647%	0.4661%	0.7394%	0.8077%	0.3065%	0.2715%	0.1585%	0.1334%	0.0607%	0.0040%	0.0059%	0.0005%	0.0003%	0.0003%	0.0002%
47	Virginia	0.0229%	0.0295%	0.0861%	0.0629%	0.0337%	0.0126%	0.0171%	0.0030%	0.0001%	0.0001%	0.0002%	0.0003%	0.0009%	0.0005%	0.0002%
48	Kansas	0.0000%	0.0001%	0.0001%	0.0001%	0.0001%	0.0022%	0.0009%	0.0000%	0.0008%	0.0001%	0.0000%	0.0050%	0.0055%	0.0001%	0.0001%
49	West Virginia	0.1280%	0.0925%	0.1633%	0.1481%	0.0379%	0.0039%	0.0072%	0.0072%	0.0000%	0.0000%	0.0000%	0.0000%	0.0000%	0.0000%	0.0000%
50	Delaware	0.1887%	0.1243%	0.2006%	0.1885%	0.1289%	0.0693%	0.0373%	0.0000%	0.0000%	0.0000%	0.0000%	0.0000%	0.0000%	0.0000%	0.0000%
51	Dist of Columbia	0.4413%	0.9239%	1.4960%	1.5137%	0.6885%	0.3519%	0.3707%	0.5510%	0.2610%	0.0124%	0.0192%	0.0000%	0.0000%	0.0000%	0.0000%

There are several important observations. First, a state target of one percent incremental annual electricity savings is achievable, but on a state-by-state basis, this may be difficult to reach. In fact, in 2006, only Rhode Island and Connecticut achieved this target for the entire state.

Second, some jurisdictions (*e.g.*, the District of Columbia) achieved one percent in the 1990s, but current incremental efforts have fallen rather sharply.

Third, other states (*e.g.*, Minnesota, Massachusetts, and California) have achieved a relatively consistent and sustained incremental electricity reduction each year. These three states are in very different circumstances in terms of restructuring, politics, climates, and relative retail prices. Regardless, they have generally renewed and sustained their annual incremental efforts to provide energy efficiency. This shows that states can reasonably make a sizeable effort to expand energy efficiency in this country.

Table 6-2 shows these same energy efficiency comparisons on a cumulative basis.

Chapter 6: Relative State-by-State Comparisons Over Time

TABLE 6-2

ENERGY EFFICIENCY (EE) CUMULATIVE MWhs SAVED / TOTAL MWhs SOLD

Rank	State	1992	1993	1994	1995	1996	1997	1998	1999	2000	2001	2002	2003	2004	2005	2006
1	Connecticut	4.0897%	4.7615%	5.1241%	5.6120%	5.7241%	6.0370%	3.3424%	6.4852%	6.7508%	7.4930%	7.5868%	7.6008%	8.0085%	8.2991%	9.2630%
2	Rhode Island	4.5857%	6.6324%	5.8411%	5.2503%	4.7574%	4.9338%	5.0147%	6.3704%	0.0000%	4.9181%	5.4592%	5.6712%	5.5571%	7.3505%	7.1403%
3	Minnesota	0.9998%	1.9094%	1.9744%	2.5889%	3.2672%	3.6113%	4.4347%	4.7860%	4.9969%	5.3276%	5.6763%	5.8524%	6.3763%	6.7408%	7.0892%
4	California	1.1337%	4.1598%	4.6475%	5.5185%	5.2924%	4.4683%	3.0732%	5.3204%	5.2800%	4.9012%	5.2859%	5.1493%	5.4553%	6.2310%	6.3481%
5	Washington	2.4369%	3.9428%	5.9165%	6.1399%	7.4958%	9.0050%	4.0439%	4.0233%	3.8630%	5.0285%	5.2366%	2.4510%	4.9569%	5.0027%	5.5698%
6	Dist of Columbia	0.4878%	1.3423%	3.0350%	4.9783%	6.2499%	6.7471%	6.6427%	7.2192%	7.4126%	6.4997%	4.7598%	4.6342%	4.5616%	4.5171%	4.4592%
7	Iowa	0.4039%	0.6360%	0.9919%	1.4310%	1.6866%	1.6672%	1.8198%	2.0189%	2.2586%	2.4754%	2.7330%	3.1802%	3.3879%	3.7044%	4.2387%
8	Hawaii	0.0578%	0.2003%	0.1954%	0.2091%	0.3025%	0.4881%	0.8516%	0.5762%	0.5778%	0.7093%	0.6757%	0.7606%	0.8054%	1.5215%	4.1375%
9	New Hampshire	0.3084%	0.4481%	0.4893%	0.7053%	0.8906%	1.0733%	1.3978%	1.4627%	1.6482%	1.7488%	2.0315%	2.5004%	3.0163%	3.0560%	4.0183%
10	Massachusetts	2.5022%	3.2406%	3.2881%	3.7065%	3.5521%	3.7237%	3.7550%	1.2337%	3.9074%	5.3420%	5.2386%	3.3604%	3.7444%	5.6420%	3.7518%
11	Idaho	0.9869%	1.4481%	1.8244%	2.1500%	2.1248%	2.2561%	2.3528%	2.3249%	2.3090%	2.7595%	2.9295%	2.9775%	3.0185%	3.3585%	3.6140%
12	Utah	0.5463%	0.8270%	1.1271%	1.6351%	1.9290%	2.1519%	2.2879%	2.5145%	2.5084%	2.7498%	2.9978%	2.9862%	2.9097%	3.3560%	3.4306%
13	Wisconsin	2.4720%	3.4096%	4.4749%	4.7801%	5.2193%	2.8142%	5.3709%	5.1194%	5.2879%	4.0511%	4.0341%	3.8371%	3.5288%	3.1881%	3.2325%
14	Florida	2.6545%	2.7850%	2.8852%	2.9580%	3.2713%	3.4941%	3.4554%	3.6415%	3.6117%	3.5760%	3.5783%	2.6300%	2.7226%	2.7735%	2.9018%
15	Montana	0.5097%	1.7985%	4.2001%	4.8526%	5.1318%	6.5789%	2.2779%	2.3810%	2.0745%	1.7889%	2.0209%	2.2684%	2.4756%	2.5927%	2.7540%
16	North Dakota	0.5131%	0.9860%	1.1590%	1.3428%	1.6227%	1.8153%	2.0365%	1.9909%	2.0976%	2.0746%	2.1591%	2.1414%	2.3056%	2.3987%	2.4810%
17	Wyoming	0.4097%	0.6182%	0.8476%	1.2282%	1.4389%	1.6076%	1.7205%	1.8774%	1.8444%	1.9941%	2.1269%	2.1198%	2.0551%	2.3270%	2.3461%
18	Oregon	0.9624%	1.7164%	2.5396%	3.2628%	3.6915%	4.0300%	3.5486%	3.5228%	3.6716%	3.8765%	4.0702%	2.1511%	2.1576%	2.2021%	2.2744%
19	South Dakota	0.4344%	0.8077%	0.7733%	1.1641%	1.2129%	1.3465%	1.5497%	1.6825%	1.7777%	1.7630%	1.8719%	1.8675%	1.9862%	2.0381%	2.1305%
20	Maryland	0.7164%	1.0766%	1.9043%	2.5176%	2.9794%	2.7628%	2.7853%	2.8425%	2.8324%	1.7027%	1.3686%	2.0748%	2.1532%	2.0437%	2.1047%
21	Colorado	0.2619%	0.5254%	0.7581%	0.5993%	0.9314%	1.1838%	1.1697%	1.1099%	1.1797%	1.1955%	1.2575%	1.2833%	1.4797%	1.6829%	1.7548%
22	New Jersey	0.2710%	0.3085%	0.4900%	1.2486%	1.6081%	2.0462%	2.6328%	2.8323%	4.1960%	2.5755%	2.9205%	3.1510%	1.5287%	1.8111%	1.5533%
23	Vermont	0.8292%	1.6295%	2.6863%	3.3774%	3.8181%	4.2060%	4.5869%	6.1126%	1.2705%	3.2640%	3.0410%	3.1043%	2.8094%	2.6456%	1.5228%
24	Nevada	0.8785%	1.6242%	1.7379%	1.8786%	0.7194%	0.8623%	0.8146%	0.0009%	0.0339%	0.1118%	0.2312%	0.1567%	0.2342%	0.6890%	1.2512%
25	Indiana	0.0777%	0.2054%	0.3900%	0.6514%	0.7168%	0.7271%	0.7655%	0.8828%	0.7920%	0.7983%	0.6981%	0.7953%	0.7834%	0.7639%	0.7828%

(Continued)

TABLE 6-2 (CONT'D)
ENERGY EFFICIENCY (EE) CUMULATIVE MWhs SAVED / TOTAL MWhs SOLD

Rank	State	1992	1993	1994	1995	1996	1997	1998	1999	2000	2001	2002	2003	2004	2005	2006
26	Texas	0.3767%	0.9476%	1.5313%	1.5310%	1.5054%	1.2940%	1.2876%	1.3190%	1.3324%	1.2435%	0.6067%	0.3688%	0.4009%	0.3877%	0.7292%
27	New York	1.8009%	2.5508%	3.7426%	4.4034%	4.6435%	3.6668%	2.7690%	0.5276%	0.6189%	0.3471%	0.3671%	0.3558%	0.3915%	0.4182%	0.5734%
28	Arizona	1.2368%	1.3694%	1.4386%	1.5203%	1.4466%	0.2310%	0.2723%	0.2357%	0.0779%	0.0583%	0.0094%	0.1637%	0.1572%	0.3651%	0.5428%
29	Kentucky	1.8664%	1.8423%	1.8446%	0.9946%	1.0050%	1.0198%	1.0526%	1.0443%	1.0250%	0.2906%	0.3296%	0.3697%	0.4064%	0.4503%	0.4148%
30	Alabama	1.8182%	1.7864%	1.7429%	0.5640%	0.5571%	0.5686%	0.5510%	0.5593%	0.5630%	0.1967%	0.2185%	0.1763%	0.5389%	0.3516%	0.3682%
31	Mississippi	1.1319%	1.1134%	1.0899%	0.6001%	0.5859%	0.5996%	0.5780%	0.5466%	0.5897%	0.2185%	0.2426%	0.2622%	0.2821%	0.3004%	0.3180%
32	Ohio	0.0784%	0.1612%	0.2087%	0.5054%	0.7783%	0.6449%	0.7191%	0.6007%	0.3927%	0.2386%	0.2296%	0.2267%	0.2193%	0.2003%	0.2148%
33	Georgia	0.1144%	0.2502%	0.3505%	0.3597%	0.3649%	0.2940%	0.2784%	0.2445%	0.2671%	0.5955%	0.2214%	0.2348%	0.2246%	0.2202%	0.2033%
34	New Mexico	0.1167%	0.3167%	0.3325%	0.2890%	0.2038%	0.0734%	0.0865%	0.0389%	0.1391%	0.1503%	0.0988%	0.1379%	0.1833%	0.1783%	0.1855%
35	Tennessee	1.3206%	1.3082%	1.2612%	0.6425%	0.6035%	0.6151%	0.5850%	0.5858%	0.5915%	0.0791%	0.1146%	0.1262%	0.1394%	0.1514%	0.1688%
36	Virginia	0.2574%	0.2930%	0.3398%	0.6847%	0.5358%	0.4390%	0.4422%	0.4180%	0.3973%	0.0516%	0.3367%	0.1551%	0.1516%	0.1473%	0.1476%
37	South Carolina	0.8329%	0.9405%	0.9680%	0.9911%	1.0199%	0.6816%	1.5381%	0.1262%	0.6114%	0.1375%	0.1342%	0.1348%	0.1330%	0.1173%	0.1240%
38	Illinois	0.0079%	0.0403%	0.0256%	0.0592%	0.0813%	0.0769%	0.1061%	0.1247%	0.0722%	0.0628%	0.0627%	0.0760%	0.0824%	0.0891%	0.1052%
39	Michigan	0.2540%	0.6132%	0.7786%	0.6937%	0.7495%	0.0344%	0.3541%	0.1915%	0.2483%	0.1661%	0.1182%	0.1013%	0.0933%	0.0814%	0.0923%
40	Alaska	0.0039%	0.0426%	0.0583%	0.0820%	0.0976%	0.1412%	0.1436%	0.1428%	0.1467%	0.1408%	0.1400%	0.0572%	0.0516%	0.0611%	0.0681%
41	Arkansas	0.0739%	0.0705%	0.0736%	0.0697%	0.0849%	0.0899%	0.0849%	0.0791%	0.0650%	0.0642%	0.0767%	0.0783%	0.0774%	0.0628%	0.0640%
42	Nebraska	0.0757%	0.0889%	0.1123%	0.1215%	0.0691%	0.0606%	0.0643%	0.0853%	0.0904%	0.0226%	0.0730%	0.1973%	0.2147%	0.0432%	0.0625%
43	Louisiana	0.0286%	0.0289%	0.1568%	0.0124%	0.0208%	0.0250%	0.0269%	0.0261%	0.0254%	0.0286%	0.0327%	0.0433%	0.0405%	0.0520%	0.0556%
44	North Carolina	1.6671%	1.7815%	1.7500%	1.7351%	1.7342%	0.9902%	0.3221%	0.0322%	0.0313%	0.0171%	0.0326%	0.0154%	0.0149%	0.0262%	0.0376%
45	Missouri	0.0193%	0.0214%	0.0215%	0.0236%	0.0252%	0.0657%	0.0703%	0.0217%	0.0233%	0.0244%	0.0248%	0.0273%	0.0284%	0.0273%	0.0320%
46	Pennsylvania	0.3509%	0.3803%	0.3510%	0.4413%	0.4312%	0.1584%	0.1694%	0.2236%	0.2208%	0.1453%	0.0670%	0.0124%	0.0104%	0.0096%	0.0098%
47	West Virginia	1.0171%	1.1438%	1.3114%	1.4047%	1.4268%	0.1865%	0.1932%	0.1952%	0.1825%	0.1609%	0.1564%	0.0088%	0.0087%	0.0085%	0.0088%
48	Oklahoma	0.2849%	0.3009%	0.2977%	0.2942%	0.2796%	0.2682%	0.2425%	0.2331%	0.2082%	0.1969%	0.1793%	0.1751%	0.1636%	0.0109%	0.0050%
49	Kansas	0.0805%	0.0006%	0.0006%	0.0006%	0.0006%	0.0014%	0.0002%	0.0005%	0.0019%	0.0011%	0.0006%	0.0009%	0.0001%	0.0006%	0.0049%
50	Maine	3.4444%	3.6626%	3.8914%	4.3846%	4.4729%	4.5018%	0.4811%	0.6684%	0.7141%	0.2345%	0.2546%	0.2187%	0.0376%	0.0420%	0.0000%
51	Delaware	0.2507%	0.3595%	0.5679%	0.6959%	0.8984%	0.9457%	0.9565%	0.0000%	0.0000%	0.0000%	0.0000%	0.0000%	0.0000%	0.0000%	0.0000%

The three states of Minnesota, Massachusetts, and California are in the top ten of the 51 jurisdictions. They join the still high-performing top two (Connecticut and Rhode Island), as well as Washington, the District of Columbia, Iowa, Hawaii, and New Hampshire.

No state has yet to achieve the 10 percent electricity (MWh) savings goal that the previous chapter discussed for IOUs. Nevertheless, such a goal is within the realm of possibility. In addition, these data represent the years through 2006. In the years subsequent to 2006, energy efficiency has become a core political objective of both major political parties.

Table 6-3 shows "actual" capacity savings on the same state-by-state basis. Three states (Nebraska, Utah, and Alabama) have beaten the one percent per year capacity savings target in 2006. Other states have done so over the past 15 years.

TABLE 6-3
LOAD MANAGEMENT (LM) INCREMENTAL ACTUAL PEAK REDUCTION / PEAK LOAD

Rank	State	1992	1993	1994	1995	1996	1997	1998	1999	2000	2001	2002	2003	2004	2005	2006
1	Nebraska	0.9263%	0.7131%	0.3390%	1.7320%	0.1795%	0.5003%	0.7087%	0.4286%	0.7416%	0.4987%	0.6262%	0.7781%	1.0206%	1.3750%	1.3750%
2	Utah	0.1804%	0.1344%	0.0216%	0.0039%	0.0171%	0.1559%	0.0128%	6.6296%	1.7127%	0.0175%	0.0989%	0.0481%	0.0000%	0.0000%	1.0969%
3	Alabama	0.0140%	0.0074%	0.0146%	0.0225%	0.0277%	0.0344%	0.0332%	0.0313%	0.0818%	0.7236%	0.7052%	0.0000%	0.3462%	0.0000%	1.0907%
4	Wyoming	0.0100%	0.0151%	0.0137%	0.0020%	0.0000%	0.1669%	0.0479%	5.4664%	1.4225%	0.0000%	0.0000%	0.0000%	0.0000%	0.0000%	0.7734%
5	Connecticut	0.0679%	0.2094%	0.0466%	0.1020%	0.0135%	0.0000%	0.0000%	0.0000%	0.0000%	0.0000%	0.1679%	0.3458%	0.2226%	0.2254%	0.7364%
6	West Virginia	0.0083%	1.1926%	0.0036%	0.0022%	0.0018%	0.0014%	0.0041%	0.0043%	0.0022%	0.0000%	0.0000%	0.0937%	0.2252%	0.0082%	0.5133%
7	New York	0.5075%	0.2144%	0.1823%	0.1258%	0.0967%	0.0799%	0.1906%	0.0092%	0.0683%	0.1483%	0.1961%	0.1884%	0.2817%	0.2400%	0.4753%
8	South Dakota	0.8159%	0.7832%	0.2658%	0.3263%	0.1679%	0.3514%	0.1608%	0.1776%	0.2003%	0.1983%	0.3759%	0.1370%	0.4277%	0.5116%	0.4713%
9	Colorado	0.0573%	0.0695%	0.3070%	0.1700%	0.3579%	1.0315%	0.3692%	0.4150%	0.2891%	0.3282%	0.5033%	0.3206%	1.4045%	0.3165%	0.4480%
10	North Dakota	2.5149%	2.6559%	2.1783%	0.4137%	0.2646%	2.1853%	2.1324%	2.1722%	2.2126%	2.3268%	0.3879%	0.3727%	0.5822%	0.2185%	0.4179%
11	Minnesota	1.4522%	1.2726%	1.5684%	2.2362%	2.1721%	0.4109%	0.4535%	0.5379%	0.5491%	0.6126%	1.2932%	0.4328%	0.6688%	0.6746%	0.4043%
12	Oregon	0.0567%	0.0530%	0.0005%	0.0000%	0.0533%	0.0000%	0.0000%	2.1753%	0.4745%	0.4416%	0.0000%	0.0000%	0.0029%	0.0199%	0.3660%
13	Indiana	0.0461%	0.3453%	0.0802%	0.2506%	0.1623%	0.0297%	0.0485%	0.0763%	0.0035%	0.0000%	0.0000%	0.0841%	0.0076%	1.3731%	0.3394%
14	Ohio	0.0484%	0.4414%	0.0273%	0.1111%	0.3679%	0.0761%	0.0355%	0.0041%	0.3013%	0.0003%	0.0241%	0.0606%	0.3420%	0.0698%	0.3139%
15	Virginia	0.3141%	0.5077%	0.2326%	0.7242%	0.0284%	0.1343%	0.0371%	0.0335%	0.0516%	0.0281%	0.0320%	0.0490%	0.1517%	0.1035%	0.2462%
16	Idaho	0.0000%	0.0000%	0.0000%	0.0000%	0.0000%	0.0000%	0.0000%	1.0712%	0.2761%	5.0547%	0.0000%	0.0000%	0.1068%	0.8124%	0.2379%
17	Iowa	1.0426%	2.4993%	0.7204%	0.5187%	0.0668%	0.1625%	0.1868%	0.1422%	0.1274%	0.2165%	3.8268%	0.1072%	0.8674%	0.1643%	0.2358%
18	Georgia	1.5331%	0.7400%	0.7453%	0.3851%	0.2350%	0.0657%	0.0853%	0.0727%	0.1262%	0.0238%	0.1006%	0.0930%	0.1292%	0.0758%	0.1989%
19	Wisconsin	0.1705%	0.2883%	0.1343%	0.1110%	0.1790%	0.0289%	0.0311%	0.0393%	0.0952%	0.1654%	0.0908%	0.0891%	0.1275%	0.1671%	0.1921%
20	California	1.0705%	1.1153%	0.0876%	1.0289%	1.0641%	0.9629%	0.9940%	0.0317%	0.0888%	0.1592%	0.0218%	0.0122%	0.0169%	0.0145%	0.1571%
21	Montana	0.0045%	0.0150%	0.0176%	0.0311%	0.0234%	0.0245%	0.0180%	0.0224%	0.0057%	0.0000%	0.0385%	0.0298%	0.0283%	0.0550%	0.1513%
22	Arkansas	0.4239%	0.1306%	0.3201%	0.0524%	0.0571%	0.0273%	0.0343%	0.0568%	0.0481%	0.0217%	0.0437%	0.0829%	0.0775%	0.0619%	0.1222%
23	Kansas	0.1551%	0.1778%	0.0982%	0.0279%	0.4319%	0.0529%	0.1427%	0.1042%	0.1148%	0.0307%	0.0729%	0.2242%	0.1029%	0.1157%	0.1052%
24	Nevada	1.2027%	0.2624%	0.6462%	0.0095%	0.0000%	0.0000%	0.0000%	0.0000%	0.0000%	0.0189%	0.0712%	0.1352%	0.1190%	0.0968%	0.0953%
25	Kentucky	0.0021%	0.1948%	0.0024%	0.0155%	0.0070%	0.0038%	0.0020%	0.0014%	0.0016%	0.0444%	0.0286%	0.0389%	0.0857%	0.1060%	0.0888%
26	Delaware	0.8926%	0.2863%	0.0847%	0.2368%	0.0286%	0.0270%	0.0311%	0.0767%	0.1348%	0.0664%	0.1621%	0.1314%	0.1664%	0.1446%	0.0847%

TABLE 6-3 (CONT'D)
LOAD MANAGEMENT (LM) INCREMENTAL ACTUAL PEAK REDUCTION / PEAK LOAD

Rank	State	1992	1993	1994	1995	1996	1997	1998	1999	2000	2001	2002	2003	2004	2005	2006
27	Michigan	0.0391%	0.1904%	0.0086%	2.4999%	2.4636%	0.0463%	0.0169%	0.0008%	0.0166%	0.0011%	0.0003%	0.0124%	0.0724%	0.1409%	0.0773%
28	Florida	0.1024%	0.4035%	0.3112%	0.4151%	0.3324%	0.2903%	0.2168%	0.1812%	0.1780%	0.1029%	0.1047%	0.2342%	0.0321%	0.1757%	0.0561%
29	Washington	0.1209%	0.0245%	0.0656%	0.1072%	0.0085%	0.0587%	0.0034%	0.2612%	0.0163%	0.0066%	0.0068%	0.0000%	0.0000%	0.0000%	0.0554%
30	Arizona	0.2385%	0.2560%	0.2118%	0.2541%	0.1040%	0.0262%	0.0258%	0.0292%	0.0259%	0.0808%	0.1104%	0.1153%	0.1419%	0.0649%	0.0477%
31	Texas	0.2275%	0.2372%	0.0247%	0.0885%	0.2238%	0.0389%	0.0245%	0.0676%	0.1030%	0.0649%	0.0704%	0.0188%	0.0241%	0.0330%	0.0426%
32	South Carolina	0.3336%	0.3061%	0.1191%	0.3846%	0.0766%	0.0554%	0.0327%	0.0101%	0.0500%	0.0059%	0.0098%	0.0093%	0.0386%	0.0404%	0.0413%
33	Oklahoma	0.5140%	1.0748%	0.0514%	0.3505%	0.0499%	0.3494%	0.5305%	0.0847%	0.1003%	0.0748%	0.0000%	0.0000%	0.0000%	0.0324%	0.0314%
34	Louisiana	0.0035%	0.0000%	0.1242%	0.0908%	0.0000%	0.0048%	0.0065%	0.0220%	0.1327%	0.0000%	0.1166%	0.0000%	0.0914%	0.0000%	0.0236%
35	Massachusetts	0.3092%	0.0590%	0.1228%	0.2868%	0.0742%	0.4438%	0.3281%	0.0239%	0.0969%	0.0132%	0.0997%	0.1153%	0.1086%	0.0260%	0.0218%
36	Illinois	0.2669%	0.0870%	0.0584%	0.1412%	0.1271%	0.3670%	2.4073%	0.9965%	0.2585%	0.6828%	0.0933%	0.0218%	0.0467%	0.0399%	0.0184%
37	Pennsylvania	0.7066%	0.6354%	0.1154%	0.2221%	0.0750%	0.0013%	0.0052%	0.0359%	0.0127%	0.1105%	0.0498%	0.0000%	0.0002%	0.0188%	0.0183%
38	Missouri	0.0200%	0.1133%	0.0732%	0.0702%	0.0474%	0.0294%	0.1427%	0.0569%	0.0378%	0.1073%	0.0667%	0.0291%	0.0474%	0.0306%	0.0161%
39	Maryland	0.8187%	0.1372%	0.0615%	0.0767%	0.1354%	0.0895%	0.0757%	0.0545%	0.0736%	0.0795%	0.0081%	0.0070%	0.0144%	0.0128%	0.0095%
40	Dist of Columbia	0.1842%	0.2954%	0.0707%	0.0659%	0.1117%	0.0510%	0.0792%	0.0675%	0.1573%	0.0000%	0.0000%	0.0000%	0.0177%	0.0160%	0.0078%
41	North Carolina	0.3553%	0.3086%	0.3642%	0.7588%	0.2598%	0.1113%	0.0440%	0.0318%	0.0316%	0.0105%	0.0070%	0.0073%	0.0467%	0.0067%	0.0040%
42	New Mexico	0.5855%	0.1795%	0.0006%	0.3472%	0.8458%	0.0194%	0.0048%	0.0052%	0.0005%	0.0009%	0.0000%	0.0000%	0.0022%	0.0028%	0.0022%
43	New Jersey	0.9368%	1.9499%	1.1626%	1.4538%	0.1629%	0.0276%	0.0242%	0.0326%	0.0404%	0.2568%	0.0154%	0.0000%	0.0021%	0.0018%	0.0009%
44	Tennessee	0.0000%	0.0000%	0.0000%	0.0000%	0.0000%	0.0000%	0.0000%	0.0000%	0.0000%	0.0365%	0.0011%	0.0000%	0.0000%	0.0000%	0.0000%
45	Rhode Island	0.4355%	0.0147%	0.0247%	0.0231%	0.0234%	0.1090%	0.0703%	0.2722%	0.0000%	1.1539%	7.1429%	7.1388%	7.1429%	0.0000%	0.0000%
46	New Hampshire	0.5067%	0.5129%	0.1624%	0.1652%	0.0022%	0.0007%	0.0181%	0.0152%	0.0155%	0.0000%	0.0000%	0.0000%	0.0000%	0.0000%	0.0000%
47	Hawaii	0.0009%	0.0523%	0.0000%	0.0000%	0.0000%	0.0050%	0.9110%	0.0680%	0.0713%	0.0598%	0.1285%	0.0000%	0.0000%	0.0000%	0.0000%
48	Vermont	0.6540%	0.2092%	0.0179%	0.0842%	0.0533%	0.0167%	0.0191%	0.0000%	0.0006%	0.0000%	1.2384%	0.0000%	0.0000%	0.0000%	0.0000%
49	Mississippi	0.0209%	0.0000%	0.0620%	0.0904%	0.0000%	0.0000%	0.0000%	0.0000%	0.0000%	0.0249%	0.0008%	0.0000%	0.0000%	0.0000%	0.0000%
50	Alaska	0.0000%	0.0000%	0.0000%	0.0000%	0.0000%	0.0000%	0.0000%	0.0000%	0.0000%	0.0000%	0.0000%	0.0000%	0.0000%	0.0000%	0.0000%
51	Maine	0.3434%	0.0400%	0.0430%	0.0152%	0.0007%	0.0110%	0.0112%	0.0102%	0.0000%	0.0000%	0.0000%	0.0000%	0.0000%	0.0000%	0.0000%

Table 6-4 ranks the potential incremental LM capacity savings over 15 years. Here, Connecticut, Wyoming, and Oregon have also achieved the one percent per year goal and join these other three states (Nebraska, Utah, and Alabama). About one of eight political jurisdictions in the U.S. reduced their capacity requirements about one percent incrementally in 2006 through utility-sponsored LM activities, including interruptible service tariffs.

TABLE 6-4
LOAD MANAGEMENT (LM) INCREMENTAL POTENTIAL PEAK REDUCTION / PEAK LOAD

Rank	State	1992	1993	1994	1995	1996	1997	1998	1999	2000	2001	2002	2003	2004	2005	2006
1	Connecticut	0.1340%	0.2094%	0.0466%	0.1020%	0.0135%	0.0000%	0.0000%	0.0000%	0.0000%	0.0693%	0.8114%	1.1816%	1.4544%	0.7026%	3.0090%
2	Utah	0.1804%	0.2546%	0.0315%	0.0060%	5.5491%	0.1609%	0.1746%	41.8642%	8.0461%	0.0175%	0.0989%	0.0481%	0.0000%	2.4202%	2.9075%
3	Wyoming	0.0182%	0.0184%	0.0154%	0.0041%	4.3300%	0.1802%	0.0705%	34.2285%	6.4330%	0.0000%	0.0000%	0.0000%	0.0000%	1.8739%	2.2939%
4	Nebraska	2.0295%	1.2155%	0.4892%	1.8106%	0.3756%	0.8623%	0.9694%	0.6731%	1.4583%	0.8977%	0.9582%	1.2094%	1.6750%	1.8067%	1.9269%
5	Alabama	0.9365%	0.1321%	0.3645%	0.3394%	1.0291%	0.7369%	0.0517%	0.1952%	0.0988%	0.7437%	0.7172%	0.0000%	0.3501%	0.1133%	1.0796%
6	Oregon	0.1316%	0.1232%	0.0005%	0.0000%	1.7304%	0.0000%	0.0000%	13.7003%	1.6576%	0.6950%	0.2892%	0.0000%	0.0029%	0.8788%	0.9967%
7	New York	1.0230%	0.2929%	0.2164%	0.1286%	0.1194%	0.0953%	0.1977%	0.0116%	0.0702%	0.1547%	0.4828%	0.2492%	1.1266%	0.8620%	0.7551%
8	Colorado	0.1587%	0.0795%	0.3406%	0.1856%	1.4661%	2.1635%	0.5444%	0.7088%	0.3488%	0.5528%	1.4550%	0.4049%	2.1464%	0.5447%	0.7522%
9	Indiana	1.1527%	0.6179%	0.0824%	0.2523%	0.2911%	0.0395%	0.2025%	0.5382%	0.0060%	0.4349%	0.0000%	0.0841%	0.5172%	1.3907%	0.7399%
10	Idaho	0.0000%	0.0000%	0.0000%	0.0000%	0.8582%	0.0000%	0.0000%	6.7453%	1.2850%	5.0547%	0.0000%	0.0000%	0.1068%	1.2964%	0.6252%
11	North Dakota	4.2689%	3.0393%	2.5191%	0.4602%	0.2935%	2.4401%	2.3552%	2.3689%	2.5339%	2.5546%	0.4348%	0.6634%	0.7917%	0.2459%	0.6089%
12	South Dakota	1.2539%	1.0735%	0.3338%	0.3704%	0.1853%	0.5506%	0.1981%	0.1833%	0.2268%	0.2128%	0.3883%	0.1498%	0.4744%	0.6178%	0.5738%
13	Wisconsin	1.4692%	0.6907%	0.7156%	0.4250%	0.6223%	0.1032%	0.0590%	0.0652%	0.2796%	0.2472%	0.8023%	0.1079%	0.3667%	0.2990%	0.5678%
14	Hawaii	0.0009%	0.0523%	0.0000%	0.0008%	0.2916%	0.0061%	1.3071%	0.4446%	0.4485%	0.0598%	0.1927%	0.0627%	0.0580%	0.0000%	0.5143%
15	Minnesota	3.2664%	2.2286%	1.6184%	2.3350%	2.2648%	0.5181%	0.5590%	0.5985%	0.6665%	1.6089%	1.3572%	0.5335%	0.7794%	0.7924%	0.4909%
16	California	1.4329%	1.6226%	0.4003%	1.6275%	2.1211%	1.0637%	1.1805%	1.1295%	0.7976%	0.9464%	0.8878%	0.0972%	1.3093%	0.2252%	0.4019%
17	Louisiana	0.5198%	0.0000%	0.1381%	0.1034%	0.0000%	0.0145%	0.0167%	0.0220%	0.4319%	0.1012%	0.4854%	0.1046%	0.3656%	0.0000%	0.3892%
18	West Virginia	1.0529%	1.7194%	0.0037%	0.3674%	0.1564%	0.0014%	0.0041%	0.0043%	0.0022%	0.0000%	0.0000%	0.0937%	0.3746%	0.0149%	0.3873%
19	Georgia	1.8558%	0.8920%	0.8146%	0.4803%	0.5133%	0.2136%	0.1374%	0.1119%	0.1717%	0.0952%	0.1490%	0.1302%	0.1735%	0.1171%	0.3303%
20	Iowa	1.7472%	4.1339%	1.0875%	0.5409%	0.1902%	1.1663%	0.3396%	0.2541%	0.2027%	0.2611%	3.8602%	0.1510%	0.8877%	0.2261%	0.3252%
21	Virginia	0.5410%	0.6935%	0.2584%	1.1172%	0.1438%	0.1474%	0.0491%	0.0370%	0.0701%	0.0328%	0.0540%	0.0626%	0.3085%	0.1722%	0.2651%
22	Florida	1.7007%	0.8558%	0.4762%	0.8197%	0.3813%	0.4235%	0.2286%	0.2130%	0.1811%	0.1297%	0.1971%	0.2997%	0.1983%	0.2644%	0.1629%
23	Kansas	0.6511%	0.3028%	0.1785%	0.0405%	0.5456%	0.0622%	0.1548%	0.1264%	0.1695%	0.0307%	0.0729%	0.3375%	0.1646%	0.1473%	0.1626%
24	Arkansas	0.4472%	1.5374%	1.6999%	1.2321%	0.2149%	0.0589%	0.0703%	0.1139%	0.0741%	0.0571%	0.0657%	0.2210%	0.1269%	0.0885%	0.1543%
25	Montana	0.0070%	0.0168%	0.0197%	0.0312%	0.0234%	0.0245%	0.0180%	0.0224%	0.0057%	0.0000%	0.0385%	0.0298%	0.0283%	0.0550%	0.1513%

(Continued)

TABLE 6-4 (CONT'D)
LOAD MANAGEMENT (LM) INCREMENTAL POTENTIAL PEAK REDUCTION / PEAK LOAD

Rank	State	1992	1993	1994	1995	1996	1997	1998	1999	2000	2001	2002	2003	2004	2005	2006
26	Washington	0.1246%	0.0769%	0.1165%	0.2258%	0.2623%	0.4180%	0.3327%	2.0865%	0.1468%	0.2951%	0.0203%	0.0000%	0.0000%	0.1147%	0.1364%
27	South Carolina	1.0192%	0.6985%	0.1100%	0.5467%	0.2686%	0.3998%	0.1089%	0.3902%	0.1236%	0.0099%	0.0195%	0.0416%	0.1109%	0.2200%	0.1148%
28	Michigan	0.1324%	0.2649%	0.0422%	2.9246%	2.8919%	0.0512%	0.0223%	0.0906%	0.0198%	0.0021%	0.0458%	0.0127%	0.1267%	0.1727%	0.1103%
29	Kentucky	1.5591%	0.2337%	0.4289%	0.0904%	0.0534%	0.0599%	0.0020%	0.0014%	0.0019%	0.1685%	0.0696%	0.0389%	0.0958%	0.1092%	0.1082%
30	Nevada	1.7126%	0.4508%	0.7917%	0.1539%	0.0000%	0.0000%	0.0000%	0.0000%	0.0000%	0.3231%	0.0712%	0.1352%	0.3305%	0.0969%	0.0978%
31	Delaware	1.1655%	0.4567%	0.1581%	0.5944%	0.4058%	0.4572%	0.0829%	0.0821%	0.1402%	0.0664%	0.1630%	0.1314%	0.1664%	0.2026%	0.0847%
32	Arizona	0.6944%	1.7139%	1.5342%	1.6715%	1.0271%	0.0368%	0.0375%	0.0438%	0.0312%	0.2146%	0.1306%	0.4339%	0.4743%	0.1412%	0.0595%
33	Texas	0.8235%	1.0899%	2.0325%	0.4739%	0.4877%	0.4254%	0.3925%	0.0720%	0.1306%	0.0722%	0.0941%	0.0244%	0.1272%	0.0424%	0.0553%
34	Missouri	0.0727%	0.2286%	0.1413%	0.0824%	0.0665%	0.0584%	0.1750%	0.0764%	0.0807%	0.1614%	0.0884%	0.1003%	0.0386%	0.0617%	0.0444%
35	North Carolina	1.6807%	1.0847%	0.3273%	0.8345%	0.3316%	0.2379%	0.0771%	0.0604%	0.0518%	0.0140%	0.0112%	0.0075%	0.0507%	0.0490%	0.0368%
36	Illinois	0.3457%	0.1201%	0.1077%	0.1719%	0.1674%	0.3812%	2.5913%	1.0115%	0.2778%	1.3549%	0.1888%	0.0341%	0.0714%	0.0555%	0.0355%
37	Pennsylvania	0.9771%	0.8982%	0.1228%	0.2299%	1.2093%	0.0013%	0.0057%	0.0390%	0.0133%	0.2671%	0.0965%	0.0877%	0.0002%	0.0251%	0.0335%
38	Massachusetts	0.3435%	0.4561%	0.2477%	0.3035%	0.1359%	0.5383%	0.5623%	0.0505%	0.2037%	0.1452%	0.1709%	0.1729%	0.1221%	0.0391%	0.0291%
39	Maryland	1.0327%	1.6768%	0.8602%	1.3003%	0.4337%	0.3880%	0.3197%	0.3954%	0.1963%	0.1446%	0.0120%	0.0070%	0.0144%	0.0224%	0.0223%
40	Ohio	0.7708%	0.7420%	0.6241%	0.1111%	0.5264%	0.0819%	0.2911%	0.0654%	0.0368%	0.1227%	0.0295%	0.0661%	0.3494%	0.0731%	0.0163%
41	Dist of Columbia	0.6988%	0.5735%	0.5477%	0.3549%	0.2432%	0.2760%	0.0792%	0.5905%	0.2272%	0.0000%	0.0166%	0.0000%	0.0177%	0.0241%	0.0078%
42	New Mexico	1.0223%	1.6081%	1.1593%	0.6345%	1.6541%	1.9145%	1.8156%	0.0066%	0.0011%	0.0017%	0.0049%	0.0000%	0.0032%	0.0039%	0.0033%
43	New Jersey	1.2277%	1.9721%	1.1772%	1.7442%	0.2748%	0.3393%	0.2015%	0.0765%	0.0413%	0.2568%	0.4438%	0.0854%	0.0298%	0.0027%	0.0009%
44	Tennessee	1.2209%	0.0000%	0.2456%	0.0637%	0.0000%	0.0000%	0.0000%	0.0000%	0.0000%	0.0465%	0.0011%	0.0000%	0.0000%	0.0000%	0.0000%
45	Oklahoma	0.6824%	1.1160%	0.2081%	1.0958%	0.2085%	0.4462%	0.6330%	0.1054%	0.1005%	0.1088%	0.0001%	0.0478%	0.0000%	0.1295%	0.0000%
46	Rhode Island	0.4355%	0.0147%	0.0247%	0.0231%	0.0234%	0.1090%	0.1739%	0.2722%	0.0030%	1.1539%	7.1429%	7.1388%	7.1429%	0.0000%	0.0000%
47	New Hampshire	0.5273%	0.8485%	0.1821%	0.2384%	0.0223%	0.0369%	0.0181%	0.0152%	0.0155%	0.0000%	0.0000%	0.0000%	0.0000%	0.0000%	0.0000%
48	Vermont	0.9839%	0.5887%	0.0179%	0.1080%	0.1397%	0.0297%	0.0191%	0.0000%	0.0009%	0.0000%	3.0960%	0.0000%	0.0000%	0.0027%	0.0009%
49	Mississippi	0.9314%	0.0000%	0.2512%	0.1419%	0.0000%	0.0000%	0.0000%	0.0000%	0.0000%	0.0317%	0.0008%	0.0000%	0.0000%	0.0000%	0.0000%
50	Alaska	0.0000%	0.0000%	0.0000%	0.0000%	0.0000%	0.0000%	0.0000%	0.0000%	0.0000%	0.0000%	0.0000%	0.0000%	0.0000%	0.0000%	0.0000%
51	Maine	0.3488%	0.0456%	0.0483%	0.0210%	0.0007%	0.0166%	0.0169%	0.0153%	0.0000%	0.0000%	0.0000%	0.0000%	0.0000%	0.0000%	0.0000%

Table 6-5 shows the state-by-state comparison for *actual cumulative* LM capacity savings over the 15-year period expressed as a percent of peak load. Three states (North Dakota, Nebraska, and Minnesota) have saved more than five percent of capacity cumulatively. North Dakota is the only state to surpass 10 percent. In 2006, North Dakota saved about 9.7 percent of its peak load through LM programs in the state.

TABLE 6-5
LOAD MANAGEMENT (LM) CUMULATIVE ACTUAL PEAK REDUCTION / PEAK LOAD

Rank	State	1992	1993	1994	1995	1996	1997	1998	1999	2000	2001	2002	2003	2004	2005	2006
1	North Dakota	14.8655%	15.0995%	14.6033%	15.8383%	14.5963%	12.0171%	12.9976%	13.1259%	12.6400%	5.9258%	14.1301%	12.9607%	6.6665%	6.3814%	9.6700%
2	Nebraska	1.9037%	0.6872%	0.6667%	5.0501%	3.9378%	1.1119%	6.5189%	5.7312%	5.5487%	7.2827%	6.6749%	6.0455%	7.2528%	6.6762%	7.1904%
3	Minnesota	3.8442%	4.9099%	5.7068%	6.4641%	6.8996%	5.0936%	4.7665%	4.9982%	5.4290%	6.1476%	6.9123%	5.0063%	5.6311%	5.8315%	5.4723%
4	Iowa	2.3214%	5.1654%	4.6052%	8.2061%	3.6993%	2.9116%	3.4286%	3.2992%	3.2369%	3.3740%	2.2895%	5.1065%	3.7071%	5.4621%	4.8448%
5	Wisconsin	0.4858%	1.0029%	2.8964%	0.8982%	3.6839%	0.4322%	0.8927%	0.9271%	0.8935%	0.8518%	0.6586%	1.5343%	3.8814%	3.4262%	3.7012%
6	Colorado	2.1463%	1.3055%	2.6953%	2.7796%	3.9266%	4.2082%	5.1803%	2.1035%	2.1939%	2.5011%	3.4590%	2.3705%	3.0283%	2.8260%	2.8892%
7	California	1.7031%	1.8845%	1.9000%	1.8998%	1.8778%	1.3792%	2.8355%	0.9924%	2.4672%	2.4090%	1.4053%	1.5295%	2.1636%	3.2500%	2.6801%
8	Florida	1.8245%	2.7908%	2.6886%	5.1129%	5.8844%	4.1141%	4.3324%	4.8844%	3.7874%	4.7603%	3.4719%	3.4751%	2.5760%	1.9884%	2.5467%
9	South Dakota	2.1973%	2.6927%	2.0141%	2.8617%	2.2715%	2.0717%	2.0315%	2.0819%	2.0457%	2.2589%	2.5882%	1.9066%	1.5634%	2.1456%	2.1616%
10	Georgia	2.7056%	3.1794%	3.9466%	5.3986%	1.0339%	0.9490%	0.9848%	0.8734%	0.9203%	0.7692%	2.4058%	1.9869%	2.0593%	1.9318%	1.8034%
11	Arkansas	3.1241%	2.8245%	2.9865%	0.8286%	0.8599%	0.7014%	2.1475%	2.5978%	1.9501%	1.7375%	1.4901%	1.0019%	0.9584%	2.1091%	1.6038%
12	Indiana	0.0741%	0.4099%	0.2004%	0.6178%	0.6814%	0.8477%	1.1559%	1.5105%	0.3828%	1.2436%	0.3001%	0.3231%	1.1383%	2.2874%	1.5190%
13	Arizona	1.0312%	1.9665%	1.5696%	2.0164%	0.7436%	0.5832%	0.5594%	0.6118%	0.5911%	0.7708%	0.6632%	0.6525%	0.7082%	0.6303%	1.2942%
14	Alabama	0.5316%	1.9427%	1.9534%	2.0522%	2.0502%	1.7151%	1.5967%	1.6283%	0.3623%	0.8773%	0.8473%	0.3314%	0.4543%	0.1846%	1.1388%
15	Delaware	3.4239%	4.9082%	7.4584%	3.7200%	0.5113%	0.4806%	0.8920%	0.5875%	0.6471%	0.8302%	0.9081%	1.0511%	1.0882%	0.9170%	1.0632%
16	Virginia	1.8399%	1.8203%	2.5485%	1.8108%	0.9383%	1.8003%	1.7672%	1.4022%	1.4496%	0.9088%	0.9124%	1.1199%	1.0092%	0.9025%	1.0537%
17	Oklahoma	1.3919%	2.5248%	1.5536%	1.7653%	1.5159%	0.8266%	1.6019%	1.2774%	0.3446%	1.0851%	1.3563%	0.2937%	0.1121%	0.1181%	0.9142%
18	Pennsylvania	2.6891%	2.2673%	1.7821%	1.8477%	1.3588%	0.8508%	0.3998%	0.5755%	0.3692%	0.4352%	1.5477%	0.4993%	0.7715%	0.8529%	0.8328%
19	West Virginia	0.1729%	1.3636%	0.7840%	0.8538%	0.0373%	1.0892%	1.8923%	1.5149%	0.9605%	0.2285%	0.0201%	0.1138%	0.3255%	0.3194%	0.8234%
20	New Jersey	1.9066%	2.9755%	3.8852%	4.4390%	1.9976%	0.5115%	0.5381%	0.3307%	0.4097%	1.2993%	0.3973%	0.3203%	0.0021%	0.8708%	0.8046%
21	Idaho	0.0043%	0.0020%	0.0019%	0.0000%	0.0000%	0.0000%	0.0000%	1.0712%	1.2952%	6.4951%	0.0000%	0.0000%	0.1068%	0.9343%	0.7475%
22	New York	1.1361%	0.5382%	0.3442%	0.3560%	0.3027%	0.2751%	0.3590%	0.0253%	0.2531%	0.6105%	0.6155%	0.8173%	0.4305%	0.3655%	0.7473%
23	Ohio	0.1452%	1.3815%	1.0041%	0.8036%	1.4115%	0.4155%	1.1018%	0.9584%	0.4677%	0.4256%	0.0828%	0.1214%	0.4746%	0.4512%	0.7467%
24	Kentucky	0.0541%	3.3957%	3.1691%	3.1789%	3.1166%	3.4059%	2.7959%	2.5512%	0.0567%	0.1766%	0.1596%	0.1756%	0.3106%	0.4181%	0.5078%
25	Illinois	1.3421%	0.5363%	0.5516%	1.5381%	1.4969%	3.1750%	5.0799%	3.8498%	0.7551%	2.3545%	0.4921%	0.3481%	0.5781%	0.5359%	0.3789%
26	South Carolina	1.8171%	2.0754%	1.5892%	1.4085%	1.2610%	1.6386%	0.3545%	0.3520%	0.3768%	0.2423%	0.2785%	0.3099%	0.4823%	0.3817%	0.3581%

TABLE 6-5 (CONT'D)
LOAD MANAGEMENT (LM) CUMULATIVE ACTUAL PEAK REDUCTION / PEAK LOAD

Rank	State	1992	1993	1994	1995	1996	1997	1998	1999	2000	2001	2002	2003	2004	2005	2006
27	Maryland	4.0104%	1.2641%	1.5926%	1.2664%	1.6689%	1.3691%	1.3919%	1.0822%	1.1536%	2.8147%	3.1158%	0.3451%	0.3611%	0.1485%	0.3160%
28	Missouri	1.1740%	1.1220%	1.0154%	1.2126%	1.0504%	0.9060%	1.0176%	1.2772%	0.9840%	0.3917%	0.2703%	0.3996%	0.4109%	0.3110%	0.2801%
29	Montana	0.6410%	0.4370%	0.4828%	0.2607%	0.0834%	0.1058%	0.2597%	0.3737%	0.0658%	0.5352%	0.2305%	0.0600%	0.0642%	0.0554%	0.2778%
30	Nevada	2.8830%	0.9624%	1.7785%	0.0577%	0.0000%	0.0000%	0.0000%	0.0000%	0.0000%	0.0189%	0.0712%	0.1352%	0.1190%	0.2655%	0.2625%
31	Michigan	0.0265%	0.2025%	0.2986%	3.2370%	3.3789%	0.1012%	0.1753%	0.2181%	0.0339%	0.1443%	0.0301%	0.0431%	0.1939%	0.2108%	0.2420%
32	Wyoming	0.0735%	0.0909%	0.0523%	0.1254%	0.0491%	0.2324%	0.2495%	5.7359%	6.7140%	6.5125%	0.2764%	0.2647%	0.2721%	0.2140%	0.2023%
33	Texas	0.4649%	0.9253%	0.7932%	0.8159%	0.9135%	0.1503%	0.1328%	0.1772%	0.1129%	0.1648%	0.1134%	0.0576%	0.0251%	0.1429%	0.1888%
34	Kansas	0.0370%	0.5673%	1.0396%	0.8225%	2.6376%	2.5848%	2.5666%	0.6655%	0.5066%	0.5390%	0.4341%	0.5460%	0.4795%	0.5956%	0.1050%
35	North Carolina	2.2706%	2.2442%	2.3624%	2.6823%	2.9552%	3.7789%	0.6120%	0.4461%	0.3998%	0.3275%	0.1525%	0.0554%	0.1561%	0.0440%	0.0512%
36	Mississippi	0.6141%	2.2869%	2.1057%	1.7607%	1.9632%	2.0579%	1.9076%	1.3879%	0.0155%	0.0864%	0.0623%	0.0605%	0.0545%	0.5033%	0.0466%
37	Louisiana	0.1979%	0.1850%	0.2803%	0.0908%	0.0906%	0.1401%	0.1377%	0.2388%	0.1565%	0.0316%	0.1488%	0.0308%	0.1341%	0.4168%	0.0413%
38	Tennessee	0.0000%	2.6961%	2.2956%	2.3215%	2.1550%	2.1487%	2.1093%	2.0045%	0.0001%	0.0963%	0.0588%	0.0570%	0.0497%	0.0441%	0.0399%
39	Oregon	0.1366%	0.0594%	0.0606%	0.0000%	0.0000%	0.0000%	0.0000%	2.1752%	1.6680%	5.3346%	0.0000%	0.0000%	0.0029%	0.0229%	0.0184%
40	Dist of Columbia	2.4883%	2.2767%	2.5795%	2.5645%	2.9028%	2.7246%	2.6175%	2.6320%	2.8841%	2.5081%	6.5745%	0.0000%	0.0177%	0.0160%	0.0078%
41	Washington	0.7428%	0.3656%	0.3707%	0.1907%	0.1429%	0.1587%	0.1576%	0.2660%	0.0873%	0.5051%	0.0068%	0.0045%	0.0042%	0.0042%	0.0043%
42	New Mexico	0.6053%	0.3232%	0.3623%	0.6779%	1.1642%	0.0256%	0.0140%	0.0211%	0.0097%	0.2420%	0.0001%	0.0001%	0.0049%	0.0056%	0.0038%
43	Massachusetts	0.4053%	0.4671%	0.6885%	1.4566%	0.4547%	0.9996%	0.8736%	0.4498%	0.5392%	0.0924%	0.1994%	0.1874%	0.1629%	0.0000%	0.0000%
44	Utah	0.2526%	0.1705%	0.2687%	0.1776%	0.2565%	0.2297%	0.0058%	6.6285%	8.0351%	6.9803%	0.1781%	0.0481%	0.0816%	0.0924%	0.0000%
45	Connecticut	0.1716%	0.8510%	0.9058%	1.0289%	2.3040%	0.5399%	0.5562%	0.4965%	0.5451%	0.0693%	0.1679%	0.4035%	0.5788%	0.6098%	0.0000%
46	Rhode Island	0.4844%	0.0274%	0.0247%	0.0231%	0.0234%	0.1090%	0.5118%	0.2722%	0.0000%	1.1539%	7.1429%	7.1388%	7.1429%	0.0000%	0.0000%
47	New Hampshire	1.2717%	1.6671%	0.6827%	0.3148%	0.2351%	0.1637%	0.1827%	0.1978%	0.1313%	0.3016%	0.1449%	0.1464%	0.1505%	0.1362%	0.0000%
48	Hawaii	0.1700%	0.0523%	0.2661%	0.0996%	0.4865%	0.4509%	1.8728%	1.3425%	1.4059%	1.3160%	0.2569%	0.1880%	0.1740%	0.0000%	0.0000%
49	Vermont	5.4993%	5.5268%	0.7243%	0.7506%	0.7121%	1.0181%	0.5330%	0.6029%	0.0000%	0.0000%	1.2384%	1.1983%	1.1765%	0.0000%	0.0000%
50	Alaska	0.6062%	0.5650%	0.1940%	0.1817%	0.3674%	0.3511%	0.3010%	0.2536%	0.2790%	0.2516%	0.3167%	0.3271%	0.1668%	0.2054%	0.0000%
51	Maine	0.7057%	0.9894%	1.2153%	1.2722%	1.3026%	1.2582%	0.0381%	0.0343%	0.0293%	0.0321%	6.7225%	0.0482%	0.0000%	0.0000%	0.0000%

Table 6-6 shows the corresponding cumulative potential capacity savings for the same 51 jurisdictions. The top eight performers exceeded five percent cumulative capacity savings in 2006 and North Dakota exceeded the 10 percent goal that same year, achieving a savings of about 11.1 percent.

TABLE 6-6
LOAD MANAGEMENT (LM) CUMULATIVE POTENTIAL PEAK REDUCTION / PEAK LOAD

Rank	State	1992	1993	1994	1995	1996	1997	1998	1999	2000	2001	2002	2003	2004	2005	2006
1	North Dakota	18.7592%	15.6446%	15.2821%	16.9341%	15.8152%	13.6143%	15.1438%	15.0226%	14.8372%	15.4815%	16.3944%	15.0327%	8.0641%	7.6927%	11.1007%
2	Nebraska	2.9319%	1.4184%	1.0490%	7.5705%	8.2904%	1.9037%	9.7167%	8.9998%	9.1106%	9.6280%	9.8650%	9.1420%	11.0410%	9.0147%	9.7007%
3	Minnesota	6.1029%	6.2936%	6.7191%	7.6455%	8.2828%	6.4705%	6.0813%	6.2755%	6.8919%	7.5159%	7.6902%	5.6747%	6.2456%	6.4625%	6.1558%
4	Iowa	5.0865%	8.9973%	9.1916%	8.7695%	8.7297%	8.9954%	9.8654%	8.2292%	8.7412%	8.0055%	3.7995%	6.7141%	4.7052%	7.0505%	5.6805%
5	Alabama	5.1575%	7.8077%	7.3301%	7.6723%	8.3180%	8.8632%	7.9885%	7.9502%	4.7489%	5.9106%	6.1485%	0.4833%	5.9459%	5.7039%	5.6158%
6	Arkansas	4.3088%	12.0984%	13.3257%	10.5527%	9.6401%	5.7291%	8.7902%	5.9030%	5.3179%	5.4209%	5.7062%	5.1740%	5.1841%	5.5504%	5.3935%
7	Florida	6.3655%	6.6766%	6.9571%	7.0650%	7.0496%	6.0052%	5.4072%	5.4793%	5.4979%	5.1081%	5.0241%	4.9738%	4.2870%	4.4766%	5.2201%
8	Wisconsin	3.0657%	3.3954%	6.0525%	6.4633%	6.3903%	2.3925%	5.5256%	5.0372%	5.3825%	4.5232%	4.7629%	4.8033%	5.2535%	4.8141%	5.2026%
9	Illinois	2.2958%	1.7997%	1.8084%	1.9803%	2.1598%	3.2743%	5.3536%	4.3682%	4.7043%	5.3923%	5.9299%	5.0538%	5.9729%	5.2238%	4.2571%
10	California	4.8370%	5.2910%	5.5244%	5.8402%	6.4616%	5.9906%	5.7798%	5.8283%	6.2315%	5.5794%	3.3399%	2.6171%	4.0599%	4.9449%	4.0289%
11	Indiana	1.4483%	1.5726%	0.5877%	1.4482%	2.4863%	1.8190%	1.9200%	2.1111%	2.5836%	4.8895%	4.2381%	3.5703%	3.8740%	3.5780%	3.9851%
12	Colorado	3.0915%	3.3509%	3.7988%	3.6930%	6.4864%	7.3006%	9.7617%	3.1651%	2.4872%	2.8842%	4.7644%	2.7962%	3.4403%	3.3906%	3.6053%
13	South Carolina	4.9482%	4.9975%	5.8198%	4.1516%	4.1003%	4.5110%	3.4940%	3.3582%	3.2046%	2.5663%	2.5535%	2.7053%	2.9868%	2.9235%	2.7061%
14	South Dakota	4.2276%	3.2351%	2.2863%	3.5743%	2.8083%	2.5889%	2.4324%	2.4294%	2.7075%	2.6448%	3.0717%	2.4352%	1.9549%	2.5615%	2.5600%
15	Kansas	5.1240%	4.5330%	5.2527%	5.3504%	5.1122%	4.5854%	4.4919%	3.1207%	2.7328%	1.0353%	2.4143%	2.3642%	2.4578%	2.1212%	2.3876%
16	Georgia	3.5095%	3.8346%	4.6730%	6.1574%	4.0398%	1.8590%	1.5054%	1.2730%	1.2440%	1.0983%	2.7782%	2.4818%	2.5833%	2.2520%	2.0730%
17	Maryland	5.6046%	7.4930%	9.9471%	9.1693%	8.5548%	22.1968%	8.7421%	7.7220%	7.7192%	7.1823%	5.9254%	4.7975%	4.0032%	1.9004%	1.8308%
18	Vermont	5.6912%	6.3599%	1.1367%	1.3135%	1.8607%	1.6700%	0.8364%	1.1524%	0.0546%	0.0000%	3.0960%	2.9958%	2.9412%	1.1635%	1.8067%
19	Alaska	0.6062%	0.5650%	0.1940%	0.1817%	0.9726%	0.6894%	0.8126%	0.6522%	0.7158%	0.6289%	0.7918%	0.8177%	0.4170%	1.0786%	1.7487%
20	North Carolina	7.6676%	7.2369%	6.9451%	5.9404%	6.8304%	7.4669%	4.4718%	3.6032%	3.1298%	2.5987%	2.4837%	2.0962%	2.1110%	1.9522%	1.7020%
21	Delaware	8.0430%	7.8170%	8.2174%	8.1394%	8.6946%	7.2645%	7.1830%	0.8370%	0.8970%	1.0628%	1.1685%	1.3138%	1.3515%	1.2611%	1.4267%
22	Arizona	3.1511%	2.8780%	3.2250%	3.6454%	2.2992%	1.0336%	0.6682%	0.7373%	2.3152%	2.1976%	1.6832%	1.5917%	1.6302%	1.4249%	1.3241%
23	Virginia	2.9606%	2.4230%	2.9515%	2.7174%	1.7899%	2.5522%	2.4590%	1.9198%	1.9347%	1.3191%	1.5691%	1.8080%	1.6094%	1.3429%	1.2573%
24	Oklahoma	4.1868%	4.3473%	3.6612%	4.2780%	3.3380%	1.8518%	2.9650%	1.8594%	0.4913%	1.7005%	1.6356%	1.5522%	0.8418%	1.5744%	1.1872%
25	West Virginia	2.4122%	2.3399%	1.5549%	2.5850%	1.0917%	2.4538%	2.5895%	2.3612%	1.5130%	1.2475%	1.2707%	1.1699%	1.1695%	0.7968%	1.1729%

(Continued)

TABLE 6-6 (CONT'D)
LOAD MANAGEMENT (LM) CUMULATIVE POTENTIAL PEAK REDUCTION / PEAK LOAD

Rank	State	1992	1993	1994	1995	1996	1997	1998	1999	2000	2001	2002	2003	2004	2005	2006
26	Ohio	2.2188%	2.4580%	2.9280%	3.0583%	3.3800%	1.1949%	1.3186%	1.3585%	1.3651%	2.0374%	1.5708%	1.3430%	1.5782%	1.2105%	1.1697%
27	Hawaii	1.0234%	0.4865%	0.6342%	0.4409%	1.2618%	1.5222%	3.6104%	3.0717%	3.5125%	2.9310%	0.7065%	0.7519%	0.6961%	0.3544%	0.9143%
28	Idaho	0.5428%	0.5361%	0.6312%	0.6275%	0.8582%	0.0000%	0.0000%	6.7453%	7.7025%	13.6205%	8.1572%	7.7692%	0.1246%	0.9749%	0.9048%
29	New York	1.2447%	0.6898%	0.3874%	0.3667%	0.3402%	0.2988%	0.3745%	0.0545%	0.3581%	0.8651%	1.1657%	1.2279%	1.2788%	1.0101%	0.8582%
30	Pennsylvania	4.6502%	3.7196%	4.3282%	4.3371%	4.3164%	0.9912%	1.3145%	1.6856%	1.2479%	0.6164%	1.8372%	0.9882%	0.9625%	0.9986%	0.8570%
31	New Jersey	2.2370%	4.2027%	4.2923%	5.6265%	3.0693%	2.7331%	2.9208%	2.8751%	2.4977%	2.8042%	2.0051%	1.6230%	0.9923%	0.8717%	0.8046%
32	Kentucky	3.4265%	6.5212%	6.5806%	6.4163%	6.5368%	7.0457%	6.0006%	5.4555%	0.1975%	0.4792%	0.4053%	0.3949%	0.4044%	0.4466%	0.5531%
33	Michigan	0.2472%	0.4176%	0.4981%	3.8814%	3.9845%	0.2687%	0.4652%	0.5241%	0.5365%	0.4718%	0.5122%	0.5318%	0.5563%	0.4684%	0.4758%
34	New Hampshire	1.3917%	2.2726%	0.7312%	0.8880%	0.9580%	0.7731%	0.2172%	0.2455%	0.2633%	0.6535%	0.5797%	0.5368%	0.5517%	0.4993%	0.3998%
35	Missouri	1.7412%	1.5297%	1.6580%	1.5659%	1.4375%	1.2854%	1.3862%	1.6875%	2.2972%	1.1129%	1.0652%	1.1671%	1.2629%	0.9916%	0.3957%
36	Wyoming	2.7091%	2.5561%	3.1117%	3.2001%	4.3977%	0.2379%	0.2981%	34.5511%	38.5450%	37.5205%	36.9875%	35.4238%	0.4003%	0.3791%	0.3543%
37	Montana	3.5367%	2.5249%	2.5393%	2.4176%	0.1064%	0.1084%	0.2621%	0.3769%	0.0756%	0.6835%	0.3501%	0.1356%	0.1360%	0.1323%	0.3418%
38	Nevada	6.2369%	2.4410%	3.9028%	0.1987%	0.0000%	0.0000%	0.0000%	0.0000%	0.0000%	0.3231%	0.0712%	0.1352%	0.3306%	0.2655%	0.2626%
39	Mississippi	2.3850%	4.1104%	4.0443%	3.6339%	3.7906%	3.8335%	3.5895%	3.2755%	0.2141%	0.3064%	0.2716%	0.2612%	0.2549%	0.2432%	0.2360%
40	Texas	4.0569%	4.9185%	4.5788%	4.4516%	3.8051%	0.6445%	0.7474%	0.4650%	0.4704%	0.4507%	0.1513%	0.0720%	0.0800%	0.1570%	0.2073%
41	Washington	2.0043%	0.8316%	0.8067%	0.9268%	0.5193%	0.5798%	0.5597%	2.1921%	0.4758%	3.4175%	3.2379%	2.0524%	0.0042%	0.1094%	0.1106%
42	Louisiana	0.7733%	0.6758%	0.2943%	0.1990%	0.1032%	0.4485%	0.4377%	0.2845%	0.4669%	0.1455%	0.5305%	0.1477%	0.4204%	0.0535%	0.0413%
43	Tennessee	2.3092%	5.2116%	4.8115%	4.9455%	4.5908%	4.5774%	4.5701%	4.3429%	0.0001%	0.1063%	0.0588%	0.0570%	0.0497%	0.0441%	0.0399%
44	Massachusetts	0.5219%	1.1496%	0.9664%	1.7687%	1.1837%	1.3354%	1.3981%	0.6554%	1.3083%	0.3565%	0.3703%	0.3171%	0.2850%	0.0651%	0.0364%
45	Oregon	1.3365%	1.0035%	1.2535%	1.2509%	1.6032%	0.0000%	0.0000%	13.6973%	9.1721%	29.7930%	17.8471%	16.0051%	0.0034%	0.0239%	0.0223%
46	Dist of Columbia	5.7158%	5.7178%	6.7668%	7.3035%	8.1581%	7.4002%	7.1982%	7.6430%	7.6560%	6.4332%	6.6525%	3.2596%	0.5599%	0.1133%	0.0222%
47	New Mexico	2.5269%	1.7570%	1.5255%	0.9697%	1.9766%	1.9269%	1.8325%	0.0309%	3.0259%	0.4830%	0.0196%	0.0204%	0.0252%	0.0089%	0.0066%
48	Utah	3.5914%	3.2900%	4.2619%	4.0869%	5.8762%	0.3526%	0.1454%	41.8511%	47.6278%	41.1480%	46.5458%	22.6455%	0.1468%	0.1664%	0.0000%
49	Connecticut	0.3721%	0.8510%	0.9058%	1.0289%	2.3040%	0.5399%	0.5562%	0.4965%	0.5451%	0.1386%	0.8114%	1.3402%	1.4544%	0.6363%	0.0000%
50	Rhode Island	0.4844%	0.0274%	0.0247%	0.0231%	0.0234%	0.1090%	0.8418%	0.2722%	0.0000%	1.1539%	7.1429%	7.1388%	7.1429%	0.0000%	0.0000%
51	Maine	0.9576%	1.2586%	1.2571%	1.3140%	1.3454%	1.2997%	0.0795%	0.0723%	0.0293%	0.0343%	6.7234%	0.0482%	0.0000%	0.0000%	0.0000%

Regardless of the aggregated focus, there are two reasonable goals, if the nation seeks realistic targets for energy efficiency or demand-side management. The first of these two goals is for IOUs and their state regulators to seek savings equal to 10 percent of total sales volume on a cumulative basis. The second goal is for governors and legislators to establish a 10 percent cumulative savings goal for all the electric utilities in their respective states. The reason to consider both types of targets is that most state regulation has no or only limited control over electricity that investor-owned utilities do not supply. There are also often some limitations on state regulatory control over electricity markets in restructured states.

The incremental goals for both energy efficiency and LM may be a bit more aggressive. Nevertheless, achieving one percent for both incremental energy (MWh) and capacity (MW) savings has been achieved at both the IOU and state levels. Therefore, a one percent incremental target seems to be a stretch, but still an achievable goal.

The next chapter reviews a second performance measure. This chapter shifts to measures of effort. This is measured in terms of the share of revenue requirements that is invested in energy efficiency and load management.

CHAPTER 7
BENCHMARKING EFFORT AND PERFORMANCE

The previous chapters ranked utilities and states in terms of energy and capacity savings as percentages of the volume sold and peak demand. These rankings show performance or savings results in one dimension. The discussion in this chapter adds a second ranking of effort, based upon the amount the IOUs and states have spent to achieve their various savings.

Relative Effort

Table 7-1 ranks the top 100 IOUs in terms of their incremental spending on energy efficiency stated as a percentage of total revenue requirements. The top 20 IOUs spent about one percent of their total revenues on incremental or additional energy efficiency for each of the last three years for which data is available. The top four spent more than two percent. Forty-four of the top 100 IOUs in terms of 2006 sales did not spend any measurable amount of money on energy efficiency over the three years from 2004 through 2006.

TABLE 7-1
EE INCREMENTAL COST / TOTAL REVENUE

Rank	Utility Name	2004	2005	2006
1	Public Service Electric & Gas Co	3.468%	3.286%	2.919%
2	Western Massachusetts Electric Co	2.679%	2.826%	2.542%
3	United Illuminating Co	2.067%	2.726%	2.378%
4	Massachusetts Electric Co	2.909%	2.454%	2.216%
5	Narragansett Electric Co	1.909%	1.934%	1.783%
6	Puget Sound Energy Inc	1.583%	1.747%	1.763%
7	Interstate Power & Light Co	2.268%	1.816%	1.759%
8	Pacific Gas & Electric Co	0.994%	1.549%	1.587%
9	Avista Corp	0.758%	0.878%	1.480%
10	Atlantic City Electric Co	1.379%	1.328%	1.413%
11	Idaho Power Co	0.491%	0.662%	1.391%
12	Connecticut Light & Power Co	2.063%	1.825%	1.390%
13	MidAmerican Energy Co	1.006%	1.164%	1.360%

(Continued)

TABLE 7-1 (CONT'D)
EE INCREMENTAL COST / TOTAL REVENUE

Rank	Utility Name	2004	2005	2006
14	San Diego Gas & Electric Co	0.000%	3.195%	1.281%
15	Northern States Power Co MN	1.510%	1.317%	1.265%
16	Public Service Co of NH	1.344%	1.370%	1.204%
17	Wisconsin Power & Light Co	1.492%	1.069%	1.202%
18	Central Maine Power Co	1.567%	1.620%	1.192%
19	Southern California Edison Co	0.729%	1.601%	1.016%
20	NorthWestern Energy LLC	0.898%	1.165%	1.008%
21	Minnesota Power Inc	0.741%	0.869%	0.877%
22	PacifiCorp	0.803%	1.014%	0.850%
23	Nevada Power Co	0.373%	0.411%	0.834%
24	Jersey Central Power & Light Co	1.254%	0.953%	0.709%
25	Northern States Power Co WI	0.755%	0.637%	0.694%
26	Hawaiian Electric Co Inc	0.629%	0.568%	0.672%
27	Arizona Public Service Co	0.000%	0.000%	0.598%
28	Tucson Electric Power Co	0.316%	0.237%	0.514%
29	Sierra Pacific Power Co	0.227%	0.343%	0.488%
30	Florida Power & Light Co	0.496%	0.466%	0.405%
31	Gulf Power Co	0.504%	0.408%	0.381%
32	Wisconsin Electric Power Co	0.006%	0.147%	0.300%
33	Louisville Gas & Electric Co	0.224%	0.322%	0.271%
34	Progress Energy Florida Inc	0.551%	0.304%	0.260%
35	Southwestern Public Service Co	0.535%	0.312%	0.243%
36	Public Service Co of Colorado	0.635%	0.643%	0.243%
37	Tampa Electric Co	0.260%	0.211%	0.200%
38	Pennsylvania Electric Co	0.174%	0.191%	0.185%
39	Kentucky Power Co	0.174%	0.172%	0.175%
40	Southwestern Electric Power Co	0.153%	0.183%	0.164%
41	Metropolitan Edison Co	0.171%	0.175%	0.163%
42	Consolidated Edison Co-NY Inc	0.000%	0.037%	0.157%
43	Duke Energy Indiana Inc	0.164%	0.153%	0.142%
44	Kentucky Utilities Co	0.135%	0.085%	0.132%
45	Indianapolis Power & Light Co	0.012%	0.081%	0.125%
46	Kansas City Power & Light Co	0.000%	0.016%	0.117%
47	Entergy Gulf States Inc	0.126%	0.122%	0.109%
48	El Paso Electric Co	0.096%	0.018%	0.085%

TABLE 7-1 (CONT'D)
EE INCREMENTAL COST / TOTAL REVENUE

Rank	Utility Name	2004	2005	2006
49	New York State Electric & Gas Corp	0.000%	0.038%	0.049%
50	Duke Energy Ohio Inc	0.113%	0.043%	0.047%
51	Georgia Power Co	0.000%	0.000%	0.029%
52	Progress Energy Carolinas Inc	0.000%	0.000%	0.024%
53	Aquila Inc	0.020%	0.025%	0.024%
54	Mississippi Power Co	0.001%	0.001%	0.006%
55	Alabama Power Co	0.000%	0.000%	0.000%
56	Appalachian Power Co	0.000%	0.000%	0.000%
57	Baltimore Gas & Electric Co	0.000%	0.000%	0.000%
58	Boston Edison Co	0.000%	2.311%	0.000%
59	Central Hudson Gas & Electric Corp	0.000%	0.000%	0.000%
60	Central Illinois Public Service Co	0.000%	0.000%	0.000%
61	Cleco Power LLC	0.000%	0.000%	0.000%
62	Cleveland Electric Illuminating Co	0.000%	0.000%	0.000%
63	Columbus Southern Power Co	0.000%	0.000%	0.000%
64	Commonwealth Edison Co	0.000%	0.000%	0.000%
65	Commonwealth Electric Co	1.701%	1.469%	0.000%
66	Consumers Energy Co	0.000%	0.000%	0.000%
67	Dayton Power & Light Co	0.000%	0.000%	0.000%
68	Delmarva Power & Light Co	0.000%	0.000%	0.000%
69	Detroit Edison Co	0.000%	0.000%	0.000%
70	Duke Energy Carolinas, LLC	0.000%	0.000%	0.000%
71	Duquesne Light Co	0.000%	0.000%	0.000%
72	Entergy Arkansas Inc	0.000%	0.000%	0.000%
73	Entergy Louisiana Inc	0.000%	0.000%	0.000%
74	Entergy Mississippi Inc	0.000%	0.000%	0.000%
75	Entergy New Orleans Inc	0.000%	0.000%	0.000%
76	Illinois Power Co	0.000%	0.000%	0.000%
77	Indiana Michigan Power Co	0.000%	0.000%	0.000%
78	Kansas Gas & Electric Co	0.000%	0.000%	0.000%
79	Monongahela Power Co	0.000%	0.000%	0.000%
80	Niagara Mohawk Power Corp	0.000%	0.000%	0.000%
81	Northern Indiana Public Service Co	0.000%	0.000%	0.000%
82	Ohio Edison Co	0.000%	0.000%	0.000%
83	Ohio Power Co	0.000%	0.000%	0.000%

(Continued)

TABLE 7-1 (CONT'D)
EE INCREMENTAL COST / TOTAL REVENUE

Rank	Utility Name	2004	2005	2006
84	Oklahoma Gas & Electric Co	0.000%	0.000%	0.000%
85	Orange & Rockland Utilities Inc	0.000%	0.000%	0.000%
86	PECO Energy Co	0.000%	0.000%	0.000%
87	Portland General Electric Co	0.000%	0.000%	0.000%
88	Potomac Electric Power Co	0.000%	0.000%	0.000%
89	PPL Electric Utilities Corp	0.000%	0.000%	0.000%
90	Public Service Co of NM	0.000%	0.000%	0.000%
91	Public Service Co of Oklahoma	0.000%	0.000%	0.000%
92	Rochester Gas & Electric Corp	0.000%	0.000%	0.000%
93	South Carolina Electric & Gas Co	0.000%	0.000%	0.000%
94	The Potomac Edison Co	0.000%	0.000%	0.000%
95	Toledo Edison Co	0.000%	0.000%	0.000%
96	Union Electric Co	0.000%	0.000%	0.000%
97	Virginia Electric & Power Co	0.000%	0.000%	0.000%
98	West Penn Power Co	0.000%	0.000%	0.000%
99	Westar Energy Inc	0.000%	0.000%	0.000%
100	Wisconsin Public Service Corp	0.000%	0.000%	0.000%

Note: Indirect costs were allocated to EE in proportion to direct costs of EE and LM.

Table 7-2 adds the additional incremental amount the largest 100 IOUs spent on load management to the energy efficiency amounts they spent. With this addition, the top 10 spent about two percent or more. The top 26 spent about one percent or more. There were 68 IOUs that spent measurable amounts on either energy efficiency or load management during the three years from 2004 through 2006. There were 32 that effectively had no appreciable demand-side spending.

TABLE 7-2
EE & LM INCREMENTAL COST / TOTAL REVENUE

Rank	Utility Name	2004	2005	2006
1	Interstate Power & Light Co	4.565%	4.066%	3.753%
2	Northern States Power Co MN	3.782%	3.345%	3.198%
3	United Illuminating Co	2.239%	3.161%	3.182%
4	Public Service Electric & Gas Co	3.474%	3.390%	3.000%
5	Western Massachusetts Electric Co	2.679%	2.826%	2.542%
6	Massachusetts Electric Co	2.927%	2.472%	2.225%
7	MidAmerican Energy Co	1.951%	2.023%	2.171%
8	Southern California Edison Co	1.816%	2.983%	2.136%
9	Connecticut Light & Power Co	2.076%	1.897%	1.985%
10	Pacific Gas & Electric Co	1.039%	1.896%	1.977%
11	Idaho Power Co	0.661%	1.004%	1.804%
12	Narragansett Electric Co	1.935%	1.940%	1.787%
13	Puget Sound Energy Inc	1.583%	1.747%	1.763%
14	San Diego Gas & Electric Co	0.000%	3.712%	1.566%
15	Northern States Power Co WI	1.393%	1.337%	1.538%
16	Avista Corp	0.758%	0.878%	1.480%
17	Atlantic City Electric Co	1.379%	1.328%	1.413%
18	Progress Energy Florida Inc	1.909%	1.677%	1.395%
19	Florida Power & Light Co	1.746%	1.582%	1.268%
20	Wisconsin Power & Light Co	1.538%	1.081%	1.232%
21	Public Service Co of NH	1.348%	1.373%	1.207%
22	Central Maine Power Co	1.567%	1.620%	1.192%
23	PacifiCorp	0.945%	1.178%	1.166%
24	Gulf Power Co	1.318%	1.176%	1.116%
25	NorthWestern Energy LLC	0.898%	1.165%	1.008%
26	Nevada Power Co	0.502%	0.599%	0.999%
27	Hawaiian Electric Co Inc	0.636%	0.766%	0.982%
28	Minnesota Power Inc	0.741%	0.869%	0.877%
29	Jersey Central Power & Light Co	1.351%	1.020%	0.762%
30	Alabama Power Co	0.729%	0.750%	0.742%
31	Tampa Electric Co	1.003%	0.935%	0.739%
32	Baltimore Gas & Electric Co	0.021%	0.660%	0.686%
33	Louisville Gas & Electric Co	0.651%	0.666%	0.663%
34	Kansas City Power & Light Co	0.038%	0.106%	0.646%
35	Arizona Public Service Co	0.000%	0.000%	0.598%

(Continued)

TABLE 7-2 (CONT'D)
EE & LM INCREMENTAL COST / TOTAL REVENUE

Rank	Utility Name	2004	2005	2006
36	Public Service Co of Colorado	1.015%	1.028%	0.581%
37	Tucson Electric Power Co	0.316%	0.237%	0.514%
38	Sierra Pacific Power Co	0.227%	0.343%	0.488%
39	Duke Energy Carolinas, LLC	0.507%	0.491%	0.458%
40	Indianapolis Power & Light Co	0.294%	0.370%	0.421%
41	Kentucky Utilities Co	0.460%	0.332%	0.375%
42	Georgia Power Co	0.348%	0.272%	0.363%
43	Duke Energy Indiana Inc	0.386%	0.356%	0.325%
44	Wisconsin Electric Power Co	0.193%	0.190%	0.300%
45	South Carolina Electric & Gas Co	0.392%	0.416%	0.283%
46	Consolidated Edison Co-NY Inc	0.076%	0.070%	0.275%
47	Southwestern Public Service Co	0.535%	0.312%	0.243%
48	Commonwealth Edison Co	0.243%	0.220%	0.225%
49	Pennsylvania Electric Co	0.174%	0.191%	0.188%
50	Kentucky Power Co	0.174%	0.172%	0.175%
51	Metropolitan Edison Co	0.171%	0.175%	0.165%
52	Southwestern Electric Power Co	0.153%	0.183%	0.164%
53	Entergy Gulf States Inc	0.126%	0.122%	0.109%
54	Virginia Electric & Power Co	0.125%	0.110%	0.106%
55	Westar Energy Inc	0.205%	0.222%	0.090%
56	Kansas Gas & Electric Co	0.000%	0.000%	0.087%
57	El Paso Electric Co	0.096%	0.018%	0.085%
58	New York State Electric & Gas Corp	0.000%	0.038%	0.049%
59	Duke Energy Ohio Inc	0.114%	0.046%	0.049%
60	Progress Energy Carolinas Inc	0.000%	0.000%	0.024%
61	Aquila Inc	0.020%	0.025%	0.024%
62	Mississippi Power Co	0.001%	0.001%	0.006%
63	PECO Energy Co	0.000%	0.002%	0.004%
64	Duquesne Light Co	0.014%	0.019%	0.001%
65	Central Illinois Public Service Co	1.036%	0.001%	0.000%
66	Union Electric Co	0.000%	0.001%	0.001%
67	Illinois Power Co	0.000%	0.001%	0.000%
68	Appalachian Power Co	0.000%	0.000%	0.000%
69	Boston Edison Co	0.000%	2.311%	0.000%
70	Central Hudson Gas & Electric Corp	0.000%	0.000%	0.000%
71	Cleco Power LLC	0.000%	0.000%	0.000%
72	Cleveland Electric Illuminating Co	0.000%	0.000%	0.000%

TABLE 7-2 (CONT'D)
EE & LM INCREMENTAL COST / TOTAL REVENUE

Rank	Utility Name	2004	2005	2006
73	Columbus Southern Power Co	0.000%	0.000%	0.000%
74	Commonwealth Electric Co	1.701%	1.469%	0.000%
75	Consumers Energy Co	0.000%	0.000%	0.000%
76	Dayton Power & Light Co	0.000%	0.000%	0.000%
77	Delmarva Power & Light Co	0.000%	0.000%	0.000%
78	Detroit Edison Co	0.000%	0.000%	0.000%
79	Entergy Arkansas Inc	0.000%	0.000%	0.000%
80	Entergy Louisiana Inc	0.000%	0.000%	0.000%
81	Entergy Mississippi Inc	0.000%	0.000%	0.000%
82	Entergy New Orleans Inc	0.000%	0.000%	0.000%
83	Indiana Michigan Power Co	0.000%	0.000%	0.000%
84	Monongahela Power Co	0.000%	0.000%	0.000%
85	Niagara Mohawk Power Corp	0.000%	0.000%	0.000%
86	Northern Indiana Public Service Co	0.000%	0.000%	0.000%
87	Ohio Edison Co	0.000%	0.000%	0.000%
88	Ohio Power Co	0.000%	0.000%	0.000%
89	Oklahoma Gas & Electric Co	0.000%	0.000%	0.000%
90	Orange & Rockland Utilities Inc	0.000%	0.000%	0.000%
91	Portland General Electric Co	0.000%	0.000%	0.000%
92	Potomac Electric Power Co	0.028%	0.000%	0.000%
93	PPL Electric Utilities Corp	0.000%	0.000%	0.000%
94	Public Service Co of NM	0.000%	0.000%	0.000%
95	Public Service Co of Oklahoma	0.000%	0.000%	0.000%
96	Rochester Gas & Electric Corp	0.000%	0.000%	0.000%
97	The Potomac Edison Co	0.000%	0.000%	0.000%
98	Toledo Edison Co	0.000%	0.000%	0.000%
99	West Penn Power Co	0.000%	0.000%	0.000%
100	Wisconsin Public Service Corp	0.000%	0.000%	0.000%

Note: Indirect costs were allocated to EE in proportion to direct costs of EE and LM.

Table 7-3 ranks the 51 state-level jurisdictions in terms of the jurisdictional spending on energy efficiency as a percentage of the amount spent on electricity. The states are ranked based upon their relative 2006 ranking. The relative percentages are shown for the entire 15-year period that EIA published this data: 1992 through 2006.

TABLE 7-3
INCREMENTAL EE SPENDING / TOTAL REVENUE

Rank	State	1992	1993	1994	1995	1996	1997	1998	1999	2000	2001	2002	2003	2004	2005	2006
1	Iowa	0.755%	0.867%	1.343%	1.520%	1.014%	1.044%	1.143%	1.008%	0.885%	1.010%	1.073%	1.246%	1.108%	1.033%	1.087%
2	Washington	2.065%	4.718%	3.731%	2.520%	2.043%	1.248%	0.625%	0.600%	0.546%	1.060%	0.670%	0.640%	0.660%	0.724%	0.891%
3	California	1.625%	1.489%	1.655%	1.419%	1.361%	0.967%	0.792%	1.306%	1.290%	1.102%	0.949%	0.368%	0.556%	1.042%	0.826%
4	New Jersey	0.224%	0.336%	0.631%	0.905%	0.721%	0.749%	1.367%	1.392%	2.110%	1.178%	1.449%	1.361%	1.075%	0.819%	0.694%
5	Minnesota	0.813%	0.862%	1.376%	1.897%	1.856%	0.921%	0.983%	0.802%	0.686%	0.740%	0.733%	0.780%	0.669%	0.613%	0.604%
6	Connecticut	2.142%	1.965%	1.643%	1.590%	1.268%	1.262%	1.068%	1.086%	2.823%	2.529%	1.478%	0.787%	0.922%	0.763%	0.582%
7	Hawaii	0.086%	0.126%	0.093%	0.074%	0.438%	0.645%	0.841%	0.753%	0.765%	0.675%	0.658%	0.733%	0.554%	0.562%	0.578%
8	Nevada	0.460%	0.400%	0.309%	0.179%	0.060%	0.047%	0.018%	0.000%	0.014%	0.087%	0.185%	0.239%	0.241%	0.281%	0.531%
9	Wisconsin	3.286%	3.028%	2.128%	1.617%	1.153%	0.839%	1.368%	1.270%	1.257%	1.060%	0.861%	0.410%	0.425%	0.367%	0.449%
10	Arizona	0.331%	0.447%	0.445%	0.548%	0.483%	0.125%	0.118%	0.160%	0.104%	0.059%	0.022%	0.045%	0.051%	0.122%	0.429%
11	New York	2.340%	2.164%	1.474%	0.919%	0.531%	0.541%	0.087%	0.385%	0.360%	0.256%	0.133%	0.259%	0.233%	0.115%	0.366%
12	Massachusetts	3.142%	3.151%	3.207%	2.364%	1.955%	1.857%	2.295%	4.555%	2.694%	1.563%	1.251%	0.639%	0.623%	0.688%	0.316%
13	Utah	0.411%	0.727%	0.618%	1.014%	0.268%	0.123%	0.131%	0.148%	0.103%	0.331%	0.348%	0.323%	0.276%	0.342%	0.309%
14	Idaho	0.288%	0.641%	0.438%	0.307%	0.186%	0.151%	0.125%	0.134%	0.124%	0.258%	0.137%	0.125%	0.144%	0.182%	0.297%
15	Florida	0.494%	0.697%	0.726%	0.747%	0.830%	0.922%	0.657%	0.592%	0.541%	0.491%	0.511%	0.430%	0.385%	0.317%	0.288%
16	New Hampshire	0.410%	0.446%	0.427%	0.709%	0.671%	0.486%	0.479%	0.488%	0.501%	0.264%	0.388%	0.695%	0.386%	0.289%	0.222%
17	Rhode Island	3.515%	4.423%	3.903%	3.041%	1.957%	1.935%	2.134%	2.913%	0.000%	0.864%	0.675%	0.584%	0.370%	0.268%	0.218%
18	Colorado	0.126%	0.365%	0.448%	0.575%	0.678%	0.185%	0.198%	0.125%	0.121%	0.134%	0.296%	0.222%	0.334%	0.339%	0.180%
19	Oregon	0.768%	1.120%	1.145%	1.313%	0.795%	0.638%	0.415%	0.417%	0.366%	0.869%	0.400%	0.209%	0.149%	0.156%	0.171%
20	Kentucky	0.354%	0.458%	0.565%	0.535%	0.169%	0.283%	0.180%	0.168%	0.131%	0.267%	0.282%	0.245%	0.217%	0.131%	0.167%
21	Texas	0.274%	0.271%	0.307%	0.334%	0.259%	0.218%	0.181%	0.141%	0.128%	0.136%	0.196%	0.092%	0.084%	0.115%	0.159%
22	Vermont	1.764%	3.001%	2.023%	1.275%	0.947%	0.743%	0.814%	0.928%	C.155%	0.102%	0.143%	0.178%	0.170%	0.153%	0.147%
23	Montana	0.170%	0.717%	0.804%	0.604%	0.367%	0.210%	0.125%	0.062%	0.103%	0.070%	0.102%	0.138%	0.098%	0.133%	0.118%
24	North Dakota	0.111%	0.121%	0.195%	0.242%	0.246%	0.131%	0.185%	0.164%	0.173%	0.154%	0.107%	0.120%	0.113%	0.098%	0.098%
25	Nebraska	0.058%	0.114%	0.030%	0.052%	0.033%	0.042%	0.022%	0.018%	0.018%	0.061%	0.015%	0.001%	0.408%	0.296%	0.095%
26	Wyoming	0.140%	0.244%	0.197%	0.340%	0.092%	0.034%	0.038%	0.003%	0.026%	0.105%	0.110%	0.103%	0.088%	0.109%	0.092%

TABLE 7-3 (CONT'D)
INCREMENTAL EE SPENDING / TOTAL REVENUE

Rank	State	1992	1993	1994	1995	1996	1997	1998	1999	2000	2001	2002	2003	2004	2005	2006
27	South Carolina	0.173%	0.263%	0.309%	0.332%	0.169%	0.113%	0.060%	0.066%	0.062%	0.063%	0.048%	0.049%	0.049%	0.050%	0.057%
28	Indiana	0.464%	0.870%	1.048%	0.998%	0.477%	0.125%	0.044%	0.064%	0.041%	0.055%	0.058%	0.064%	0.039%	0.048%	0.054%
29	Alabama	0.242%	0.250%	0.326%	0.117%	0.043%	0.097%	0.045%	0.847%	0.042%	0.171%	0.187%	0.092%	0.114%	0.048%	0.047%
30	Maine	1.707%	1.491%	1.056%	1.199%	1.497%	2.311%	0.022%	1.223%	2.010%	0.459%	0.355%	0.205%	0.104%	0.073%	0.044%
31	South Dakota	0.072%	0.074%	0.098%	0.105%	0.090%	0.048%	0.045%	0.036%	0.040%	0.042%	0.045%	0.046%	0.041%	0.037%	0.037%
32	Mississippi	0.169%	0.273%	0.178%	0.117%	0.013%	0.054%	0.025%	0.029%	0.030%	0.071%	0.082%	0.062%	0.048%	0.033%	0.031%
33	Missouri	0.005%	0.005%	0.007%	0.022%	0.048%	0.059%	0.275%	0.008%	0.011%	0.017%	0.015%	0.007%	0.018%	0.023%	0.029%
34	Louisiana	0.035%	0.034%	0.028%	0.001%	0.008%	0.007%	0.009%	0.003%	0.003%	0.014%	0.027%	0.037%	0.030%	0.030%	0.029%
35	Alaska	0.044%	0.077%	0.047%	0.130%	0.050%	0.054%	0.051%	0.059%	0.045%	0.038%	0.036%	0.027%	0.028%	0.029%	0.028%
36	New Mexico	0.035%	0.030%	0.037%	0.045%	0.034%	0.037%	0.054%	0.001%	0.042%	0.003%	0.036%	0.039%	0.050%	0.030%	0.028%
37	Tennessee	0.207%	0.333%	0.148%	0.094%	0.008%	0.055%	0.023%	0.027%	0.026%	0.071%	0.077%	0.058%	0.044%	0.024%	0.026%
38	Georgia	0.166%	0.716%	0.654%	0.422%	0.034%	0.050%	0.024%	0.012%	0.021%	0.025%	0.014%	0.017%	0.017%	0.008%	0.026%
39	Illinois	0.030%	0.032%	0.079%	0.052%	0.027%	0.010%	0.049%	0.063%	0.014%	0.025%	0.018%	0.022%	0.015%	0.014%	0.016%
40	Kansas	0.000%	0.000%	0.000%	0.000%	0.000%	0.000%	0.095%	0.001%	0.002%	0.001%	0.001%	0.001%	0.001%	0.002%	0.015%
41	Pennsylvania	0.248%	0.266%	0.285%	0.291%	0.164%	0.056%	0.031%	0.001%	0.037%	0.026%	0.026%	0.021%	0.018%	0.016%	0.015%
42	Arkansas	0.007%	0.007%	0.001%	0.010%	0.001%	0.008%	0.004%	0.003%	0.004%	0.004%	0.006%	0.001%	0.006%	0.012%	0.007%
43	North Carolina	0.161%	0.230%	0.317%	0.371%	0.246%	0.172%	0.001%	0.001%	0.002%	0.002%	0.002%	0.000%	0.001%	0.000%	0.004%
44	Ohio	0.087%	0.470%	0.247%	0.376%	0.205%	0.122%	0.080%	0.065%	0.034%	0.019%	0.007%	0.006%	0.008%	0.005%	0.004%
45	Michigan	0.397%	0.674%	0.197%	0.182%	0.149%	0.004%	0.036%	0.033%	0.068%	0.021%	0.003%	0.000%	0.000%	0.002%	0.004%
46	Oklahoma	0.037%	0.033%	0.007%	0.006%	0.003%	0.003%	0.009%	0.006%	0.007%	0.006%	0.009%	0.006%	0.009%	0.007%	0.001%
47	Virginia	0.092%	0.141%	0.213%	0.125%	0.048%	0.051%	0.002%	0.001%	0.000%	0.000%	0.000%	0.001%	0.001%	0.000%	0.001%
48	Maryland	0.701%	1.347%	2.040%	1.710%	1.033%	0.752%	0.402%	0.175%	0.035%	0.008%	0.005%	0.001%	0.000%	0.000%	0.000%
49	West Virginia	0.013%	0.047%	0.108%	0.082%	0.021%	0.012%	0.016%	0.007%	0.000%	0.000%	0.000%	0.000%	0.000%	0.000%	0.000%
50	Delaware	0.053%	0.152%	0.281%	0.293%	0.306%	0.182%	0.202%	0.000%	0.000%	0.000%	0.000%	0.000%	0.000%	0.000%	0.000%
51	District of Columbia	0.706%	1.543%	2.562%	2.747%	1.381%	0.704%	0.512%	0.429%	0.082%	0.021%	0.008%	0.000%	0.000%	0.000%	0.000%

Note: Indirect costs were allocated to EE in proportion to direct costs of EE and LM.

During this period, every jurisdiction had specific years in which utility money was spent to increase energy efficiency. About half the states spent money fairly evenly over these 15 years. Some of the others spent early in the 15-year period, then they eased back in later years. From 2003 through 2006, four jurisdictions did not spend any measurable percentage of electric utility revenue on energy efficiency. Two of these (Maryland and the District of Columbia) had fairly significant spending before 2000.

Table 7-4 adds incremental load management spending to the incremental amount spent on energy efficiency. This increases the number of states with measurable spending performance and demonstrates that some jurisdictions adjusted their demand-side conservation efforts between energy efficiency and load management. On such a combined basis, only two jurisdictions had no measurable spending for just the last two years the data covers.

TABLE 7-4
INCREMENTAL EE & LM SPENDING / TOTAL REVENUE

Rank	State	1992	1993	1994	1995	1996	1997	1998	1999	2000	2001	2002	2003	2004	2005	2006
1	Iowa	1.407%	1.692%	2.419%	2.181%	1.773%	1.799%	1.813%	1.703%	1.567%	1.570%	1.549%	1.775%	2.345%	2.243%	2.219%
2	Minnesota	1.446%	1.269%	1.697%	2.208%	2.196%	1.219%	1.413%	1.089%	1.030%	1.027%	0.964%	1.009%	1.579%	1.440%	1.396%
3	California	1.873%	1.677%	1.864%	1.631%	1.431%	0.997%	0.833%	1.352%	1.327%	1.144%	0.984%	0.412%	0.876%	1.523%	1.262%
4	Florida	2.027%	2.530%	2.722%	2.498%	2.391%	2.244%	2.119%	2.024%	1.937%	1.622%	1.674%	1.473%	1.328%	1.188%	0.992%
5	Washington	2.333%	5.130%	4.099%	2.912%	2.208%	1.389%	0.625%	0.600%	0.547%	1.060%	0.670%	0.640%	0.667%	0.731%	0.900%
6	Connecticut	2.189%	1.988%	1.664%	1.618%	1.291%	1.262%	1.068%	1.086%	2.823%	2.602%	1.850%	0.855%	0.994%	0.821%	0.817%
7	Hawaii	0.086%	0.126%	0.093%	0.074%	0.438%	0.645%	0.949%	0.753%	0.765%	0.675%	0.658%	0.733%	0.558%	0.684%	0.770%
8	New Jersey	1.056%	1.141%	1.306%	1.186%	1.103%	1.039%	1.610%	1.625%	2.321%	1.307%	1.566%	1.422%	1.088%	0.847%	0.715%
9	Nevada	0.891%	0.713%	0.605%	0.210%	0.060%	0.047%	0.027%	0.009%	0.014%	0.132%	0.271%	0.322%	0.304%	0.371%	0.613%
10	Wisconsin	3.637%	3.577%	2.559%	2.072%	1.616%	1.051%	1.535%	1.409%	1.392%	1.182%	0.993%	0.534%	0.587%	0.489%	0.593%
11	Alabama	1.047%	1.037%	1.219%	1.428%	1.382%	1.342%	1.266%	1.188%	1.084%	0.594%	0.605%	0.114%	0.491%	0.471%	0.458%
12	Arizona	0.352%	0.512%	0.483%	0.554%	0.587%	0.130%	0.121%	0.162%	0.105%	0.066%	0.030%	0.058%	0.059%	0.125%	0.446%
13	Utah	0.413%	0.729%	0.620%	1.016%	0.268%	0.124%	0.135%	0.153%	0.110%	0.331%	0.348%	0.323%	0.323%	0.395%	0.416%
14	Nebraska	0.945%	0.441%	0.650%	0.862%	0.722%	0.632%	0.527%	0.465%	0.624%	0.647%	0.575%	0.521%	0.588%	0.469%	0.411%
15	New York	2.596%	2.308%	1.518%	0.981%	0.611%	0.559%	0.104%	0.386%	0.362%	0.273%	0.174%	0.306%	0.275%	0.132%	0.406%
16	Idaho	0.288%	0.641%	0.439%	0.307%	0.186%	0.151%	0.125%	0.134%	0.124%	0.263%	0.137%	0.134%	0.173%	0.240%	0.366%
17	Colorado	0.147%	0.407%	0.495%	0.609%	0.727%	0.283%	0.256%	0.173%	0.166%	0.156%	0.365%	0.350%	0.519%	0.538%	0.364%
18	Kentucky	1.011%	1.081%	0.937%	0.960%	0.417%	0.407%	0.300%	0.274%	0.192%	0.332%	0.401%	0.404%	0.390%	0.276%	0.322%
19	Massachusetts	3.275%	3.307%	3.415%	2.531%	2.104%	1.989%	2.440%	0.463%	2.797%	1.575%	1.259%	0.645%	0.626%	0.691%	0.317%
20	North Dakota	0.328%	0.391%	0.422%	0.457%	0.434%	0.307%	0.295%	0.270%	0.287%	0.271%	0.297%	0.275%	0.346%	0.300%	0.301%
21	Georgia	0.282%	1.052%	1.096%	0.804%	0.462%	0.147%	0.092%	0.045%	0.071%	0.079%	0.308%	0.252%	0.237%	0.174%	0.232%
22	New Hampshire	0.472%	0.579%	0.561%	0.794%	0.742%	0.534%	0.528%	0.523%	0.527%	0.265%	0.391%	0.701%	0.389%	0.289%	0.224%
23	Rhode Island	3.702%	4.548%	3.903%	3.041%	1.957%	1.951%	2.161%	2.954%	0.003%	0.877%	0.700%	0.587%	0.375%	0.269%	0.219%
24	Oregon	0.790%	1.158%	1.179%	1.340%	0.808%	0.642%	0.415%	0.420%	0.370%	0.944%	0.400%	0.209%	0.171%	0.170%	0.195%
25	South Carolina	0.640%	0.748%	0.730%	0.751%	0.559%	0.403%	0.289%	0.270%	0.246%	0.238%	0.171%	0.217%	0.198%	0.193%	0.189%
26	Texas	0.396%	0.399%	0.422%	0.435%	0.319%	0.240%	0.186%	0.152%	0.144%	0.181%	0.209%	0.234%	0.092%	0.127%	0.176%

(Continued)

TABLE 7-4 (CONT'D)
INCREMENTAL EE & LM SPENDING / TOTAL REVENUE

Rank	State	1992	1993	1994	1995	1996	1997	1998	1999	2000	2001	2002	2003	2004	2005	2006
27	Vermont	1.834%	3.543%	2.553%	1.747%	1.327%	1.065%	1.084%	1.126%	0.200%	0.102%	0.146%	0.182%	0.182%	0.162%	0.156%
28	Indiana	0.596%	1.028%	1.197%	1.114%	0.524%	0.196%	0.090%	0.116%	0.094%	0.200%	0.096%	0.132%	0.154%	0.153%	0.156%
29	Wyoming	0.141%	0.248%	0.198%	0.340%	0.092%	0.036%	0.039%	0.046%	0.034%	0.108%	0.114%	0.105%	0.105%	0.128%	0.129%
30	Montana	0.224%	0.812%	0.880%	0.670%	0.391%	0.214%	0.128%	0.064%	0.104%	0.072%	0.432%	0.140%	0.099%	0.133%	0.118%
31	Missouri	0.477%	0.320%	0.328%	0.302%	0.336%	0.276%	0.483%	0.241%	0.072%	0.069%	0.054%	0.035%	0.055%	0.065%	0.113%
32	North Carolina	1.008%	0.902%	0.976%	1.035%	0.878%	0.586%	0.238%	0.178%	0.165%	0.137%	0.132%	0.111%	0.112%	0.112%	0.108%
33	South Dakota	0.124%	0.127%	0.128%	0.131%	0.117%	0.070%	0.063%	0.052%	0.061%	0.060%	0.064%	0.061%	0.103%	0.093%	0.090%
34	Illinois	0.115%	0.106%	0.144%	0.124%	0.144%	0.152%	0.214%	0.472%	0.495%	0.257%	0.183%	0.191%	0.138%	0.089%	0.085%
35	Virginia	0.448%	0.513%	0.709%	0.689%	0.527%	0.327%	0.201%	0.169%	0.142%	0.019%	0.087%	0.066%	0.074%	0.072%	0.071%
36	Kansas	0.141%	0.153%	0.166%	0.119%	0.129%	0.095%	0.206%	0.050%	0.049%	0.039%	0.039%	0.037%	0.032%	0.036%	0.070%
37	Maryland	1.249%	2.272%	2.695%	2.180%	1.628%	1.311%	1.068%	0.756%	0.536%	0.111%	0.240%	0.161%	0.028%	0.069%	0.064%
38	Maine	1.933%	1.631%	1.086%	1.238%	1.529%	2.364%	0.025%	1.225%	2.022%	0.459%	0.362%	0.205%	0.104%	0.074%	0.045%
39	Mississippi	0.289%	0.306%	0.271%	0.253%	0.056%	0.071%	0.041%	0.044%	0.044%	0.085%	0.096%	0.073%	0.057%	0.041%	0.038%
40	Tennessee	0.335%	0.358%	0.297%	0.287%	0.057%	0.073%	0.040%	0.043%	0.043%	0.088%	0.094%	0.071%	0.056%	0.037%	0.037%
41	Arkansas	0.138%	0.128%	0.098%	0.086%	0.078%	0.058%	0.041%	0.045%	0.028%	0.025%	0.035%	0.026%	0.030%	0.037%	0.036%
42	Alaska	0.076%	0.105%	0.096%	0.150%	0.072%	0.069%	0.066%	0.072%	0.050%	0.039%	0.038%	0.041%	0.036%	0.033%	0.033%
43	Louisiana	0.038%	0.035%	0.034%	0.013%	0.011%	0.008%	0.011%	0.009%	0.008%	0.016%	0.029%	0.038%	0.031%	0.031%	0.029%
44	New Mexico	0.065%	0.068%	0.062%	0.064%	0.048%	0.040%	0.055%	0.001%	0.042%	0.003%	0.036%	0.039%	0.050%	0.030%	0.028%
45	Pennsylvania	0.397%	0.404%	0.401%	0.341%	0.208%	0.057%	0.049%	0.066%	0.052%	0.044%	0.098%	0.022%	0.024%	0.022%	0.020%
46	Ohio	0.508%	0.670%	0.398%	0.430%	0.259%	0.141%	0.090%	0.071%	0.042%	0.037%	0.012%	0.007%	0.012%	0.006%	0.007%
47	Oklahoma	0.587%	0.570%	0.405%	0.444%	0.384%	0.189%	0.045%	0.045%	0.042%	0.033%	0.030%	0.020%	0.013%	0.013%	0.005%
48	Delaware	0.852%	0.766%	0.828%	0.713%	0.698%	0.566%	0.536%	0.000%	0.000%	0.000%	0.001%	0.000%	0.000%	0.006%	0.004%
49	Michigan	0.504%	0.736%	0.216%	0.206%	0.189%	0.006%	0.046%	0.037%	0.072%	0.024%	0.006%	0.004%	0.005%	0.003%	0.004%
50	West Virginia	0.029%	0.054%	0.109%	0.083%	0.021%	0.012%	0.016%	0.013%	0.000%	0.000%	0.000%	0.000%	0.000%	0.000%	0.000%
51	District of Columbia	1.210%	2.008%	3.096%	3.184%	1.841%	1.028%	0.894%	0.817%	0.403%	0.218%	0.090%	0.035%	0.003%	0.000%	0.000%

Benchmarking

Utilities and regulators may target savings percentages, but the focus in regulation is typically the amount spent. Consider a utility that arranges for one high efficiency compact fluorescent light (CFL) bulb to be installed to replace an incandescent light bulb. The amount saved, assuming similar use, would be the same across every utility and state jurisdiction. Population size, of course, matters. This is why the various savings are expressed in terms of the amount of kWhs sold or kW of peak demand.

Spending also matters. Suppose one utility gave away the CFL and another combined with retailers to provide a forty percent discount to encourage consumers to purchase a CFL bulb. The give-away might cost the utility $2.50, for example. However, the discount or incentive might only cost the utility forty percent of the price, or about $1.00. This is the type of issue regulators would sink their teeth into when they consider energy efficiency.

Utilities and jurisdictions that use customer incentives are, in effect, able to spend less to save the same "X" kWhs; or they can save some multiple of more kWhs for a given $Y invested in demand-side programs. Programs that require the utility to pay the full cost of energy efficiency or load management would, all else the same, be less cost effective.

There are some complications with respect to how "the" costs of demand-side programs are interpreted and assigned over the multi-year life of different utility-sponsored programs. These will be explained and analyzed below. Here, the analysis is simplified, when cumulative energy savings data are discussed for benchmarking purposes.

Chart 7-1 takes the top 100 IOUs in terms of sales revenue in 2006 and omits the IOUs with no meaningful energy efficiency savings in that same year. The two dashed lines divide the sample in terms of the mean value of direct energy efficiency spending as a percentage of sales revenue (vertical axis), and the mean value of the percent of energy efficiency savings in 2006 relative to the MWhs sold that year.

A straight-line regression is fitted to the data in Chart 7-1. The better performing IOUs are to the right and at or below the regression line. These IOUs are outperforming the others in terms of the relative incremental amount spent per incremental MWh saved.

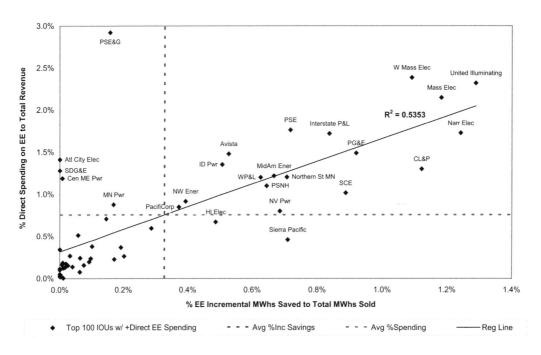

Chart 7-2 replaces incremental results with cumulative results, or the total amount saved in any given year. The year selected for cumulative savings is 2006. The horizontal axis shows the total percentage saved from all prior years cumulatively in 2006. Since these savings in 2006 reflect prior years' spending, the vertical axis shows the direct spending on energy efficiency summed over the last ten years stated as a percentage of 2006 sales revenues.

Again, the best performers shown in Chart 7-2 are at or below the regression line and to the right. These are the IOUs that have saved more over time and cumulatively spent relatively less to do so.

CHART 7-2
PERCENT OF CUMULATIVE (10 YEARS) DIRECT EE SPENDING TO REVENUE AND PERCENT OF CUMULATIVE EE MWhs SAVED TO TOTAL MWhs SOLD FOR 2006
(Top 100 IOUs by 2006 Revenue)

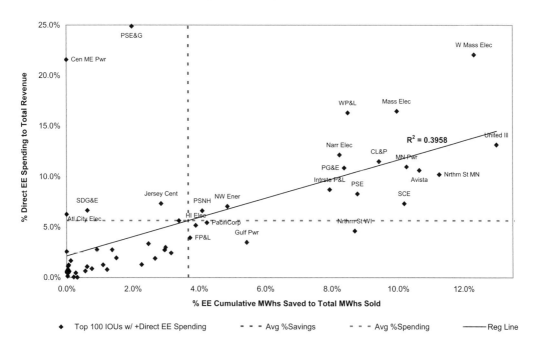

Charts 7-3 and 7-4 show the same cumulative benchmarking results for the potential and actual load management savings using direct spending on load management over the last ten years expressed as a percent of 2006 sales revenue on the vertical axis.

CHART 7-3
PERCENT OF CUMULATIVE (10 YEARS) DIRECT LM SPENDING TO REVENUE AND PERCENT OF CUMULATIVE LM POTENTIAL PEAK REDUCTION TO PEAK LOAD FOR 2006
(Top 100 IOUs by 2006 Revenue)

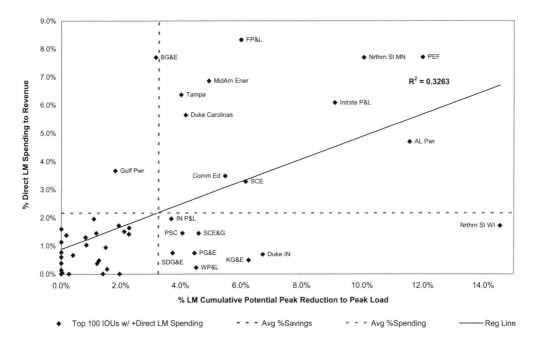

Chapter 7: Benchmarking Effort and Performance

CHART 7-4
PERCENT OF CUMULATIVE (10 YEARS) DIRECT LM SPENDING TO TOTAL REVENUE AND PERCENT OF CUMULATIVE LM ACTUAL PEAK REDUCTION TO PEAK LOAD FOR 2006
(Top 100 IOUs by 2006 Revenue)

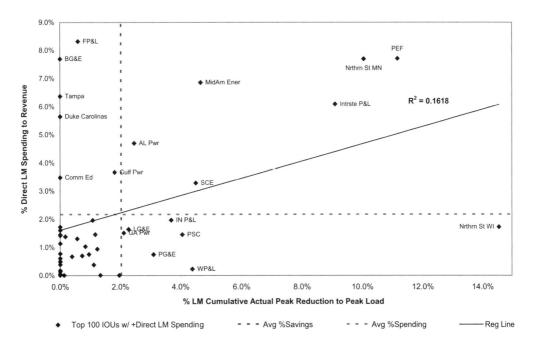

103

Similar charts can be used to benchmark states, as well, over time. These corresponding state benchmarking data are shown in Charts 7-5 through 7-8. These cumulative charts are particularly important if states seek to combine energy efficiency and demand response policies as part of any state renewable portfolio standards (RPS) target.

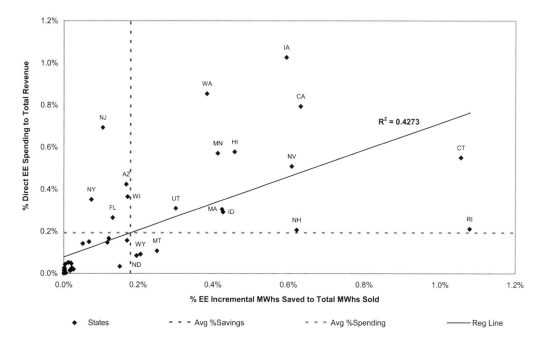

CHART 7-5
PERCENT OF INCREMENTAL DIRECT EE SPENDING TO TOTAL REVENUE AND PERCENT OF INCREMENTAL EE MWhs SAVED TO TOTAL MWhs SOLD FOR 2006
(50 States & District of Columbia)

Chapter 7: Benchmarking Effort and Performance

CHART 7-6
PERCENT OF CUMULATIVE (10 YEARS) DIRECT EE SPENDING TO TOTAL REVENUE AND PERCENT OF CUMULATIVE EE MWhs SAVED TO TOTAL MWhs SOLD FOR 2006
(50 States & District of Columbia)

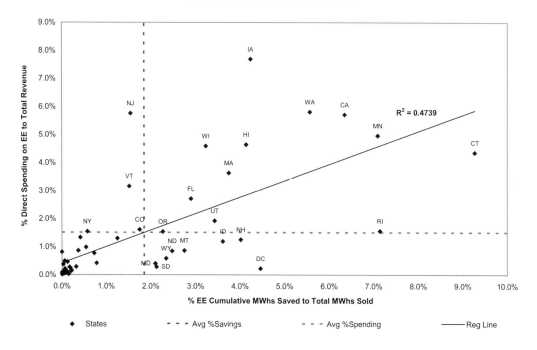

105

Going Green and Getting Regulation Right

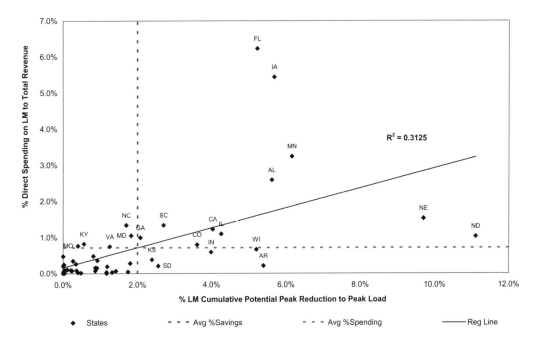

**CHART 7-7
PERCENT OF CUMULATIVE (10 YEARS) DIRECT LM SPENDING TO TOTAL REVENUE AND
PERCENT OF CUMULATIVE LM POTENTIAL PEAK REDUCTION TO PEAK LOAD FOR 2006
(50 States & District of Columbia)**

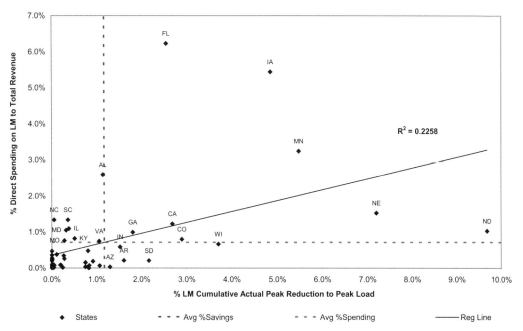

**CHART 7-8
PERCENT OF CUMULATIVE (10 YEARS) DIRECT LM SPENDING TO TOTAL REVENUE AND PERCENT OF CUMULATIVE LM ACTUAL PEAK REDUCTION TO PEAK LOAD FOR 2006
(50 States & District of Columbia)**

Regardless, there are two conclusions. First, the cost-effectiveness of demand-side management programs matter. Second, there are very significant differences across utilities in terms of their achievements (savings as a percent of a utility's size) and relative effectiveness (percent spending relative to the amount saved).

Finally, states and utilities could adopt "best practice" targets of "one percent" incremental savings and "10 percent" cumulative savings. The "one percent" incremental goal would reduce most states' growth to about their respective population growth. The "10 percent" cumulative goal would equal about half the high end of the RPS "20 percent" targets being considered around the nation generally and as already established in California.

SECTION III: REGULATING ENERGY EFFICIENCY

This section explains the economic theory of regulating energy efficiency. Relative marginal costs and prices matter. So does the value of externalities.

This discussion brings nonparticipating customers into sharper focus. This is important, because these are the customers that might object to utility-sponsored demand-side management.

CHAPTER 8
AN OVERVIEW OF ENERGY EFFICIENCY REGULATIONS

The Practice

The previous discussion introduced the "participant" and "nonparticipant" issue. There were also some introductory remarks related to energy services' unbundling and tariff design. This chapter discusses the regulatory aspects of these concepts in greater detail. As with most regulatory discussions, the focus quickly shifts to cost allocation and tariffs.

Most state regulators do not serve long terms as commissioners. Accordingly, issues previously debated and sometimes resolved need to be revisited periodically by successive waves of state regulators. The nation seeks more energy efficiency. There are no two states that approach this task in the same way. There are conceptually two extreme regulatory approaches.

The Pure Market Approach

- The goal of regulation is to design tariffs that communicate economically efficient price signals for energy, such as time-sensitive or real-time tariffs.

- Utilities should provide information to customers about energy use (audits) and propose energy saving solutions to customers.

- Utilities could help inform and police the market through appliance programs, such as EnergyStar®, and help verify contractor installation performance.

- Ultimately, the customers choose and pay and the utilities simply spend money to provide information.

The Societal Approach

- The goal of regulation is least-cost energy supply.

- Regulators should use integrated planning analyses to determine the least cost mix of build, buy, and demand-side or energy efficiency choices.

- Utilities should install, or pay others to install, these least-cost demand-side, as well as traditional supply-side, choices.

- The customer, other than their time to choose, would not pay directly. The utility's demand-side program costs would be assigned to all customers. Regulators would seek to achieve the least total revenue requirements for all customers.

The Reality

There are no major programs that occupy either the extreme market or societal approaches. The vast middle ground is where one would find most utility-sponsored energy efficiency programs. The most important regulatory questions relate to how much direct participants in energy efficiency or load management would pay directly, with the difference between these direct participant payments and the actual cost shifted to the utility's revenue requirement that all customers, including nonparticipants, would pay.

There is a closely related secondary regulatory question. Shareholders expect an opportunity to earn a just and reasonable return "of" and "on" their investments in regulated utilities. Selling less energy raises questions concerning fixed-cost recovery and lost operating margins under rate base cost-of-service (COS) regulation. This problem is more severe, when investor-owned utilities retain generation as a regulated cost recovery mechanism. Nevertheless, even utilities that have wholesale competition and recover a portion of their income through wires charges are not immune to the potential lost revenue or lost margin effects of increased energy efficiency. These states also often have forms of performance-based regulation (PBR) that make prices and revenue the focus.

There is also the regulatory treatment of other financial incentives. Few, if any, would invest in a business that cannot earn money, when it provides useful things and services, such as energy efficiency and load management. This pushback would likely increase, if a business cannot earn, when it convinces its core customers *not* to consume its products. Customers and shareholders, along with other interested parties, debate both lost margins and the role of direct utility financial incentives to encourage energy efficiency and load management. Regulators are required to sort out these matters and to design sensible compromises to allocate cost responsibility and tariffs between participants and nonparticipants.

Low-income consumers, particularly people living in rental units, present yet another regulatory challenge. Most cannot afford their utility bills, let alone spend more money in the short run to save later, particularly when someone else owns the housing unit. Most states and utilities carve out this problem and address it separately. This works best when community groups, government agencies, utilities, and regulators work together. This may sound easy. In practice, it is not always that straightforward or quite so simple.

The second major regulatory decision is to determine the best regulatory and market/business design for a particular state or set of circumstances. There are three quite different types of approaches across the nation. Starting with the least utility involvement in programmatic decisions and operations, these are:

Utilities as Bankers

- Regulators assess public benefits charges, as quasi-taxes, through utility tariffs.

- Third parties, often nonprofit entities, design, operate, and distribute the utility-financed energy efficiency programs.

- There is often a higher degree of micro-managing than in other approaches to ensure that specific programs are implemented.

- Monitoring and verification of results might be required, but this is not always the case.

A Regulatory Performance Focus

- Regulators specify the amount of money utilities will spend and/or how much energy and capacity savings utilities need to target and achieve.

- Regulators often include some form of revenue decoupling or specifically otherwise address lost margins or utility income.

- Regulators often allocate indirect utility cost and/or direct financial incentives to reward utility performance with respect to energy and capacity savings.

- Again, regulators may mandate specific programs, such as third-party run programs.

- Monitoring and verification is likely, if the focus is on utility performance in terms of spending or estimated savings. Actual savings verification is somewhat sporadic, if there is a different focus.

Market Approaches

- The utility would design a new profit-based energy efficiency business.

- Regulators would directly approve the financial incentive system, cost/benefit allocations, and revenue recovery mechanisms.

- Verification and monitoring are more essential.

- Regulators need to take steps to ensure reasonable and fair competition with nonutility vendors.

These three program design approaches are mixed with conceptual regulatory and cost recovery approaches. The various regulatory jurisdictions in the U.S. and Canada currently, and over the past couple of decades, are using or have tried various aspects of these approaches.

Some regulators approach these matters as pro-active participants. More often, they serve as judges listening to others' proposals. In almost every case, regulators rely on a systematic review of the pros and cons of the various proposals, as well as the opinions of others.

The Theory

In 1988, Charles Cicchetti, William Hogan, and Irwin M. Stelzer provided comments on the FERC's proposed rulemaking on "Bidding Programs" (RM 88-5-000).[1] They focused particular attention on questions related to utility-sponsored energy efficiency and how to compare demand-side bidding alternatives to traditional supply-side alternatives. This discussion focused on both traditional COS rate regulation and, at the time, states contemplating using nonutility generator (NUG) bidding systems. The FERC's goal seemingly was to advance the concepts of least cost and economic efficiency, which means increasing energy efficiency when marginal benefits exceed marginal costs. These economists reached three conclusions with respect to utility-sponsored "conservation programs and demand-side bidding."[2]

(1) The primary objective of utility conservation programs should be to achieve economic efficiency. This means that selecting the least-cost

---

[1] Cicchetti, Charles J., Irwin M. Stelzer, and William Hogan. (1988, July 18). Comments Before the Federal Energy Regulatory Commission, Comments of the Energy and Environmental Policy Center, RE: Regulations Governing Bidding Programs, Docket No. RM88-5-000.

[2] Today, the term most used is energy efficiency. This includes conservation, demand response, and customer premises renewable energy.

alternative for society is still the primary objective. Regulators, politicians, and consumer advocates must not mistake this objective as being synonymous with a utility achieving the lowest price, or lowest average revenue, for all of the utility's consumers …

(2) … the revenue consequence of utility-sponsored conservation is important for the utility's shareholders and its customers. Accordingly, the state regulatory authority that attempts to rely upon conservation when it is the efficient choice must take into account the unique tariff implications of this policy … the conservation proposals should be unbundled from the energy services. In addition, state regulators should explicitly describe how they intend to recover the cost of demand-side choices from the various retail customers …

(3) … Rate-base regulation was the traditional method for recovering costs, assigning risks, and arranging for returns to investors … it is a separate issue that should receive increased attention.

The authors, some twenty years earlier, went on to explain that "the FERC and some of the … state regulatory authorities were in the midst of revolutionizing the nation's electric generating choices and institutions." Written nearly two decades ago, this statement anticipated the emergence of organized competitive, wholesale markets, and retail choice. The comments explained how both traditionally regulated and newly deregulated energy markets could unbundle energy services and promote energy efficiency.

These observations apply today and can be converted to regulatory principles that are relevant to all regulated regimes. "Economic efficiency" or achieving socially optimal outcomes is the core reason for utility-sponsored energy efficiency. This suggests two "tests."[3] The first reflects cost effectiveness.

(1) Utility Cost of Service Test

Least-cost regulations would replace traditional electricity and natural gas supply-side choices when the marginal cost of energy efficiency (MC_{EE}) is less than the marginal cost of electricity (MC_E). In brief, utilities should expand energy efficiency when MC_E exceeds MC_{EE}.

---

[3] Sedano, Richard, Rick Weston, and Gordon Dunn. (2006, January 27). "Energy Efficiency Workshop." Kansas Corporation Commission.

(2) Societal Test (Economic Efficiency)

The second test broadens the objective and includes external costs and benefits, which are also important for achieving economic efficiency. The Societal Test expands the Utility Cost of Service Test to include the social costs and benefits that utilities do *not* internalize. Under the Societal Test, direct utility costs, as well as external benefits and costs, matter.

This means that energy efficiency should expand to a socially optimal level when:

$$(MC_E - MC_{EE}) + MEC_E + MEB_{EE} > 0$$

Where:

MEC_E represents the external costs of supply-side energy (*e.g.*, noninternalized environmental pollution costs), and

MEB_{EE} represents the marginal benefits of energy efficiency (*e.g.*, reduced dependence on foreign energy and related macroeconomic benefits).

This equation is somewhat overly simplistic, because new supply-side choices may, in fact, also reduce pollution, if older less efficient units are replaced. Similarly, some energy efficiency choices may have external costs, such as windmills affecting aesthetic and land use values. Regardless, after "netting" or adjusting for such secondary effects, the basic accounting of direct utility costs and benefits, as well as external societal effects, should be quantified. This would lead to a comprehensive assessment and regulatory determination of the socially optimal mix of supply-side and energy efficiency choices.

Two decades ago there were also, as there are today, regulatory concerns related to the "unique tariff implications" of utility-sponsored energy efficiency.[4] One important aspect of this regulatory concern is how much "participants" in utility-sponsored conservation should pay for energy efficiency. In addition, regulators remain concerned with how much "nonparticipants" should pay in terms of both direct subsidies and potentially higher retail prices for the energy they purchase from regulated utilities.

Several states have or are considering revenue decoupling. This approach removes the perverse incentive of COS regulation, where conservation (*i.e.*, selling less energy) would cause utilities to lose revenue and, more importantly, earnings margins when the retail price exceeds the marginal operating costs. Decoupling does not, however, remove the potential ad-

---

[4] Cicchetti, Charles J., Irwin M. Stelzer, and William Hogan. (1988, July 18). Comments Before the Federal Energy Regulatory Commission, Comments of the Energy and Environmental Policy Center, RE: Regulations Governing Bidding Programs, Docket No. RM88-5-000.

verse effect on nonparticipants, because the utility must spread its fixed-cost recovery across a smaller volume of energy sales. Other things being equal, this would cause regulated retail prices to increase.

There are two potential offsets to potentially higher retail prices. First, participants could, and from an economist's perspective should, pay at least a portion of the cost of energy efficiency, because participants would reduce their energy purchases and save money. Second, if the marginal cost of energy exceeds regulated prices, regulators should also use this difference to reduce the cost of energy efficiency for participating customers.

There are two economic and regulatory principles for establishing how much nonparticipants should pay for energy efficiency. First, societal benefits (*i.e.*, economic efficiency) would increase, if the alternative with the lower marginal cost replaces the other. Therefore, paying participants up to ($MC_E - MC_{EE}$) to encourage energy efficiency would improve societal welfare. There would, however, be some unsavory tariff implications, if a utility simply pays some customers (*i.e.*, participants) this marginal cost difference to install energy efficiency measures at their premises. Unless modified, such a program would, in effect, provide participating customers with all the tariff savings benefits of energy efficiency without necessarily paying a reasonable share for the services the utility provides.

Accordingly, regulators are often urged to consider a second principle to accomplish the least-cost outcome in a fair manner. This approach seeks to make all customers better off. This would happen when nonparticipants pay a subsidy for energy efficiency that equals the difference between the marginal cost of energy (MC_E) and the regulated price of energy (P_E), when marginal cost exceeds regulated prices. This is known as the "No Losers Test."

Amory Lovins and others have referred to this approach as a "hardly any winners test"[5] and seem to prefer requiring utility companies to simply choose the least-cost approach. Regulators would assign lost income and costs to all regulated customers in their cost allocation and tariff design. This begs the question of how much various customers should pay for both traditional utility services and energy efficiency. Consider the least controversial circumstances when energy prices are increasing and energy efficiency costs less than regulated prices:

$MC_E > P_E$ (Price of Energy) $> MC_{EE}$

The no losers test would limit the subsidy for energy efficiency to ($MC_E - P_E$). Such assistance would increase the amount of energy efficiency. If nothing else happens, this would not

---

[5] Lovins, Amory. (1985). Saving Gigabucks With Negabucks. *Public Utilities Fortnightly*, Vol. 115, Issue 6, p. 25.

be socially optimal because MC_{EE} is less than both P_E and MC_E. The economically efficient level of energy efficiency would occur when the amount spent equals the subsidy from nonparticipants ($MC_E - P_E$) plus the direct participating customer benefits ($P_E - MC_{EE}$), or, in other words, when society spends the difference between MC_E and MC_{EE}. This logic suggests that the costs could be shared between direct participants and all other utility customers.

The regulatory challenge is to establish a policy that works for all customers. Participants should pay when they benefit directly. Nonparticipants should not be made worse off. The challenge for regulators is how sharply to draw the tariff distinction between participating and nonparticipating customers in utility-sponsored energy efficiency. Economists generally believe that rational consumers, who are provided with sufficient information, would spend MC_{EE} to save the difference between P_E and MC_{EE}. If they did so, regulators could be reasonably content with subsidies limited to ($MC_E - P_E$) so as to avoid the cost of expanding energy supply.[6]

Regulators are sometimes not permitted and others are often unwilling to embrace the "subsidy" word to describe their decisions. They do not need to do so. There is an alternative and economically equivalent conceptual justification for establishing how much both participants and nonparticipants should pay for utility-sponsored energy efficiency. It comes in the form of unbundling new utility services.

> Consider the following example where
>
> Marginal Cost of Energy = MC_E = 10¢
>
> Price of Energy = P_E = 8¢ = Average Total Cost of Energy = ATC_E
>
> Marginal Cost of Energy Efficiency = MC_{EE} = 5¢
>
> Volume Consumed = 1,000
>
> Volume Conserved = 100

The customer would initially pay $80 (8¢ × 1,000). Economic efficiency would reduce energy use to 900 units and add 100 equivalent units of energy efficiency. Under what some would call a regulatory subsidy approach, this would mean the utility pays $2 (10¢ - 8¢) × 100, or ($MC_E - P_E$) multiplied by the volume conserved. In order to receive the "subsidy" based on the utility's reduction (or avoided change) in revenue requirements, the participat-

---

[6] Cicchetti, Charles J. and William Hogan. (1989, June). "Incentive Regulation: Some Conceptual and Policy Thoughts." Energy and Environmental Policy Center, Harvard University, Discussion Paper E-89-09.

ing customer must spend $MC_{EE} \times 100$ (5¢ × 100), or $5, on conservation to get these savings. However, this is a gross amount, not a net expense. The net cost for the participant is $3 ($5 - $2) because nonparticipants would be "no worse off" if they pay 2¢ per kWh for the energy efficiency of participants. The participating customer's bill would also decline from $80:

```
8¢ × 900        = $72.00
(5¢ - 2¢) × 100 = + $ 3.00
                = $75.00
```

A second approach that Cicchetti-Hogan[7] called "unbundling" would produce the same result. This is important because regulators could adopt it without using a "subsidy" concept, which some regulatory jurisdictions do not permit. Here, utility tariffs would reflect energy services such as lighting, cooling, etc., and not the more traditional units of energy sold through a utility's meter. Since the customer had been paying $80 (8¢ × 100), the customer's initial bill would remain at $80. However, regulators could use least-cost planning and either actual or virtual bidding to invite energy efficiency to compete against traditional supply-side alternatives.

The winning bid would reflect a price of MC_E, or the marginal cost of energy. The customer that pays $80 for energy service, directly or indirectly, would submit a virtual "bid" to sell 100 units of conservation and would be paid $10 (10¢ × 100). The cost to the customer that supplies those 100 units of efficiency would be $5 (5¢ × 100), and the net income from energy efficiency would, coincidentally in this example, also equal $5 ($10 - $5). The net effect for the conserving customer that virtually bids to sell the utility conservation would be:

Pay for Energy Service:	$80
Income from Conservation Bid:	– $ 5
Net Payments	$75

This is precisely the amount the participating customer would pay under an explicit regulatory subsidy approach. This means that regulators could direct utility companies or energy service providers to make these same sorts of energy efficiency decisions for their more passive retail consumers as they do when they regulate supply-side choices.

---

[7] Cicchetti, Charles J. and William Hogan. (1989, June). "Incentive Regulation: Some Conceptual and Policy Thoughts." Energy and Environmental Policy Center, Harvard University, Discussion Paper E-89-09; Cicchetti, Charles J. (1989, November 7). Prepared Statement Before the Committee on Energy and Natural Resources of the United States Senate Related to The Demand-Side Provisions of the Public Utility Regulatory Policies Act of 1978 (PURPA).

Regulators could treat such a least-cost energy efficiency policy as a virtual bid and guarantee that "participating" consumers would pay less when energy efficiency and traditional service are combined on their premises. Both regulatory approaches require participants to be active participants. Regulators would design tariffs to ensure that participants pay a portion of the costs of energy efficiency, which would become a new regulated energy service. Both conceptual approaches achieve the same allocation of costs and payments for energy efficiency using the difference in the respective marginal costs of energy (MC_E) and energy efficiency (MC_{EE}) to establish the socially desirable amount of energy efficiency.

The nonparticipating customer is held harmless because they would pay just a portion based upon the difference between MC_E and P_E, or the utility's avoided cost and its regulated price. The participating customer would pay the same for "energy" service. These joint payments would mean that society would spend the difference between the marginal cost of energy and the marginal of energy efficiency, $MC_E - MC_{EE}$. This is the socially optimal amount.

Using the numbers in this example, regulators could achieve the optimal energy efficiency effort using joint payments:

Nonparticipants pay	= $2 for 100 units
Participants pay	= $3 for 100 units
Society Gains ($10 - $5)	= $5 for 100 units

There is more to consider. The previous discussion has been limited to direct utility costs, prices, and revenue requirements. This represents only a partial analysis because energy efficiency likely has significant marginal external benefits. These affect both participants and nonparticipants. Accordingly, these external benefits affect both the economically efficient quantity of energy efficiency, as well as who should pay and how much. The next chapter rectifies this omission and provides another important conceptual reason for regulators to require utility-sponsored energy efficiency.

Amory Lovins may have been correct in his observation that payments equal to $MC_E - P_E$ may be "hardly" enough to induce utility customers to conserve, when the gap between retail prices and the cost of energy efficiency has not succeeded.[8] However, this observation does not explicitly include the external reasons for increasing energy efficiency. These are very important, as the body of Lovins' work shows. Society would gain when all costs and benefits, including externalities, are used to establish the socially optimal amount of energy efficiency and the respective amounts that participating and other customers should pay.

---

[8] Lovins, Amory. (1985). Saving Gigabucks With Negabucks. *Public Utilities Fortnightly*, Vol. 115, Issue 6, p. 25.

CHAPTER 9
THE ECONOMICS OF UTILITY-SPONSORED ENERGY EFFICIENCY WHEN THERE ARE EXTERNALITIES

The previous chapter suggested using a least-cost regulatory approach, in which energy efficiency would replace traditional energy supplies, when $MC_E > MC_{EE}$. When the external benefits of energy efficiency (MEB_{EE}) are significant, economic efficiency would favor even greater efforts to expand energy efficiency. More specifically, today society broadly believes that energy efficiency would relieve some of the adverse consequences of climate change, enhance national or economic security, or improve local environmental conditions. This means that regulators should not simply follow a least-cost approach for choosing the best mix of energy supply and energy efficiency, which, other things being equal, would reduce to equating the respective marginal costs of energy supply (MC_E) and the marginal cost of energy efficiency (MC_{EE}). If either specific marginal cost exceeded the other, then regulators, utilities, consumers, or society should substitute more of the lower cost choice and rely on less of the higher marginal cost option.

This changes when there are significant marginal benefits that flow from energy efficiency. This is the current consensus and that seems correct. Nevertheless, new generation or traditional supply choices may also improve external benefits.

Few debate the climate change and security advantages of "renewables," such as wind, tidal, solar, and geothermal sources. Many would assign similar external conceptual benefits to nuclear, clean-coal, and carbon sequestration. The point is there are likely external benefits when new technology on either the supply or demand side pushes aside the use of existing sources of energy supplied through generation that is less efficient and uses fossil fuels, primarily coal.

This current discussion is conceptual. To keep it simple, the term marginal external benefits of energy efficiency (MEB_{EE}) will be defined as a "net" concept that represents the additional external advantages of energy efficiency (EE) relative to the actual supply-side effect for energy (E). Accordingly, the economically efficient balance between energy and efficiency is when $MC_E = MC_{EE} - MEB_{EE}$.

This means that energy efficiency should be substituted for supply-side energy beyond the point of equivalence between their respective marginal costs. Put more programmatically, regulators and utilities should expand energy efficiency if the marginal cost of energy plus the marginal external (net) benefits of energy efficiency exceed the marginal cost of energy efficiency: $MC_E + MEB_{EE} > MC_{EE}$.

There are two cases to consider.

Case I: The Marginal Cost of Energy Equals the Regulated Price

In the 1980s and 1990s, the marginal cost of supply-side electricity (MC_E) likely exceeded regulated utility prices (P_E). When $MC_E > P_E$ (= ATC_E), it is easy for regulators to justify aggressive utility-sponsored energy efficiency programs, because all customers would benefit and should be willing to pay a portion of the cost of energy efficiency. Other traditionally regulated jurisdictions would not necessarily be in this position, because prices (P_E) average total cost (ATC_E) and marginal energy costs (MC_E) may be approximately equal.

These other states might have less existing inefficient supply-side options or other expensive supply sources. If such states add new technically more efficient generating stations and/or other cost-reducing supply-side choices, they might be able to reduce, or at least maintain, current regulated prices. This would be possible, because not all states have the same political, environmental, and regulatory constraints. These states might reasonably have $ATC_E = P_E$ and regulated prices approximately equal to or even less than MC_E. There would be little or no utility cost-of-service reason for such regulators to require utility-sponsored programs that encouraged energy efficiency, because efficiency would not reduce the utility's cost of service and regulated prices would likely increase under cost-of-service regulation.

Suppose, however, that the marginal cost of energy efficiency (MC_{EE}) is less than the marginal cost of electricity (MC_E). Society would gain if lower cost energy efficiency replaces electricity and natural gas. Regardless, there would be resistance to utility-sponsored energy efficiency that nonparticipants subsidize, when the average total cost of electricity (ATC_E) is either not affected over time, or more likely, increases in the near term, because energy efficiency reduces sales and some utility costs are fixed.

A similar resistance to energy efficiency would exist in restructured states, because the prices consumers pay for energy are conceptually equal to the marginal cost of energy. In such states, there would still be societal gains if consumers consume more energy efficiency and less energy when $MC_E > MC_{EE}$. This means that energy service providers that "unbundled" energy efficiency and energy would add value and expect to profit, if they sell products based upon energy efficiency. This would require competitive retail choice or direct modifications to wholesale market rules.

Regulators would, other things being equal, be hard pressed to adopt policies that require *all* customers to help defer the costs of energy efficiency, if they expect higher prices for some customers. Accordingly, viewed narrowly as a utility matter, unless $MC_E > ATC_E = P_E$, there would be regulatory and political resistance to mandatory utility-sponsored energy efficiency programs. This negative or neutral conclusion changes, however, when the marginal external benefits of energy efficiency are added to the public policy analyses of social benefits and costs. Consider Figure 9-1.

FIGURE 9-1
SOCIAL LOSS EQUALS ABC

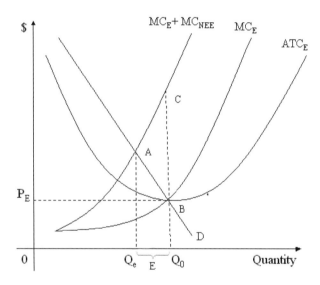

States in which $P_E = ATC_E = MC_E$ would have little or no traditional utility cost-of-service reason to mandate utility-sponsored energy efficiency. The optimal quantity of energy sold and consumed would equal Q_O. This changes when external factors are important. Define MC_{NEE} as the lost marginal external benefits when society foregoes energy efficiency that has lower marginal costs than electricity or natural gas. Therefore, MC_{NEE} equals the negative value of the marginal external benefits of energy efficiency, or $(-MEB_{EE})$.

This resistance dissipates when external factors are quantified and compared to the costs of energy and energy efficiency. This is shown in Figure 9-1, when the marginal external benefits of energy foregone (MEC_{NEE}) are added to the utility's marginal cost of energy (MC_E). This forms the marginal social cost of energy. The socially optimal level of energy consumption falls to Q_e. The triangle ABC represents the gain for society, when energy efficiency is mandated up to an amount equal to E, which is defined as Q_O minus Q_e. This same triangle ABC also represents the external benefits gained when the utility achieves E units of energy efficiency. The vertical difference between the marginal social cost ($MC_E + MC_{NEE}$) and marginal cost (MC_E), or MC_{NEE}, represents the amount of external benefits *all* utility customers would be willing to pay to avoid in order to achieve additional amounts ($Q_e - Q_O$) of energy efficiency.

Case II. The Marginal Cost of Energy Exceeds the Regulated Price

There are two reasons to increase energy efficiency when a utility also had a marginal cost of energy (MC_E) that exceeded regulated prices ($P_E = ATC_E$). First, regulators should recognize that the utility would be selling more energy than the economically efficient amount, excluding external factors. Consider Figure 9-2.

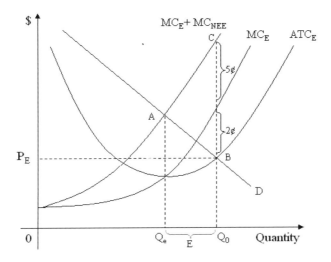

FIGURE 9-2
SOCIAL LOSS ALSO REFLECTS EXCESS CONSUMPTION
WITHOUT ENERGY EFFICIENCY

Second, regulators should also consider the societal losses due to external cost when energy consumption initially equals to Q_O. This can be shown when MC_{NEE} is added to MC_E. The two economically inefficient effects (average cost pricing and external costs) mean that consumption should be reduced E units to Q_e.

For illustrative purposes, Figure 9-2 shows a 2¢ per kWh difference between MC_E and P_E (= ATC_E). In addition, at a consumption of Q_O, Figure 9-2 shows that the marginal external benefits foregone (MC_{NEE}) would equal 5¢ per kWh.

In these circumstances, regulators should be willing to spend at the margin 2¢ plus 5¢ per kWh, or 7¢ per kWh to increase energy efficiency. All customers would gain and all customers, even nonparticipants, should pay to achieve both effects. To the extent participating customers pay more for energy efficiency, the utility would find less latent resistance to its utility-sponsored energy efficiency programs. More could be spent to achieve the socially optimal mix of energy and energy efficiency, when externalities are added and participants pay part of the costs related to energy efficiency.

As a general matter, regulators would likely continue to recognize that spreading the consumed energy's fixed costs over fewer sales (E) would raise regulated prices. Participating customers would almost certainly pay less, when the effect of less consumption and higher prices are combined. However, nonparticipants would pay more because their regulated price increase would have no offsetting reduced sales volume. While true, this conclusion omits the very important external benefit reasons for expanding energy efficiency. A simple equation shows how regulators could reasonably protect nonparticipants, if this is their focus.

Consider the following. Regardless of the relationship between MC_E and P_E, let:

ΔP_{NP} = Increase in Nonparticipating Customer Prices

Q_{NP} = Volume a Nonparticipating Customer Consumes

E_{NP} = Quantity or Share of Energy Efficiency the Nonparticipant Would Subsidize.

The nonparticipant would be willing to subsidize energy efficiency and pay higher prices equal to ΔP, if:

$$MEB_{EE} * E_{NP} \geq \Delta P_{NP} * Q_{NP}$$

Accordingly, regulators can determine the break-even value of MEB_{EE} for nonparticipating customers using:

$$\text{Break-Even MEB}_{EE} = \frac{\Delta P_{NP} * Q_{NP}}{E_{NP}}$$

The value of MEB_{EE} affects the value of the social benefits of energy efficiency in any Societal Test, as well as how to hold harmless nonparticipants. That said, MEB_{EE} is most important, because it would also help regulators determine how much nonparticipants would be willing to pay to increase the energy efficiency of society. The flip side is determining how much participating customers should pay, because any such participant payments would reduce ΔP_{NP}. This conclusion is always true. However, as shown above, when marginal energy costs (MC_E) are increasing, this would also reduce ΔP_{NP}, because MC_E exceeds $P_E = ATC_E$. Quantifying external benefits and energy efficiency costs would also help regulators answer the basic regulatory questions: (1) How much energy efficiency? and (2) Who should pay?

A Conceptual Approach

This can be reduced to a mathematical approach showing the effect for regulated retail prices, with and without an energy efficiency program. One purpose is to determine the break-even value attributable to external factors, such as clear air, climate change, and national security. The second purpose is to determine the value nonparticipants would be willing to pay so that other customers participate in energy efficiency and reduce their consumption.

The conceptual formulation is as follows:

Define

TC_0	= Total Initial Cost
Q_0	= Initial Electricity Volume
ΔQ	= Change in Electricity Volume with and without Energy Efficiency
MC_E	= Marginal Cost of Electricity
K	= Percent of Efficiency that Participants Pay
MEB_{EE}	= Marginal External Benefits of Energy Efficiency
P_N	= Price without Energy Efficiency

$$P_N = \frac{TC_0 + MC_E \Delta Q}{Q_0 + \Delta Q}$$

The Efficiency Budget (EB) equals:

$$EB = MC_{EE} * E = MC_{EE} * \Delta Q$$

where

MC_{EE} = Marginal Cost of Energy Efficiency

E = Number of Efficiency Units to Offset ΔQ, or the Growth in Electricity that Efficiency Displaces

Consider why nonparticipants would be willing to pay for others to become more efficient. First, they would consider the change in regulated prices, ΔP, which equal:

$$\Delta P = P_N - P_0$$

where P_0 = average total cost = $\dfrac{TC_0}{Q_0}$.

Therefore,

$$\Delta P = \frac{TC_0 + MC_E \Delta Q}{Q_0 + \Delta Q} - \frac{TC_0}{Q_0}$$

When energy efficiency is less expensive (MC_{EE}) than expanding traditional energy supplies, and MC_E exceeds $ATC_0 = P_0$, nonparticipants would be willing to pay some amount to avoid a rate increase based upon the difference between the marginal cost of electricity and the current regulated price ($MC_E - P_0$) to defer the cost of energy efficiency.

There is a second reason why nonparticipants would be willing to pay for other customers to become more efficient. This occurs when there are external benefits (MEB_{EE}) for energy efficiency relative to electricity production. Assume for simplicity that all nonparticipants are alike. They would formulate a new net willingness to pay after subtracting credits equal to (MEB_{EE}), based upon the values they assign to energy efficiency for reducing negative externalities.

Define (P_N^*) as:

$$P_N^* = \frac{TC_0 + k_1 MC_E \Delta Q - k_2 MEB_{EE} \Delta Q}{Q_0 + k_1 \Delta Q}$$

where k_1 equals the percentage of growth that energy efficiency does *not* replace (if $E = \Delta Q$, then $k_1 = 0$) and k_2 equals the percent of the energy efficiency budget ($EB = MC_{EE} * E$) that nonparticipants pay. For simplicity, assume $k_1 = 0$ and therefore, $E = \Delta Q$. This assumes that regulators and utilities expand energy efficiency to the point of canceling all system growth. This also presumes that the MC_{EE} is less than MC_E, at least up to this point.

The break-even value of MEB_{EE} is the amount nonparticipants would pay in higher rate riders based on these two sources of benefits: (1) avoiding expansion in traditional supply when $MC_E > P_E > MC_{EE}$; and (2) reducing negative externalities with a value of MEB_{EE}.

For nonparticipants, when there is no system growth, because energy efficiency replaces all ($k_1 = 0$) growth, the break-even value of MEB_{EE} can be determined as follows:

$$P_N = P_N^*$$

$$\frac{TC_0 + MC_E \Delta Q}{Q_0 + \Delta Q} = \frac{TC_0 - k_2 MEB_{EE} \Delta Q}{Q_0}$$

Cross multiplying

$$Q_0(TC_0 + MC_E \Delta Q) = Q_0 TC_0 - Q_0 k_2 MEB_{EE} \Delta Q + TC_0 \Delta Q - k_2 MEB_{EE} \Delta Q^2$$

$$MC_E \Delta Q = \frac{\Delta Q}{Q_0} TC_0 - k_2 MEB_{EE} \Delta Q - \frac{k_2 MEB_{EE} \Delta Q^2}{Q_0}$$

$$MC_E = \frac{TC_0}{Q_0} - k_2 MEB_{EE} - \frac{k_2 MEB_{EE} \Delta Q}{Q_0}$$

$$MC_E = P_0 - k_2 MEB_{EE} \left(1 - \frac{\Delta Q}{Q_0}\right)$$

$$MEB_{EE} = \frac{P_0 - MC_E}{k_2\left(1 - \frac{\Delta Q}{Q_0}\right)} = \left(\frac{P_0 - MC_E}{k_2}\right)\left(\frac{Q_0}{Q_0 + \Delta Q}\right)$$

In other words, the nonparticipants would determine how much more they would pay when growth is supplied traditionally and they pay P_N. When efficiency displaces traditional growth, nonparticipants effectively pay P_N^*, after subtracting their implicit credits for avoiding externalities. The size of the break-even value for MEB_{EE} depends upon three factors:

(1) the growth avoided as a percent of the initial sales of energy $\left(\dfrac{Q_0}{Q_0-\Delta Q} = \dfrac{Q_0}{Q_0-E}\right)$;

(2) the difference between the marginal cost of energy (MC_E) and the initial regulated price P_0; and (3) the share of total spending on energy efficiency (EB) the nonparticipant pays (k_2).

An additional complication is the share of growth that energy efficiency replaces. This can be accounted for either by adding k_1 back into the analysis or calculating a new baseline initial price based upon a new volume of electricity supplied traditionally. When k_1 exceeds zero, energy efficiency does not account for enough savings to avoid all supply-side costs. The optimal mix of electricity and energy efficiency, and thus the optimal value of k_1, does not depend upon the effective difference in prices that the participants and nonparticipants would pay as a result of a utility-sponsored energy efficiency program. In other words, this assumes all utility customers and their regulators use objective reasoning such as integrated resource planning or its equivalent to establish the targets for energy efficiency and steel in the ground, or in algebraic form, the target value for k_1. Regulators would also need to continue to perform cost assignments based upon those factors they deem to be "just and reasonable."

The basic conclusions are what are important. These are:

- Nonparticipants would benefit from more energy efficiency, if existing regulated prices are below the marginal cost of new energy supplies.

- Nonparticipants also benefit and would, therefore, be willing to pay some of the costs of other customers' energy efficiency, if these programs reduce negative externalities and increase marginal benefits.

- Nonparticipants also would benefit and would be willing to pay, when the savings between the marginal cost of energy efficiency and the marginal cost of energy increase and exceed regulated prices.

- Other things being equal, the more participants pay directly to improve energy efficiency, the greater the amount of energy savings, and nonparticipants would pay less. The tension is to make certain that energy efficiency increases up to the point where the marginal cost of energy equals the marginal cost of energy efficiency net of the positive external benefits of energy efficiency.

Much of the regulatory debate concerning questions related to nonparticipant payments to support energy efficiency could be addressed rather definitively with a quantitative assessment of the benefits and costs of energy efficiency. This would answer the traditional regulatory question of whether a subsidy is justified or not. Of course, it is important to quantify the "size" or cost of any nonparticipant payments (higher regulated prices). This would depend on the answer to four questions: (1) How much do direct participants pay for their own energy efficiency improvements? (2) What is the size of the gap between the avoided cost of electricity and current regulated prices that nonparticipants pay? (3) What is the utility's cost for energy efficiency and how does this affect revenue requirements? and (4) Have nonparticipants already invested in their own energy efficiency without utility financial support?

The next two chapters explain how regulators could quantify externalities.

CHAPTER 10
EXTERNAL BENEFITS AND ENERGY EFFICIENCY

Broadly speaking, society would achieve external benefits from increased energy efficiency. Specifically, on the other side of the ledger, regulators would need to recognize that there are three categories of energy efficiency benefits that reduce the negative externalities of traditional supply-side utility service. These are:

- *Clean Air*: reductions in sulfur dioxide (SO_2), nitrogen oxide (NO_X), and other emissions.

- *Climate Change*: reductions in carbon dioxide (CO_2) emissions.

- *National and Economic Security*: reductions in U.S. energy consumption and displacing imported oil.

Quantifying societal benefits and cost is something of an "art" that relies on statistical analyses of data. One aspect of such comparisons is that "costs" are typically more readily determined and quantified through market data and/or engineering analyses. The very nature of externalities is that these are matters not reflected in either market prices or internalized in the production of goods that are sold and consumed.

There are various names given to the analytic methods used to resolve the typical asymmetry in the data and information on the "benefit" and "cost" sides of the societal comparisons of externalities and analyses. Start with the more obvious notion that benefits should exceed costs. This can be expressed as either:

$$\frac{\text{Benefits}}{\text{Costs}} \geq 1$$

or

$$\text{Benefits} - \text{Costs} \geq 0$$

Regardless of the expression chosen, either of these result in the same algebraic expression of the notion that if benefits exceed costs, it is worth doing. The typical conditions are that "costs" can typically be quantified and that some "benefits" are actually cost savings.[1] This can simplify the analysis using avoided costs to represent the benefits of energy as:

---

[1] *See* Krutilla, John V. and Charles J. Cicchett. (1972). Evaluating Benefits of Environmental Resources With Special Application to the Hells Canyon. *Natural Resources Journal*, Vol. 12, No. 1.

Avoided Costs of Energy + External Benefits ≥ Costs of Energy Efficiency

This also means that:

External Benefits ≥ Costs of Energy Efficiency − Avoided Costs of Electricity

One conceptual approach is to determine the "break-even" value of external benefits. This is sometimes more formally called a "shadow price." This is the minimum value that the non-quantified, portion of external benefits, in monetary terms, would need to be in order for benefits to just equal costs. Policy makers, such as regulators, would then be able to use their judgment to determine if the nonquantified difference between benefits (avoided cost) and energy efficiency costs is "big enough" for them to expand energy efficiency and to require nonbeneficiaries to pay higher prices in order to expand energy efficiency.

This particular regulatory approach is what the Federal Power Commission (FPC) adopted in 1970[2] in determining the relative value of nonquantifiable future preservation benefits of the Hells Canyon project. The analyses relied upon the near-term development net benefits of a hydroelectric dam that would convert a wild river into a lake that would produce electricity. Conventional coal-fired power costs were deemed to be the avoided cost benefits of development. When the costs of the hydro project were compared to coal, the direct comparison demonstrated that hydro was lower cost and the benefits of a dam exceeded its costs.

The licensing of the Hells Canyon dam was a foregone conclusion before the Sierra Club brought externalities to federal court. The environmental argument was that preservation benefits had not been included. In effect, the environmentalists averred that the external benefits of preservation needed to be compared to the costs of coal minus the costs of hydroelectricity.

The FPC took evidence that showed, over time and with reasonable discounting methods, the amount of current preservation benefits would need to be relatively small to determine that coal development would be a better choice than a new dam in Hells Canyon, despite a dam being less expensive than a new coal-fired generating station. This was particularly evident when the FPC converted the aggregate dollar amount necessary for external preservation benefits to just break even with the net development (supply-side) benefits per unit (*e.g.*, number of current wilderness users and the amount they currently spend to experience a wild and scenic river left in its undeveloped condition).

The lesson here is that state regulators can compare the quantified net benefits and costs of energy efficiency, while also recognizing additional factors that have not been internalized or

---

[2] *See* Krutilla, John V. and Charles J. Cicchetti, *op cit.*

quantified. A break-even value can be readily attached to the omitted external benefit values. These can be compared to data that seem to reasonably represent what society is willing to pay to improve air quality, address climate change, and increase economic and national security.

There is also a converse analysis that is sometimes applicable in the current energy efficiency regulatory debate. Some negative externalities are being internalized using "market-like" prices. The policy approach is broadly called "cap and trade." Under this approach, the government would establish the target levels of pollution and assign or sell property rights to utilities that limit their combined emissions to the target levels, or the "cap." The market aspect of this policy approach takes over at this point, because utilities are encouraged to trade these permitted property rights in a market that establishes the per-unit value of the permitted emissions.

The supply-side of electricity generation is very much understood given its engineering focus and emphasis. Therefore, it is possible to estimate and reasonably determine the costs of internalizing through avoiding the need to purchase pollution permits or to derive value through selling permits not needed, when an electric generator alters supply-side operations and/or investments.

In addition, the policy of using market-like approaches to address negative externalities is not typically adopted or implemented simultaneously throughout the electricity generating industry. These real world conditions combine to provide regulatory policy makers with information about the likely future cost/value of internalizing current externalities. This information can be used in a similar break-even fashion to establish how much the utility, society, nonparticipants, and participants should be willing to pay for energy efficiency, in order to reduce the future market-based costs of internalizing externalities such as greenhouse gases. Experience in Europe, for example, can help Americans determine the potential per-unit cost of future compliance because the supply side of these industries has similar engineering.

The current debate concerning "cap and trade" is not done. Some advocate a more direct tax on pollution approach. In such debates, per-unit taxes are often discussed. These debates, therefore, are another source of information as to the likely range of the per-unit cost/value for internalizing current externalities related to greenhouse gases.

Regulators can convert the per-unit compliance costs of internalizing externalities to a break-even analysis to establish how much energy efficiency to pursue and how much should reasonably be paid per unit of projected electricity savings. Consider the previous benefit-to-cost comparison:

External Benefits ≥ Costs of Energy Efficiency − Avoided Costs of Electricity

This was used to infer how the break-even amount of external benefits could be established. When future "cap and trade" or pollution taxes can help to quantify external benefits, regulators could use this information and "benefit-to-cost" logic to establish how much to spend on energy efficiency. In broad terms, the logic would be to increase energy efficiency when:

$$\text{Estimated Market Value of Pollution Permits or Taxes} + \text{Avoided Costs of Electricity Supply} \geq \text{Costs of Energy Efficiency}$$

This reformulation means that regulators and utilities should seek to expand conservation and energy efficiency, when the all-inclusive supply-side future avoided societal costs exceed the costs of energy efficiency.

There are, at this relatively high level, some missing concepts. First, these concepts of benefits and costs were not very precisely defined. The discussion was mostly vague as to whether these were aggregate concepts or incremental or marginal. The theoretically correct interpretation and regulatory approach should be more precise, and marginal, not aggregate, concepts should be used. That said, this may not always work in practice. Using an aggregate break-even approach may conceptually cause regulators to go beyond the economically efficient target. Since many jurisdictions are not very far advanced in adopting energy efficiency, this conceptual risk may not be very great in practice.

A second factor is timing and discounting. The first observation is obvious; the time value of money is important, when energy efficiency investments occur in the present and the savings take place over time. Just as with supply-side analyses, demand-side program evaluations should accurately reflect the time value of benefits and costs. This observation will be discussed further below.

The difference between regulated prices (P_E) and the marginal cost of energy efficiency (MC_{EE}) plus the marginal external benefits of energy efficiency (MEB_{EE}) determine the socially optimal amount of utility-sponsored energy efficiency, as well as the amount nonparticipants could reasonably be assigned under regulation. The next chapter explains how to quantify the relevant externalities that would cause regulators to increase energy efficiency and load management.

CHAPTER 11
SOME QUANTITATIVE EVIDENCE TO ANALYZE EXTERNAL BENEFITS

Energy efficiency benefits include reductions in air emissions per MWh. Throughout much of the nation, coal-fired generation would be the fuel most likely to be displaced with energy efficiency programs. Table 11-1 uses average national data for coal-fired power stations in 2006 to quantify air pollution emissions. This is conservative, because marginal reductions are likely greater and these would typically be reduced first. Older, less efficient generating stations would emit more, while newer units would emit less than average. There is, however, a particular reason to use average coal-fired generation. This is because some forms of energy efficiency may be spread across the utility's load curve. Therefore, assigning the benefits of emission reductions to energy efficiency would require a more detailed utility load curve analysis. The average reductions would be a better estimate of how the entire load curve might react to energy efficiency.

The data on the supply side for air pollution emissions is very extensive. The Environmental Protection Agency (EPA) publishes data[1] for each electric generating station over time. Therefore, regulators could select the best assumption and likely find the data needed to quantify the physical reduction in pollution. These data include electricity output, as well as air pollution emission data in the following categories: (1) SO_2 emissions; (2) NO_X emissions; and (3) CO_2 emissions.

Table 11-1 shows the most recent EPA emission data in three important categories for average coal-fired electricity generators in the U.S. in 2006.

---

[1] All utilities are required to report their data using electronic data reporting (EDR) to the EPA. The EPA compiles and publishes these data on its Website.

TABLE 11-1 AIR EMISSIONS PER MWh FROM COAL-FIRED ELECTRICITY GENERATION	
	National Emissions EPA 2006[1]
CO_2 (tons/MWh)	1.0269
NO_x (lbs/MWh)	3.1895
SO_2 (tons/MWh)	0.0045

[1] U.S. EPA Clean Air Markets Program/Unit Level Emissions.
http://camddataandmaps.epa.gov/gdm/index.cfm?fuseaction=emissions.wizard&EQW_datasetSelection=

There are two other extreme air pollution cases to consider, when estimating the external benefits of energy efficiency. On the low end, energy efficiency would be assumed to replace new electricity generation, when a utility satisfies growth in demand from either new generation or energy efficiency. In this case, the external benefits of energy efficiency equal the avoided external costs of new generation. On the high end, there would be no need for new generation. In this case, energy efficiency would likely cause the retirement or sharp reduction in use of old, inefficient, and more often coal-fired generation that have relatively high emissions. Also in this case, the external benefits of energy efficiency would equal the pollution reduction and other savings related to these older units, not new generation.[2]

Accordingly, if there is no "new" generating station requirements, the external benefits of reduced air pollution are likely greater. In addition, there would be no offsets related to new, potentially much cleaner generation coming on line. This likely would also be the case when there is little or no difference between the marginal cost of electricity (MC_E) and the regulated price (P_E). Therefore, when new generation is needed, the external benefits related to an air pollution reduction would be less, because the new generator would be required to internalize the costs of air pollution reduction. This new power station would also displace the older more polluting generation stations. The need for energy efficiency would not necessarily falter, because new generation typically would mean that regulators would confront a new

---

[2] This assumes energy efficiency is less expensive than new generation.

round of rate increases. Accordingly, any lower external benefits due to new technology would be offset by additional economic and finance benefits. The offset would not necessarily be evenly felt across the various customer categories.

In most cases, external benefits and their support for energy efficiency reflect some combination of the two extremes and a mix of economic, finance, and external factors. The bottom line is that with this particular set of likely offsetting factors, using a national or utility average, air pollution emission effects per kWh may be more than a fairly reasonable assumption. If specific facts or conditions require more definitive measurements, the available data would make this possible. The specific combination may also vary over the life of the energy efficiency improvement.

The quantification of the external benefits for reduced air pollution emissions also requires a reasonable estimate of the dollars saved for specific types of emissions. Table 11-2 uses estimates of the market values for "cap and trade" programs to establish these per-unit values. Regardless of their relative dollars-per-unit values, they are conceptually conservative, because they represent estimates of the cost of internalizing these externalities. This is conceptually the same as determining what society pays. This is equivalent to "value in exchange." Economists recognize that society might be willing to pay a greater amount, known as "value in use," rather than to do without a good that has public benefits, such as externality-reducing energy efficiency programs.

Table 11-2 shows a low and high estimate of the per-unit cost of air emissions for these same three air emission categories. The CO_2 values shown reflect future estimates of the per-unit costs of "cap and trade." These likely reflect relatively higher per-unit values, because, unlike SO_2 and NO_X, CO_2 has mostly not been internalized in the selection, design, and operation of generating stations.

No one can know with certainty what the future will bring in terms of carbon taxes or cap-and-trade markets. Nevertheless, these per-unit CO_2 cost estimates represent the per-unit amounts that, as shown in the reported references, could be avoided, if utilities were to cut CO_2 emissions. Regulators using these estimates would, in effect, expect that there would be a relatively serious effort to reduce greenhouse gases (GHG). Regulators that are less aggressive would employ lower per-unit values.

The NO_X and SO_2 per-unit values shown reflect both past experience and future estimates. These per-unit values are far less than they would have been even twenty years ago. All these data reflects a combination of regulations to reduce emissions, emission trading prices, and potential carbon taxes. The actual marginal damages or costs of these emissions could exceed the per-unit values, because these trades would reflect the level of pollution control that governments establish for "cap" and "trade" purposes and not the likely greater social costs or value in use equivalents of pollution reduction.[3]

TABLE 11-2
AIR EMISSIONS PER UNIT OF VALUE

	Low	High
CO_2	$40/ton[1]	$80/ton[2]
NO_X	$1.20/lb[3]	$5.00/lb[4]
SO_2	$473/ton[5]	$900/ton[6]

[1] Sweden trades at $40 per ton. See Komanoff, Charles and Dan Rosenblum, *Carbon Taxes First* (April 24, 2007).

[2] The MIT Modeling group models three cases for U.S. permit prices starting at about $20-$50 (2005$) per ton in 2015 and increasing to $30-$100 per ton by 2030. William Nordhaus of Yale finds the optimal price to rise to more than $80 per ton by 2050. See the discussion of these studies In Aldy, Joseph E. *Assessing the Cost of Domestic Regulatory Proposals*, Resources for the Future (May 2007). Komanoff and Rosenblum also show a CO_2 allowance for cap and trade for the U.S. beginning at $37 and damages as high as $370 per ton.

[3] EIA's reference case shown in its Energy Outlook for 2007 uses $2,400 per ton, or $1.20 per pound.

[4] Historic NO_X prices in the west have been at or above $5/lb.

[5] U.S. EPA Office of Air and Radiation presentation "Cap and Trade Programs An Update" (May 7, 2007).

[6] Cambridge Energy Research Associates (CERA) puts the floor at $800 to $1,000 per ton based on the cost of Flue Gas Desulphurization (DFG).

The marginal external benefits when air emissions are reduced can be estimated for a typical utility by multiplying the values shown in Tables 11-1 and 11-2. Since regulators often use kWhs, the results shown in Table 11-3 convert MWhs to kWhs and dollars to cents. The estimated per-unit savings from reducing air emissions range from 4.7¢ to 10.2¢ per kWh. Since SO_2 has been addressed historically to a considerable extent, the marginal benefits related to additional SO_2 emission reductions are relatively small. The opposite is true for climate change (*i.e.*, CO_2 emissions), which are just beginning to be internalized. Therefore, there is relatively more incremental benefit associated with reducing CO_2 emissions. NO_X falls in between, because outside of Southern California, NO_X markets are not very commonplace.

---

[3] *See also* Natural Resources Defense Council. "Risky Business: Hidden Environmental Liabilities of Power Plant Ownership." *http://www.nrdc.org/air/energy/rbr/chap3.asp*; Green, Kenneth P., Steven F. Hayward, and Kevin A. Hassett. (2007, June). Climate Change: Caps vs. Taxes. American Enterprise Institute for Public Policy Research. *Environmental Policy Outlook*, No. 2; and Ishii, Ken. (2005, October 26). Sempra Energy Presentation.

TABLE 11-3
EXTERNAL BENEFITS PER kWh RELATED TO REDUCING AIR EMISSIONS
Case 1: National Emissions (¢/kWh)

	Low	High
CO_2	4.1076	8.2152
NO_x	0.3828	1.5948
SO_2	0.2129	0.4050
Total for 3 Emissions	4.7033	10.2150

The second major external reason for a renewed interest in energy efficiency is economic and national security. This has grown as the nation recognizes the economic consequences of high imported energy prices, with crude oil increasing to almost $150 per barrel in the summer of 2008. The external costs increase as the nation factors in the cost of securing a flow of crude oil from parts of the world that are admittedly anti-American.

There are two other matters related to the high prices of crude oil. First, the nation is expanding its importation of liquefied natural gas (LNG). These imports come with both high commodity prices and expensive new infrastructure investments. Second, paying foreigners more dollars to meet domestic energy needs causes macroeconomic problems in the form of increased inflationary concerns, as well as recessionary pressures in the U.S.

Paying more for imports represents a drain on the macro economy similar to a tax increase. This slows economic expansion. In addition, the dollar weakens relative to other currencies around the world. The electric industry is among the casualties of a weaker dollar, because a weaker dollar causes increases in the material and component prices needed to maintain and grow electric utility systems in the U.S. Finally, a weaker dollar makes it more difficult for the Federal Reserve to help economic expansion through lower interest rates, because doing so causes further exchange rate weakening and this increases the relative, and perhaps even the absolute, price of energy imports and general inflation.

Accordingly, the external benefits of energy efficiency would also include any savings related to increased energy independence. These include both macroeconomic (*e.g.*, a stronger dollar, lower interest rates, etc.) and national security benefits from reducing energy use. The current high prices for crude oil attributable to heightened international problems in the Middle East cause both macroeconomic and national security problems for the United States.

At the low end, the benefit of reduced energy imports would be about $27 per barrel. This amount reflects the estimated per-barrel premium for imported oil that has been used for several government purposes.[4] The most recent premium for reduced oil imports was $13.58 per barrel. It was based on 2004 dollars per barrel and an imported price of $45 per barrel. Using an imported price of crude oil closer to more likely, but still conservative based on the 2008 experience, $90 per barrel would increase the estimated premium from $13.58 per barrel to a low estimate of $27 per barrel.

Nevertheless, a premium of $27 per barrel is a quite conservative estimate, because the premium reflects just two aspects of the adverse consequence of America's oil import dependence: (1) lost efficiency due to seller market power;[5] and (2) the strategic storage costs that reduce potential losses during supply disruptions. A more accurate estimate of the per-unit benefit of reduced oil dependence should also include:

- higher defense budgets and the consequence of national security on troops and their families;

- the danger and consequences when higher oil revenues finance terrorists and anti-American hatred;

- the adverse macro-economic effects of increased imports, a weakened dollar, higher inflation, and higher interest rates and their corresponding effect on the U.S. economy, businesses, and families; and

- the effect of the above on our friends and allies throughout the world.

In addition, the full adverse effect of America's oil import dependency needs to be focused through the lens of a world oil market that has virtually little spare capacity. World oil con-

---

[4] See Leiby, Paul N. (2007, February 28). Estimating The Energy Security Benefits of Reduced U.S. Oil Imports. *Oak Ridge National Laboratory*; Leiby, Paul N., Donald W. Jones, T. Randall Curlee, and Russell Lee. (1997, November). Oil Imports: An Assessment of Benefits and Costs. ORNL-6851. *Oak Ridge National Laboratory*; U.S. Department of Tranportation, NHTSA. (2006, March). "Final Regulatory Impact Analysis: Corporate Average Fuel Economy and CAFÉ Reform for MY 2008-2011 Light Trucks." Office of Regulatory Analysis and Evaluation, National Center for Statistics and Analysis; National Academy of Sciences. (2002). "Effectiveness and Impact of Corporate Average Fuel Economy (CAFÉ) Standard." Committee on the Effectiveness and Impact of Corporate Average Fuel Economy (CAFÉ) Standards, National Research Council. Washington, D.C.: National Academy Press.

[5] On an energy equivalent basis, natural gas prices average $40 per barrel less than crude oil. This suggests a much greater efficiency loss for the Organization of the Petroleum Exporting Countries' (OPEC's) monopoly power.

sumption is equal to about 85 million barrels per day. Spare capacity has fallen to as low as about one million barrels per day in the last few years.[6] This is a reserve of a little over one percent.

As recently as the 1990s, this spare capacity reserve cushion was closer to about ten percent of world production. The precipitous decline in spare crude oil production capacity and the growth in demand in China and India are relatively new phenomena, which cause much greater energy prices, unprecedented volatility in prices, and international tensions. The future, if there is no recession, seems likely to be worse, not better, in terms of oil prices and America's dependency.

Accordingly, a reasonable range for the estimated benefits of reducing oil dependence would be $27 per barrel for the low case and about $40 per barrel or more for the high case. Reducing a barrel of oil dependence would likely reduce the nation's import dependence and yield benefits to the nation in line with the $27 to $40 per-barrel premiums.

In many states, the energy saved would likely reduce coal-fired electricity generation, not oil dependence. Nevertheless, a case could be made that any energy conservation would help to increase the nation's, as well as specific businesses' and households', energy efficiency and reduce the energy intensity indices for Gross National Product (GNP).

In addition, when oil prices increase, all domestic energy prices increase. Therefore, any foreign oil dependence premium that increases crude oil prices would likely increase other primary energy prices, such as coal and natural gas. Further, complex construction engineering costs increases when oil markets tighten. This means that new power stations cost more, when crude supplies are tight and prices are high.

An alternative argument is that the U.S. has plenty of coal and many electric utilities depend upon coal. Therefore, utility-sponsored energy efficiency would not significantly affect the nation's oil dependency. A reasonable estimate might be to value the external consequences of high oil prices for power generation at some place in the middle, say 50 percent. Furthermore, the current political support for successful energy efficiency includes national security and foreign energy dependence concerns. That said, the low premium is already quite conservative, because it omits what seems to be very costly factors related to the nation's energy dependence on imports.

Table 11-4 shows the marginal national security/oil dependence benefits related to energy efficiency.

---

[6] Currently, spare capacity has increased to about two million barrels per day. This is still less than a 2.5 percent reserve.

TABLE 11-4
THE NATIONAL SECURITY AND MACRO-ECONOMIC BENEFITS OF REDUCING ENERGY CONSUMPTION
(Cents per kWh Saved)[1]

Low Case: Premium = $27 per Barrel

Oil-to-Coal Premium Ratio	
50%	0.77
100%	1.54

High Case: Premium = $40 per Barrel

Oil-to-Coal Premium Ratio	
50%	1.14
100%	2.28

[1] The approximate energy content of a barrel of crude equals about 6,000,000 BTUs. The conversion factor for 1 kWh is about 3,413 BTUs.

Each region of the country and regulatory jurisdiction would need to perform its own benefit/cost analysis of the effects that energy efficiency would have on reducing negative externalities. Facts and conceptual differences likely would vary. In addition, marginal external benefits are but one component of the full regulatory comparison of the benefits and costs of energy efficiency. The additional factors would include the marginal cost of energy efficiency, regulated prices, and the marginal cost of electricity.

Table 11-5 combines the low and high estimates of the marginal external benefits of energy efficiency.

TABLE 11-5

SOCIAL BENEFITS FOR AVOIDING CLIMATE CHANGE, AIR EMISSIONS, AND NATIONAL SECURITY AND SAVE-A-WATT COSTS

¢/kWh

	Low	High
Three Emissions	4.703¢	10.215¢
National Security	0.77¢	2.28¢
Total Benefits	5.473¢	12.495¢

As explained above, regulators should not ignore who pays. Nonparticipant break-even values can be established using these or similar per-unit estimates of external benefits to determine the size of any price increases for electricity that would make them indifferent. Without any difference between the marginal cost of energy and regulated prices, nonparticipants would attach marginal benefits easily more than about 5.5¢ per kWh. The nonparticipant break-even values would also depend on the share of energy efficiency costs they would be assigned through regulation.

Some nonparticipants may be given the opportunity to "opt out" of paying anything to support an expansion of utility-sponsored energy efficiency programs. These customers would likely be required to verify that they have previously increased energy efficiency and/or renewable energy on their customer premises. The rationale for this is quite understandable. Nevertheless, all customers benefit, if there are additional reductions in negative externalities. Nonparticipants would also be better off when the marginal cost of energy (traditional supply-side) exceeds regulated prices and energy efficiency increases, because they would avoid paying even higher future rate increases. In addition, external benefits would not be trivial for even the already very efficient nonparticipants. Finally, there may also be other reasons why nonparticipants may be quite willing to help expand energy efficiency programs.

Commercial customers might perceive value, when their potential customers have more disposable income, because their electric and natural gas bills are less. Also, if schools and other socially significant retail consumers pay less when they participate, some nonparticipants may view this as a "good" outcome that would benefit many locally. Finally, some nonparticipants may simply want to design and pay for their own energy efficiency programs and would not object to others receiving the help they need to do something similar, because it helps the nation and the environment.

The next chapter combines these quantified estimates of the break-even model discussed in Chapter 9 to identify how regulators can put the nonparticipant matter to rest.

CHAPTER 12
COMBINING DATA AND THEORY

Chapter 9 included an algebraic model that identified the break-even conditions for nonparticipating consumers. This particular issue has previously either delayed or derailed serious utility-sponsored energy efficiency programs. The following discussion uses the range of low to high per-unit estimates of the external benefits of energy efficiency in Chapter 11 to determine break-even values under different assumptions. These estimates are combined with the algebraic model in a sensitivity analysis to expand the approach that various jurisdictions might apply to this nonparticipant issue, when it raises its regulatory head.

This illustrative analysis starts with some basic assumptions to frame the sensitivity analyses. These are: (1) annual electricity sales are $2 billion; (2) the average retail price is 8¢ per kWh; and (3) annual volume is 25 billion kWhs.

The sensitivity analysis assumes that the marginal cost of electricity is some percentage greater than the regulated price of electricity. Three values (10 percent, 20 percent, and 30 percent) are analyzed. The growth in sales also matters. These are assumed to be fully offset through energy efficiency. Nevertheless, three electricity sales growth rates are analyzed (one percent, two percent, and three percent).

There are also three different assumptions related to the percentage of the cost of energy efficiency that the participant pays. These are assumed to be 25 percent, 50 percent, and 75 percent, which equal $(1-k_2)$ in Chapter 9. The most important factor is the relationship between the marginal cost of energy efficiency (MC_{EE}) and the marginal cost of electricity (MC_E). This is treated as a percentage, designated ϕ. For equality, ϕ equals 1.0. In addition, when MC_{EE} is 20 and 40 percent less, ϕ equals 0.8 and 0.6, respectively.

The break-even values for these sensitivity analyses are shown in Tables 12-1, 12-2 and 12-3.

TABLE 12-1
THE BREAK-EVEN VALUE FOR MARGINAL EXTERNAL BENEFIT

TC_0	$2,000,000,000
P_0	$0.080
Q_0	25,000,000,000
φ	1.0

Growth Rate 1% Per Year

Participant Pays	\multicolumn{3}{c}{MC_e % adder:}		
K	10%	20%	30%
25%	0.06006	0.06012	0.06018
50%	0.04004	0.04008	0.04012
75%	0.02002	0.02004	0.02006

Growth Rate 2% Per Year

Participant Pays		MC_e % adder:	
K	10%	20%	30%
25%	0.06012	0.06024	0.06035
50%	0.04008	0.04016	0.04024
75%	0.02004	0.02008	0.02012

Growth Rate 3% Per Year

Participant Pays		MC_e % adder:	
K	10%	20%	30%
25%	0.06017	0.06035	0.06052
50%	0.04012	0.04023	0.04035
75%	0.02006	0.02012	0.02017

TABLE 12-2
THE BREAK-EVEN VALUE FOR MARGINAL EXTERNAL BENEFIT

TC_0	$2,000,000,000
P_0	$0.080
Q_0	25,000,000,000
φ	0.8

Growth Rate 1% Per Year

Participant Pays	MC_e % adder:		
K	10%	20%	30%
25%	0.04686	0.04572	0.04458
50%	0.03124	0.03048	0.02972
75%	0.01562	0.01524	0.01486

Growth Rate 2% Per Year

Participant Pays	MC_e % adder:		
K	10%	20%	30%
25%	0.04692	0.04584	0.04475
50%	0.03128	0.03056	0.02984
75%	0.01564	0.01528	0.01492

Growth Rate 3% Per Year

Participant Pays	MC_e % adder:		
K	10%	20%	30%
25%	0.04697	0.04595	0.04492
50%	0.03132	0.03063	0.02995
75%	0.01566	0.01532	0.01497

TABLE 12-3
THE BREAK-EVEN VALUE FOR MARGINAL EXTERNAL BENEFIT

TC_0	$2,000,000,000
P_0	$0.080
Q_0	25,000,000,000
φ	0.6

Growth Rate 1% Per Year

Participant Pays	MC_e % adder:		
K	10%	20%	30%
25%	0.03366	0.03132	0.02898
50%	0.02244	0.02088	0.01932
75%	0.01122	0.01044	0.00966

Growth Rate 2% Per Year

Participant Pays	MC_e % adder:		
K	10%	20%	30%
25%	0.03372	0.03144	0.02915
50%	0.02248	0.02096	0.01944
75%	0.01124	0.01048	0.00972

Growth Rate 3% Per Year

Participant Pays	MC_e % adder:		
K	10%	20%	30%
25%	0.03377	0.03155	0.02932
50%	0.02252	0.02103	0.01955
75%	0.01126	0.01052	0.00977

These tables show that several variables matter quite a bit in terms of the nonparticipants' break-even value for marginal external benefits (MEB) related to energy efficiency. These are:

- As the share participants pay increases (in increments of 25 percent of the total), the nonparticipants' break-even value for MEB_{EE} declines (by about one third for the increments analyzed).

- As the ratio of the marginal cost of energy efficiency decreases relative to the marginal cost of electricity (a decrease of 20 percent in the respective cases), there is a significant decrease in the nonparticipants' break-even value for MEB_{EE} (about a 25 percent decrease for each 20 percent decrement in relative marginal cost).

- Not shown is the fact that at lower relative regulated prices, the break-even value for MEB_{EE} decreases, and *vice versa* for higher initial regulated prices.

Two of the variables analyzed in this sensitivity analysis do not particularly matter. These are the growth in the system replaced through energy efficiency and the spread between the initial marginal cost of energy and regulated prices. This is because these factors were set to approximate their significance for a relatively typical utility. Compared to an assumed $2 billion in revenue, there is not much difference in the break-even MEB_{EE} when growth is one or three percent per year or marginal cost exceeds regulated prices for this size company by 10 or 30 percent. The reason this is so for the latter comparison is the model assumes that only growth is affected by this difference between the regulated price and marginal cost. Again, these differences affect only a small portion of the total revenues.

It is also important to compare these break-even results and the quantitative estimates of MEB discussed in Chapter 11. These ranged from about 5.5¢ per kWh on the low end to about 12.5¢ per kWh on the high end. This sensitivity analysis demonstrates two things:

- If participants pay at least half the utility's cost of energy efficiency, nonparticipants would be willing to pay more when they factor in even low estimates (about 4¢ or less) of the value of the marginal external benefits of energy efficiency.

- If the marginal cost of energy efficiency is about 80 percent or less than the marginal cost of electricity, nonparticipants would be willing to pay more than the low estimate (about 4.7¢ or less) of the value of the marginal external benefits of energy efficiency.

The next section investigates the amount utilities typically spend for energy efficiency and load management.

SECTION IV:
THE COST OF ENERGY EFFICIENCY

The costs of energy efficiency and load management are the focus of this section. Regulators, utilities, and customers recognize that benefits require effort and money must be spent. Mandating utilities to perform will get results. A central question is: At what cost? The more complex questions are: Who pays? And how much do they pay? This section answers the first of these questions. The other two require regulators to make a just and reasonable allocation of these costs. Part of the "cost" answer is quantitative and uses the EIA database for the last 15 years to determine what utilities and states have spent. Some of this is rather ambiguous.

The more complex answers to these three interdependent questions are conceptual. Opportunity costs matter and affect the way regulators think about the cost of conservation. Regulated prices and consumers' bills produce inconsistent answers, because, typically, as prices increase to finance energy efficiency, some customers' bills may fall and overall revenue requirements might be less. Nevertheless, some customers could pay more for the same volume of electricity they had been consuming.

Regulation also differs across the nation and energy efficiency and demand response programs add another layer of regulatory differences, because the 51 jurisdictions directly involved in demand-side programs frequently adopt unique regulatory approaches. Since regulatory discretion is rather wide when it comes to demand-side management, the answers across the nation will also vary.

CHAPTER 13
THE COST OF ENERGY EFFICIENCY AND LOAD MANAGEMENT

The idea is simple: most people think that energy efficiency costs less than electricity. Consumers that purchase efficiency, such as a "better" light bulb, would find they would pay less over time, and service would be the same or better. Consider a simple example, based on the following assumptions.

Cost of a compact fluorescent light (CFL) bulb	= $2.50
CFL energy use	= 25 watts
Incandescent light equivalent	= 100 watts
Life	= 8,000 hours
Price of kWhs	= 10¢

The consumer would pay for 800 kWhs of electricity using incandescent bulbs (100 watts * 8,000 hours), plus the cost of more frequently replacing the light bulb. At 10¢ per kWh, the electricity cost would be $80 over 8,000 hours of use. If instead the consumer used one CFL to produce the same light for its 8,000 hour life, the electricity use would be 200 kWhs and the customer would pay $20.

Over time, the customer would probably spend more in time and money to replace several traditional incandescent bulbs, which have about 1,000 hours of useful life, than one CFL bulb, which has about 8,000 hours of useful life. At about three times the initial cost, the CFL bulb would last about eight times longer. The consumer would, therefore, also save the time and money needed to buy and replace seven incandescent light bulbs. At about one-third the cost, the incandescent purchases would cost almost three times more than the CFLs. In addition, in this example, the customer would save $60 in the amount spent on electricity for the light bulb over 8,000 hours of use.

This simple example shows that consumers would save many times over, if they substitute CFL bulbs for incandescent light bulbs. This raises the obvious question: Why would consumers need any utility-sponsored help? And if help is needed, what form should this utility assistance take and at what cost?

The answer to the first question can be abrupt: consumers are lazy, irrational, and apathetic. These are not the reasons that regulators or utility executives would likely assign to ratepayers, who are voters and customers. Instead, the explanation that is more likely forthcoming is cast in terms of a lack of information or too many, sometimes conflicting, choices. The utility is put in the position of closing the information gap and providing something of an assurance/guarantee that CFLs actually work and last as long as manufacturers claim.

Another explanation is that consumers either do not want to pay more initially or, somewhat inconsistently, do not think the monthly savings would be big enough for them to bother to change their normal purchasing and use patterns. Here, the utility role becomes something of a teacher, banker, and retail marketer.

The savings in the money consumers spend, avoided utility costs, and reduced emissions of GHG and other pollution are astonishingly huge per light bulb, per household, per utility, and nationally. Accordingly, mandating the utility to cause CFL replacement is rather a natural regulatory response, once regulators deem the current CFL bulb replacement pace is too slow.

This is where the issue of cost becomes rather complicated. Consider two extremes. First, utilities could subsidize CFLs, if the initial cost is the reason why consumers do not buy CFLs directly and save money and time, help the environment, and reduce energy dependence. Some utilities simply give away CFLs at no charge to show customers that they work. These free bulbs are often accompanied with information in terms of the economic and environmental benefits associated with CFLs. Other utilities might subsidize the purchase price of CFLs, either directly with their customers or through retailers that sell CFLs. Some utilities will rely on advertising and education programs to reduce customer confusion and point them in the right direction.

All such programs effectively address the customer barriers and reduce the effective monetary and time-related costs for utility customers. When regulated utilities engage in such CFL marketing and promotional activities, they incur costs. These costs must be recovered and/or recognized, if regulated entities that promote energy efficiency are to be able to earn a fair rate of return.

During the first five years that the EIA collected energy efficiency data, its figures showed some of those utility costs. This detail is no longer provided by the EIA. Nevertheless, the information in Chart 13-1 shows some important detail. The utilities that had active demand-side programs, in the early to mid 1990s, reported nearly 25 percent was spent on indirect costs. This means that whatever they spent directly on programs, such as subsidized CFLs, they would spend additional amounts on administrative, marketing, monitoring, financial incentives, etc. Since early load management was usually provided through interruptible service tariffs, these programs could be omitted and then indirect costs were closer to about 40 percent of just the direct energy efficiency costs.

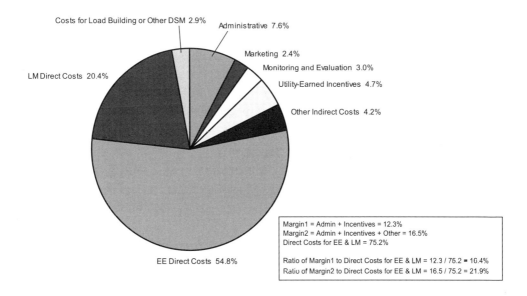

CHART 13-1
THE BREAKDOWN OF TOTAL ENERGY EFFICIENCY COSTS
(Sum of 1992-1996, For Utilities with Positive Indirect Costs)

In effect, utilities marked up the direct amount spent to cause their customers to expand energy efficiency. These markups or margins were in the form of the allocated cost recovery for administrative and general expenses, direct regulatory margins or incentives, or something euphemistically called "other." The latter could reflect a regulatory decoupling or lost margin adjustment. The panel shown in Chart 13-1 quantifies the relative size of these margins or markup for these early years.

The data show that some regulators recognized two facts. First, utilities could not simply buy or subsidize things like efficient CFLs and just give these benefits away at their direct cost. There were both additional indirect costs and financial incentives. Second, utility cost recovery involves cost allocation and returns.

There are also some important differences across the nation. Table 13-1 shows the breakdown of the various margins, defined as allocated administrative costs plus direct utility financial incentives, as a percent of direct costs for the top 100 utilities with direct costs in a particular year. Table 13-1 also shows the weighted average margins by year for IOUs with markups and direct costs, as well as across each IOU for the five-year period.

In later years, the EIA does not provide the same level of detail. Nevertheless, there is some useful detail in the information that EIA does provide. Chart 13-2 focuses on the last three years for which EIA provided data. It shows that indirect cost assignments averaged about 12.1 percent and that the combined direct costs for energy efficiency and load management averaged about 38.6 percent. During the last three years, the largest portion of the costs was for direct incentive payments paid to retail customers to subsidize their own energy efficiency spending and load management compliance. These were, respectively, 28.6 percent for energy efficiency incentive payments to customers and 20.8 percent for load management compliance. The combined total consumer incentive payment for all demand response for these three years was nearly 50 percent (49.4 percent).

TABLE 13-1
MARGIN (ADMIN COSTS + UTILITY INCENTIVES) AS A PERCENTAGE OF DIRECT COSTS FOR EE & LM
(TOP 100 INVESTOR-OWNED UTILITIES BY 2006 REVENUE)

Utility Name	1992	1993	1994	1995	1996	WGT AVG (1992-1996)
Alabama Power Co	0.00%	0.00%	3.53%	36.00%	34.82%	17.01%
Appalachian Power Co	0.00%	0.00%	50.00%	19.07%	0.00%	11.24%
Arizona Public Service Co	25.25%	30.38%	38.38%	56.17%	56.17%	40.02%
Atlantic City Electric Co	31.86%	29.60%	29.60%	0.00%		26.38%
Avista Corp	18.28%	9.45%	8.35%	46.79%	46.79%	14.08%
Baltimore Gas & Electric Co	5.80%	8.38%	7.06%	9.31%	4.28%	7.12%
Boston Edison Co	16.86%	18.74%	21.80%	16.21%	5.27%	17.64%
Central Hudson Gas & Electric Corp	11.92%	26.44%	15.06%	16.00%	2.75%	15.40%
Central Maine Power Co	1.49%	9.03%	7.74%	3.78%	3.52%	4.79%
Cleveland Electric Illuminating Co	0.00%	17.68%	35.55%	18.71%		23.13%
Columbus Southern Power Co	0.00%	14.86%	126.92%	167.19%	0.00%	82.02%
Commonwealth Edison Co	0.00%	0.00%	0.00%	0.00%	0.00%	0.00%
Commonwealth Electric Co	18.24%	17.88%	6.44%	43.37%	29.62%	18.67%
Connecticut Light & Power Co	3.52%	2.23%	3.50%	8.14%	9.48%	5.06%
Consolidated Edison Co-NY Inc	3.53%	6.65%	3.82%	19.46%	8.06%	6.66%
Consumers Energy Co	0.00%	0.00%	32.14%	20.50%	6.83%	3.16%
Dayton Power & Light Co	2.32%	25.45%		0.00%	0.00%	13.19%
Delmarva Power & Light Co	8.76%	9.32%	6.75%	0.00%	0.00%	4.92%
Detroit Edison Co	3.71%	5.50%	3.67%	3.69%	3.69%	3.93%
Duke Energy Carolinas, LLC	13.11%	22.38%	26.04%	26.73%	0.00%	18.32%
Duke Energy Indiana Inc	10.98%	0.63%	2.33%	2.90%	3.87%	3.28%

TABLE 13-1 (CONT'D)
MARGIN (ADMIN COSTS + UTILITY INCENTIVES) AS A PERCENTAGE OF DIRECT COSTS FOR EE & LM
(TOP 100 INVESTOR-OWNED UTILITIES BY 2006 REVENUE)

Utility Name	1992	1993	1994	1995	1996	WGT AVG (1992-1996)
Duke Energy Ohio Inc	0.00%	0.00%	0.00%	0.00%	1.33%	0.46%
El Paso Electric Co	63.96%	28.78%	48.17%	42.69%	32.63%	40.94%
Entergy Arkansas Inc	0.00%	0.00%	0.00%			0.00%
Entergy Gulf States Inc	4.00%	4.00%	0.00%			3.58%
Florida Power & Light Co	11.20%	8.68%	7.92%	9.37%	6.33%	8.41%
Georgia Power Co	0.00%	0.00%	0.00%	0.00%	0.00%	0.00%
Gulf Power Co	0.00%	0.00%	0.00%	0.00%	0.00%	0.00%
Hawaiian Electric Co Inc	9.38%	0.00%	47.97%		159.86%	58.85%
Idaho Power Co	0.00%	0.00%	0.00%	0.00%	16.28%	2.12%
Illinois Power Co	0.00%	0.00%				0.00%
Indiana Michigan Power Co	0.00%	0.00%	14.56%	9.89%	0.00%	5.80%
Indianapolis Power & Light Co	0.00%	0.00%	0.00%	0.00%	0.00%	0.00%
Jersey Central Power & Light Co	0.00%	0.00%	37.26%	18.75%	103.54%	29.39%
Kansas City Power & Light Co	0.00%	0.00%	0.00%	0.00%	0.00%	0.00%
Kansas Gas & Electric Co		0.00%	0.00%	0.00%	0.00%	0.00%
Kentucky Power Co	0.00%	0.00%	646.67%	0.00%	0.00%	9.47%
Kentucky Utilities Co	84.16%	93.95%	21.06%	9.71%	20.96%	27.24%
Louisville Gas & Electric Co	0.00%	0.00%	0.00%	0.00%	0.00%	0.00%
Massachusetts Electric Co	34.00%	19.71%	18.87%	21.59%	25.25%	23.44%
Metropolitan Edison Co	4.81%	6.17%	9.52%	9.20%	153.63%	15.40%
MidAmerican Energy Co				2.28%	4.99%	3.23%
Minnesota Power Inc	0.00%	0.00%	0.00%	0.00%	0.00%	0.00%
Mississippi Power Co		0.00%	0.00%	0.00%	0.00%	0.00%
Monongahela Power Co	0.00%	0.00%	0.00%	0.00%		0.00%
Narragansett Electric Co	21.09%	19.70%	24.20%	22.78%	18.51%	21.12%
Nevada Power Co	3.96%	4.31%	7.76%	11.41%	0.00%	5.75%
New York State Electric & Gas Corp	76.20%	44.11%	13.85%	0.00%	0.00%	44.22%
Niagara Mohawk Power Corp	110.16%	79.86%	54.86%	59.27%	214.59%	81.13%
Northern States Power Co MN	6.01%	0.00%	6.05%	15.15%	43.85%	15.23%
Northern States Power Co WI	12.75%	6.48%	4.54%	4.05%	3.84%	6.69%
NorthWestern Energy LLC	13.19%	31.83%	18.24%	0.00%	0.00%	13.04%
Ohio Edison Co	0.00%	0.00%	0.00%	0.00%	6.74%	0.38%
Ohio Power Co	0.00%	13.01%	172.33%	69.19%	0.00%	74.39%
Oklahoma Gas & Electric Co	1.78%	1.18%	0.00%	0.00%	0.00%	0.77%
Orange & Rockland Utilities Inc	9.82%	6.25%	6.41%	6.59%	3.04%	6.80%
Pacific Gas & Electric Co	38.00%	11.84%	14.15%	53.34%	37.09%	29.79%

(Continued)

TABLE 13-1 (CONT'D)
MARGIN (ADMIN COSTS + UTILITY INCENTIVES) AS A PERCENTAGE OF DIRECT COSTS FOR EE & LM
(TOP 100 INVESTOR-OWNED UTILITIES BY 2006 REVENUE)

Utility Name	1992	1993	1994	1995	1996	WGT AVG (1992-1996)
PacifiCorp	0.00%	0.00%	13.94%	1.24%	2.77%	3.02%
PECO Energy Co	0.00%	0.00%	0.00%	0.00%		0.00%
Pennsylvania Electric Co	0.00%	0.00%	0.00%	0.00%	166.67%	6.05%
Portland General Electric Co	0.00%	0.00%	0.00%	0.00%	8.13%	1.06%
Potomac Electric Power Co	4.41%	4.77%	7.15%	7.09%	18.47%	8.20%
PPL Electric Utilities Corp			42.85%	0.00%	0.00%	43.39%
Progress Energy Carolinas Inc	0.00%	0.00%	0.00%	0.00%	0.00%	0.00%
Progress Energy Florida Inc	4.38%	3.94%	4.63%	3.73%	2.98%	3.97%
Public Service Co of Colorado	0.00%	10.73%	28.50%	47.26%	13.47%	22.74%
Public Service Co of NM	0.00%	16.98%	11.88%	14.85%	3.44%	10.16%
Public Service Electric & Gas Co	7.78%	3.93%	5.36%	24.62%	1.98%	8.64%
Puget Sound Energy Inc	0.00%	0.09%	0.59%	5.34%	15.17%	1.05%
Rochester Gas & Electric Corp	0.00%	0.00%	0.00%	11.09%	18.07%	2.35%
San Diego Gas & Electric Co	29.64%	33.72%	29.18%	96.98%	88.17%	60.16%
Sierra Pacific Power Co	2.20%	2.84%	0.75%	1.27%		1.99%
South Carolina Electric & Gas Co	8.47%	9.57%	0.00%	0.00%	0.00%	3.54%
Southern California Edison Co	0.00%	0.00%	0.00%	0.00%	28.56%	4.68%
Southwestern Electric Power Co	22.25%	20.49%	20.07%	0.00%	0.00%	8.90%
Southwestern Public Service Co	0.00%	0.00%	0.00%	37.06%	17.21%	12.01%
Tampa Electric Co	1.08%	1.23%	2.37%	1.77%	1.23%	1.55%
The Potomac Edison Co	0.00%	29.92%	0.00%	0.00%	0.00%	1.75%
Toledo Edison Co	0.00%	23.68%	35.31%	19.76%		22.93%
Tucson Electric Power Co	0.00%	0.00%	0.00%	0.00%	0.00%	0.00%
Union Electric Co	0.00%	0.00%	0.00%	0.00%	0.00%	0.00%
United Illuminating Co	0.00%	0.00%	3.11%	4.72%	4.35%	1.93%
Virginia Electric & Power Co	0.97%	2.26%	2.25%	15.23%	10.59%	5.73%
West Penn Power Co	0.00%	0.00%	0.00%	0.00%		0.00%
Westar Energy Inc		0.00%	0.00%	0.00%	0.00%	0.00%
Western Massachusetts Electric Co	18.82%	10.03%	11.85%	10.81%	12.60%	12.95%
Wisconsin Electric Power Co	4.76%	32.34%	38.39%	94.43%	53.14%	32.89%
Wisconsin Power & Light Co	8.01%	6.80%	7.77%	2.02%	5.38%	5.77%
Wisconsin Public Service Corp	0.00%	30.50%	0.00%	0.00%	0.00%	8.42%
MEAN	9.44%	10.01%	21.73%	15.04%	19.49%	13.03%
MEAN W/O ZEROS	18.02%	17.22%	33.37%	25.07%	31.43%	16.33%

Chapter 13: The Cost of Energy Efficiency and Load Management

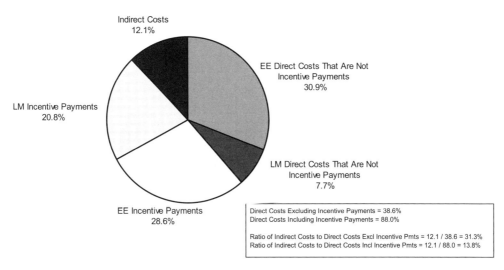

CHART 13-2
THE BREAKDOWN OF TOTAL COSTS FOR ENERGY EFFICIENCY
(Sum of 2004-2006, For Utilities with Positive Indirect Costs)

The panel in Chart 13-2 breaks out the respective margins, when these incentive payments to customers are included and when they are not included. When utilities provide money and information, they are less likely to be viewed as supplying a new demand-side product or service. Therefore, any markups or earnings margins are less likely to be significant. However, a utility that purchases or otherwise provides direct energy efficiency and load management services would very likely be treated like a producer or retailer for such demand-side products and services. The utility would likely seek and receive some additional earnings or revenue requirements, in addition to the direct costs for goods and services that it provides, packages, advertises, promotes, and sells to retail customers. The combined indirect cost was about 31 percent of the direct program cost, when consumer incentive payments are removed.

For the entire fifteen-year period, the available EIA data are limited to indirect costs and direct costs. It is not possible to remove any direct customer incentive payments prior to 2004. This means the estimated markups are understated, because any customer incentive payments would typically not be marked up, if at all, to the same extent as direct programmatic activities. Table 13-2 reflects this potential bias and shows the markup of indirect to direct costs for the entire period for the top 100 IOUs that reported any direct demand-side costs in any given calendar year. This shows that investor-owned utilities spend and invest dollars to increase energy efficiency and load management. Their customers that participate often pay some of the direct costs. Regardless, utilities also typically incur and regulators approve the additional

TABLE 13-2
INDIRECT COSTS AS A PERCENTAGE OF DIRECT COSTS FOR EE & LM
(TOP 100 INVESTOR-OWNED UTILITIES BY 2006 REVENUE)

Utility Name	1992	1993	1994	1995	1996	1997	1998	1999	2000	2001	2002	2003	2004	2005	2006	WGT AVG (1992-2006)
Alabama Power Co	17.00%	17.01%	17.98%	64.48%	91.50%	42.32%	38.89%	0.00%	32.25%	215.29%	217.61%	0.00%	200.14%	137.77%	124.67%	52.68%
Appalachian Power Co	0.00%	0.00%	58.26%	22.32%	0.00%											51.94%
Aquila Inc							0.00%	0.00%	0.00%	0.00%	0.00%	0.00%	0.00%	0.00%	0.00%	0.00%
Arizona Public Service Co	56.00%	65.05%	66.16%	106.54%	106.54%											42.25%
Atlantic City Electric Co	33.42%	38.68%	38.68%	0.00%						0.00%	0.00%	0.00%	0.00%	0.00%	0.00%	8.87%
Avista Corp	18.29%	10.83%	11.12%	47.81%	47.81%	41.83%	148.31%	0.00%	0.00%	0.00%	0.00%	0.00%	0.00%	0.00%	0.00%	10.26%
Baltimore Gas & Electric Co	9.93%	17.11%	10.73%	11.25%	5.33%	0.00%	0.00%	0.00%	0.00%	0.00%	0.00%	0.00%	0.00%	0.00%	0.00%	6.05%
Boston Edison Co	25.98%	28.00%	30.47%	26.37%	14.66%	7.55%	7.55%		30.66%	17.95%	30.41%			26.99%		24.07%
Central Hudson Gas & Electric Corp	20.09%	39.80%	36.55%	22.98%	7.12%	34.35%										26.00%
Central Illinois Public Service Co												0.00%	0.00%	0.00%	0.00%	28.88%
Central Maine Power Co	3.35%	10.83%	12.65%	4.40%	4.27%	104.02%		2.55%	1.31%	1.91%	0.00%	1.24%	0.31%	0.35%	0.33%	11.88%
Cleveland Electric Illuminating Co	0.00%	27.13%	39.63%	18.71%												27.77%
Columbus Southern Power Co	0.00%	168.58%	128.46%	169.96%	0.00%	0.00%	0.00%									94.47%
Commonwealth Edison Co	0.00%	0.00%	0.00%	0.00%	0.00%	0.00%	0.00%	0.00%	0.00%	28.97%	19.53%	75.65%	8.33%	8.55%	8.62%	9.10%
Commonwealth Electric Co	43.17%	26.63%	18.68%	50.22%	33.94%	33.94%	8.86%		30.05%	37.57%	23.37%		25.12%	27.91%		27.10%
Connecticut Light & Power Co	7.36%	9.95%	9.19%	14.29%	18.52%	5.75%	9.68%	12.92%	19.43%	17.26%	18.16%	21.16%	9.82%	10.97%	6.85%	12.68%
Consolidated Edison Co-NY Inc	5.15%	14.03%	18.43%	48.65%	17.45%	15.52%	46.26%			24.73%	6.73%	9.09%	0.00%	0.00%	0.00%	15.52%
Consumers Energy Co	1.25%	2.98%	82.54%	30.14%	9.00%	0.00%	0.00%	0.00%	0.00%							7.90%
Dayton Power & Light Co	14.31%	31.35%		0.00%	0.00%	0.00%										14.48%
Delmarva Power & Light Co	13.59%	15.69%	52.81%	26.51%	16.64%	20.60%	18.64%	0.00%	0.00%	0.00%						22.61%
Detroit Edison Co	6.90%	19.80%	11.47%	11.50%	11.50%	0.00%	0.00%	0.00%	0.00%		0.00%					8.57%
Duke Energy Carolinas, LLC	41.54%	59.69%	56.96%	78.39%	0.00%	0.00%	0.00%	0.00%	0.00%	0.00%	0.00%	0.00%	0.00%	0.00%	0.00%	23.93%

TABLE 13-2 (CONT'D)
INDIRECT COSTS AS A PERCENTAGE OF DIRECT COSTS FOR EE & LM
(TOP 100 INVESTOR-OWNED UTILITIES BY 2006 REVENUE)

Utility Name	1992	1993	1994	1995	1996	1997	1998	1999	2000	2001	2002	2003	2004	2005	2006	WGT AVG (1992-2006)
Duke Energy Indiana Inc	23.26%	30.79%	5.09%	6.31%	8.46%	2.37%	13.56%	5.77%	4.80%	26.17%	46.09%	45.41%	23.37%	3.49%	2.63%	13.92%
Duke Energy Ohio Inc	0.00%	11.09%	0.00%	0.00%	2.92%	2.64%	10.62%	13.49%	6.64%	19.61%	63.02%	35.27%	21.73%	1.94%	2.65%	6.21%
Duquesne Light Co										750.00%	0.00%	0.00%	0.00%	0.00%	0.00%	4.27%
El Paso Electric Co	132.13%	48.33%	90.17%	127.88%	76.84%	57.57%	50.00%	0.00%	0.00%	0.00%		123.14%	8.82%	16.82%	11.07%	59.01%
Entergy Arkansas Inc	0.00%	0.00%	0.00%													0.00%
Entergy Gulf States Inc	8.00%	8.00%	0.00%				0.00%			27.57%	20.30%	10.54%	10.94%	9.04%	7.13%	10.35%
Florida Power & Light Co	11.20%	8.68%	7.92%	10.82%	7.10%	6.88%	10.01%	10.68%	12.77%	8.39%	9.45%	8.65%	8.33%	8.52%	9.99%	9.22%
Georgia Power Co	0.00%	0.00%	0.00%	0.00%	0.00%	0.00%	0.00%	0.00%	0.00%	0.00%	0.00%	0.00%	0.00%	0.00%	37.10%	2.24%
Gulf Power Co	0.00%	0.00%	0.00%	0.00%	0.00%	0.00%	0.00%	0.00%	0.00%	0.00%	0.00%	0.00%	0.00%	0.00%	0.00%	0.00%
Hawaiian Electric Co Inc	23.78%	0.00%	66.22%		247.22%	0.00%	0.00%	0.00%	0.00%	0.00%	0.00%	0.00%	0.00%	0.00%	0.00%	2.97%
Idaho Power Co	0.00%	0.00%	0.00%	0.00%	16.28%	36.52%	149.87%	358.62%	507.17%	213.31%	261.38%	137.09%	38.93%	2.73%	2.77%	22.13%
Illinois Power Co	0.00%	19.72%				0.00%	0.00%	0.00%	0.00%		0.00%	0.00%	0.00%	0.00%	0.00%	18.13%
Indiana Michigan Power Co	0.00%	0.00%	14.89%	9.89%	0.00%											25.85%
Indianapolis Power & Light Co	43.75%	60.43%	20.84%	1.05%	0.00%	0.00%	0.00%	0.00%	0.00%	0.00%	0.00%	0.00%	0.00%	0.00%	0.00%	5.37%
Interstate Power and Light Co											0.00%	0.00%	0.00%	2.05%	2.34%	1.08%
Jersey Central Power & Light Co	19.62%	13.86%	64.13%	48.72%	141.77%	66.10%	22.33%	15.54%	7.03%	0.00%	0.00%	0.00%	2.18%	0.00%	0.00%	19.48%
Kansas City Power & Light Co	0.00%	0.00%	0.00%	3.83%	11.72%	6.35%	9.37%	13.86%	0.84%	0.11%	0.37%	0.77%	4.11%	3.16%	2.44%	3.36%
Kansas Gas & Electric Co	0.00%	0.00%	0.00%	0.00%	0.00%	0.00%	0.00%									0.00%
Kentucky Power Co	0.00%	0.00%	646.67%	0.00%	0.00%					7.89%	6.00%	0.00%	0.18%	12.13%	0.15%	93.19%
Kentucky Utilities Co	230.03%	258.55%	64.26%	68.65%	139.79%	5.77%	6.68%	7.82%		17.12%	6.56%	2.21%	1.37%	1.58%	1.25%	33.83%
Louisville Gas & Electric Co	0.00%	0.00%	0.00%	0.00%	0.00%	0.00%	0.00%	0.00%	0.00%	1.60%	3.20%	1.74%	1.25%	1.05%	1.02%	0.87%
Massachusetts Electric Co	48.49%	30.41%	28.20%	29.45%	31.60%	34.02%	31.03%		27.70%		121.40%	31.95%	19.52%	16.68%	3.13%	28.38%

(Continued)

TABLE 13-2 (CONT'D)
INDIRECT COSTS AS A PERCENTAGE OF DIRECT COSTS FOR EE & LM
(TOP 100 INVESTOR-OWNED UTILITIES BY 2006 REVENUE)

Utility Name	1992	1993	1994	1995	1996	1997	1998	1999	2000	2001	2002	2003	2004	2005	2006	WGT AVG (1992-2006)
Metropolitan Edison Co	6.28%	15.66%	19.50%	18.62%	379.96%	400.54%	421.69%	1481.33%	0.00%	0.00%	0.00%	0.00%	0.00%	0.00%	0.00%	35.88%
MidAmerican Energy Co				7.06%	20.37%	15.81%	16.00%	10.60%	9.35%	7.68%	7.79%	9.11%	13.47%	12.33%	11.53%	11.40%
Minnesota Power Inc	0.00%	0.00%	0.00%	0.00%	0.00%	0.00%	0.00%	0.00%	0.00%	0.00%	0.00%	0.00%	0.00%	0.00%	0.00%	0.00%
Mississippi Power Co		0.00%	0.00%	0.00%	0.00%	0.00%	0.00%	0.00%	0.00%	0.00%	0.00%	0.00%	0.00%	0.00%	0.00%	0.00%
Monongahela Power Co	0.00%	27.64%	30.54%	0.00%												16.06%
Narragansett Electric Co	31.21%	27.80%	35.72%	32.98%	26.26%	22.82%	18.08%	15.20%		20.45%	19.07%	22.87%	10.69%	9.84%	3.18%	19.24%
Nevada Power Co	3.96%	4.31%	7.76%	16.38%	0.00%	0.00%				29.03%	11.94%	17.65%	0.00%	9.09%	4.12%	7.70%
New York State Electric & Gas Corp	76.20%	46.82%	13.87%	0.00%	0.00%	0.00%	0.00%		2.68%	22.68%	9.08%	-11.73%		0.00%	0.00%	42.68%
Niagara Mohawk Power Corp	116.96%	88.84%	62.02%	70.19%	224.89%	0.00%	0.00%									88.39%
Northern States Power Co MN	6.01%	0.00%	10.50%	17.86%	43.85%	7.78%	10.84%	13.31%	12.05%	10.58%	14.80%	15.16%	0.00%	0.00%	4.93%	9.73%
Northern States Power Co WI	37.43%	35.21%	44.22%	75.03%	59.18%	35.78%	196.74%	116.81%	100.00%	36.07%	80.69%	151.21%	110.46%	112.15%	102.25%	74.96%
NorthWestern Energy LLC	21.08%	48.06%	38.05%	0.00%	42.88%	13.86%	12.30%	18.30%	15.27%	20.39%	17.96%	12.53%	4.83%	8.14%	9.89%	19.28%
Ohio Edison Co	0.00%	0.00%	8.56%	0.00%	9.32%	0.00%	0.00%	0.00%								1.76%
Ohio Power Co	0.00%	237.62%	173.77%	69.79%	0.00%											128.28%
Oklahoma Gas & Electric Co	1.78%	1.18%	0.00%	0.00%	0.00%	0.00%										0.71%
Orange & Rockland Utilities Inc	24.37%	14.56%	10.68%	8.67%	15.28%	15.70%	0.00%		0.00%	0.00%	0.00%					14.15%
Pacific Gas & Electric Co	50.48%	14.83%	21.09%	59.70%	47.43%	51.64%	24.27%	26.36%	9.76%	5.96%	3.87%		0.00%	6.71%	6.64%	20.14%
PacifiCorp	0.00%	0.00%	26.30%	3.64%	11.64%	30.01%	13.35%		0.00%	0.00%	0.00%	0.00%	0.00%	0.00%	0.00%	6.33%
PECO Energy Co	0.00%	0.00%	0.00%	0.00%												0.00%
Pennsylvania Electric Co	0.00%	0.00%	0.00%	0.00%	463.18%	400.39%	155.84%	509.03%	0.00%	0.00%	0.00%	0.00%	0.00%	0.00%	0.00%	22.81%
Portland General Electric Co	2.37%	2.05%	1.08%	1.60%	8.13%	7.07%	10.01%	7.52%	3.75%	1.40%	1.72%					3.27%
Potomac Electric Power Co	8.61%	7.23%	7.75%	7.68%	19.56%	-8.12%	-2.20%	3.53%	2.99%	1.20%	1.24%	0.27%	0.00%	0.00%		6.43%

Chapter 13: The Cost of Energy Efficiency and Load Management

TABLE 13-2 (CONT'D)
INDIRECT COSTS AS A PERCENTAGE OF DIRECT COSTS FOR EE & LM
(TOP 100 INVESTOR-OWNED UTILITIES BY 2006 REVENUE)

Utility Name	1992	1993	1994	1995	1996	1997	1998	1999	2000	2001	2002	2003	2004	2005	2006	WGT AVG (1992-2006)
PPL Electric Utilities Corp			42.85%	0.00%	0.00%											97.75%
Progress Energy Carolinas Inc	10.55%	9.56%	7.68%	5.99%	0.00%	0.00%									0.00%	5.56%
Progress Energy Florida Inc	5.85%	4.87%	4.99%	4.01%	3.57%	672.90%	52.56%	58.09%	61.17%	53.75%	58.88%	60.91%	75.74%	8.32%	9.97%	32.01%
Public Service Co of Colorado	15.10%	13.98%	50.63%	47.26%	13.47%	2.27%	8.36%	4.47%	0.67%	0.00%	0.00%	0.00%	0.00%	0.00%	7.28%	9.92%
Public Service Co of NH	0.00%	16.98%	17.66%	14.85%	7.74%	0.00%	8.46%	2.98%	3.08%	0.00%	0.00%	0.00%	13.68%	9.47%	9.50%	6.52%
Public Service Electric & Gas Co	20.43%	13.11%	17.94%	28.30%	3.24%	0.00%	0.00%	0.00%	0.00%	0.00%	0.00%	0.00%	0.00%	0.00%	0.00%	1.79%
Puget Sound Energy Inc	11.31%	4.53%	3.09%	5.95%	15.36%	0.00%	0.00%	0.00%	0.00%	0.00%	0.00%	0.00%	0.00%	0.00%	0.00%	3.62%
Rochester Gas & Electric Corp	8.92%	9.42%	6.41%	21.13%	28.73%											11.05%
San Diego Gas & Electric Co	49.64%	55.43%	53.93%	112.39%	100.96%	0.00%	30.91%	29.55%						0.00%	0.00%	41.17%
Sierra Pacific Power Co	26.17%	26.45%	8.50%	7.63%	0.00%				0.00%	50.00%	50.28%	32.03%	0.00%	14.05%	5.58%	18.57%
South Carolina Electric & Gas Co	15.43%	15.22%	14.66%	0.00%	0.00%	0.00%	18.77%	0.00%	15.66%	0.00%	0.00%	0.00%	0.00%	0.00%	0.00%	5.40%
Southern California Edison Co	29.74%	20.73%	24.55%	25.87%	47.41%	23.88%	14.08%	0.00%	0.00%	14.41%	0.00%	0.00%	0.00%	0.00%	0.00%	10.29%
Southwestern Electric Power Co	83.50%	77.23%	40.99%	0.00%	0.00%	0.00%	0.00%	0.00%	0.00%	0.00%	0.00%	0.00%	21.45%	10.32%	6.22%	10.53%
Southwestern Public Service Co	0.00%	0.00%	0.00%	57.32%	23.54%	0.32%	0.15%	0.00%	0.00%	0.00%	13.64%	0.73%	11.22%	0.00%	0.00%	5.93%
Tampa Electric Co	1.08%	1.23%	2.37%	1.77%	1.23%	1.32%	2.08%	1.40%	1.07%	1.00%	0.96%	0.92%	1.05%	1.18%	1.35%	1.34%
The Potomac Edison Co	0.00%	32.74%	0.00%	0.00%	0.00%	0.00%										1.85%
Toledo Edison Co	0.00%	38.72%	38.27%	19.76%												26.91%
Tucson Electric Power Co	0.00%	0.00%	0.00%	0.00%	0.00%	0.00%	0.00%	0.00%	0.00%	0.00%	0.00%	0.00%	0.00%	0.00%	0.00%	0.00%
Union Electric Co	0.00%	0.00%	0.00%	0.00%	0.00%	0.00%	0.00%	0.00%				0.00%	0.00%	0.00%	0.00%	0.00%
United Illuminating Co	7.92%	7.71%	11.41%	15.52%	9.98%	9.15%	31.24%	25.25%	39.04%	25.52%	27.36%	35.37%	10.29%	2.78%	2.51%	14.23%
Virginia Electric & Power Co	14.09%	16.49%	19.21%	39.12%	17.90%	43.99%	0.00%	0.00%	0.00%	0.00%	0.00%	0.00%	0.00%	0.00%	0.00%	16.66%
West Penn Power Co	0.00%	0.00%	0.00%	0.00%												0.00%

(Continued)

TABLE 13-2 (CONT'D)
INDIRECT COSTS AS A PERCENTAGE OF DIRECT COSTS FOR EE & LM
(TOP 100 INVESTOR-OWNED UTILITIES BY 2006 REVENUE)

Utility Name	1992	1993	1994	1995	1996	1997	1998	1999	2000	2001	2002	2003	2004	2005	2006	WGT AVG (1992-2006)
Westar Energy Inc		0.00%	0.00%	0.00%	0.00%	0.00%	0.00%	0.00%	0.00%	0.00%	0.00%	0.00%	0.00%	0.00%	0.00%	0.00%
Western Massachusetts Electric Co	24.68%	21.50%	22.70%	24.36%	24.92%	14.70%	14.97%	22.00%	12.16%	17.91%	14.13%	8.39%	7.49%	6.49%	6.53%	16.59%
Wisconsin Electric Power Co	35.70%	46.20%	63.06%	135.79%	63.24%		5.26%	8.56%	4.15%	2.63%	0.00%	93.77%	72.43%	29.94%	13.82%	39.29%
Wisconsin Power & Light Co	22.92%	17.76%	18.19%	10.42%	22.22%	32.49%	5.28%	0.31%	0.14%	1.48%	18.10%	0.00%	0.00%	0.00%	0.00%	8.48%
Wisconsin Public Service Corp	25.35%	30.50%	55.55%	67.73%	44.87%	137.26%	481.24%	117.15%	11.98%							50.59%
MEAN	19.68%	24.66%	31.92%	24.65%	36.31%	35.59%	31.61%	52.27%	16.99%	28.03%	18.83%	15.62%	11.70%	7.95%	6.41%	19.92%
MEAN W/O ZEROS	29.52%	34.05%	42.23%	36.13%	54.12%	62.97%	55.74%	97.57%	32.85%	49.65%	36.52%	34.04%	25.42%	16.64%	13.01%	22.59%

Chapter 13: The Cost of Energy Efficiency and Load Management

recovery "of" and "on" these direct utility costs for energy efficiency and load management. Depending upon the program, the utilities' costs for the light bulbs, appliance replacement, heating and cooling improvements, etc. will increase as indirect and shareholder incentives increase, and will decline as participating customers pay a larger portion of the improvement.

Chart 13-3 combines these various margin calculations for the different time periods, when EIA provided the data to make such calculations possible.

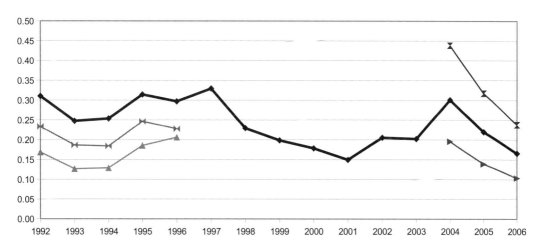

CHART 13-3
ENERGY EFFICIENCY COST MARKUP
(For Utilities with Positive Indirect Costs)

— ▲ — Ratio of Margin1 (Admin + Incentives Earned By Utilities) to Direct Costs
— ✕ — Ratio of Margin2 (Admin + Incentives Earned By Utilities + Other Indirect Costs) to Direct Costs
— ✱ — Ratio of Total Indirect Costs to Direct Costs Minus Incentive Payments
— ● — Ratio of Total Indirect Costs to Direct Costs Including Incentive Payments 2004-2006
— ◆ — Ratio of Total Indirect Costs to Direct Costs 1992-2003; Avg of [Total Indirect Cost / (Direct Costs w/ Incentive Pmts)] and [Total Indirect Costs / (Direct Costs w/o Incentive Pmts)] 2004-2006

The generic observation is that markups or margins of about 25 percent (.25) are added to direct costs, when regulators recognize that some incentives and additional cost recovery is necessary. There is an additional matter called revenue or income decoupling that regulators often explicitly consider. The data discussed here is ambiguous, because some of the markups used might have been adjusted upwards to offset any expected utility losses due to selling less electricity. In other cases, any such losses due to reduced sales would have been addressed separately, if at all.

CHAPTER 14
THE PER-UNIT COSTS OF ENERGY EFFICIENCY AND LOAD MANAGEMENT

The EIA collects two types of "savings" variables: incremental and annual. These reflect, respectively, additional "new" savings in a calendar year, and the cumulative "annual" savings in the same calendar year. The EIA also produces data for the incremental amount utilities spend each calendar year for demand-side management (DSM), with separate estimates for energy efficiency, load management, and the combined indirect related costs. These annual costs are akin to investment dollars and reflect both the various demand-side programs direct and indirect costs. EIA also has altered the detail it provides for these cost estimates over the period the data have been reported from 1992 through 2006.

The EIA does not collect information on the economic or physical life of these DSM investments. Some demand-side management programs, such as efficient light bulbs (CFLs), may last several years, depending on how many hours they are used each year. Others, such as high-efficiency air conditioning could last for decades. Regardless, it seems likely that end-use customers, after experiencing the improvements of a new technology, such as high efficiency light bulbs, would seek to replace them with similar light bulbs, when they need to be replaced at their own expense. If so, this would increase the savings at no cost to the utility. This also suggests that the benefits of a utility's DSM efforts could extend beyond the physical life of its specific efforts. Similarly, neighbors may learn of the savings and also purchase CFLs, or the participating customer may just decide to add more CFLs at his/her own expense. Regardless, these additional energy efficiency gains would be rather directly due to the utility's initial market stimulation. These so-called "halo effects" would increase and extend the savings per dollar spent and reduce the utility's unit cost for DSM.

This present discussion reviews what utilities have spent to increase energy efficiency. This approach omits so-called halo effects. In addition, two other cost factors are omitted. First, this discussion does not reflect the amount participating customers pay to increase energy efficiency. For example, if the utility pays half the cost of $2.50 per CFL bulb and the customer pays the other $1.25 per bulb, this discussion, which is based on EIA data, reflects the utility's half and omits the customer's share.

Second, this discussion also omits any utility losses that might result from lost income, fixed cost recovery, or sales margins due to reduced energy sales. These opportunity costs would typically represent utility losses, until the next general rate case, due to regulatory lag. If a utility regulator anticipated such "losses" due to energy efficiency and/or had some type of sales or revenue adjustment mechanism, the utility would tend to be made at least partially whole and would avoid some of these opportunity costs. Regardless, the cost data discussed in this chapter do not explicitly include any such lost sales or lost utility income effects. These adjustment mechanisms and potential losses will, however, be addressed in the following chapters.

For purposes of calculating annual unit costs, it is reasonable to match DSM annual cost recovery to future annual benefits. This would mean comparing prior incremental spending for DSM to a time period that reflects annual cumulative benefits.

IOUs and their regulators are well aware of the fact that money is often invested in lump sums, with the associated benefits and cost recovery spread over a number of years. Energy efficiency (EE) and load management (LM) programs also require money to be spent or invested in lump sums, with the resulting savings spread over a number of years. The conceptual treatment of both time and this per-unit cost determination is not different for IOU supply-side and demand-side curtailing efforts.

Table 14-1 shows two different methods to estimate the cost per MWh saved using EE savings in 2006. These vary considerably and neither is conceptually correct. The high estimate sums the current and previous years' incremental spending and divides this sum by the cumulative "annual" savings each year. This high estimate of 21.4¢ per kWh assumes two things that increase the per-unit cost estimate. First, all the prior years' demand-side programs are still saving energy each year. This means all prior costs remain in the numerator. Second, in effect, there are no future savings related to the more recent years' programs. This decreases the denominator. For example, 2006's incremental spending only adds savings for calendar year 2006, when there would be additional out-year savings and a lower per-unit cost. Obviously, this high end estimate is "too high."

TABLE 14-1
HIGH & LOW ESTIMATES OF COST PER kWh OF EE SAVINGS
(1992-2006)

Year	LOW: Weighted Average of Incremental Direct EE Costs per kWh of Cumulative EE Savings	HIGH: Weighted Average of Incremental Direct EE Costs per kWh of Incremental EE Savings
1992	$0.047	$0.225
1993	$0.041	$0.189
1994	$0.032	$0.197
1995	$0.025	$0.178
1996	$0.018	$0.165
1997	$0.016	$0.191
1998	$0.016	$0.230
1999	$0.017	$0.270
2000	$0.018	$0.285
2001	$0.021	$0.250
2002	$0.019	$0.277
2003	$0.017	$0.274
2004	$0.017	$0.201
2005	$0.020	$0.201
2006	$0.020	$0.236
TOTAL (1992-2006)	$0.022	$0.214

Note: Includes all utilities that filed EIA Form 861.

At the low end, the incremental spending or current year's investment amount is divided by the cumulative annual savings each year. These savings include all the prior (*i.e.,* pre-existing) energy efficiency programs still producing savings in a given year. This low per-unit cost estimate for energy efficiency is conceptually flawed and significantly biased downward, because it only includes one year's incremental cost, while capturing savings from prior years' spending on energy efficiency.

One reason for showing this second conceptually flawed case is that some analysts incorrectly suggest that energy efficiency costs the typical utility about 2¢ per kWh. This misconception seems to have been caused when the amount that EIA calls "annual" savings is divided by a particular year's incremental costs. This is not reasonable, because these "annual" savings are actually the cumulative savings that were caused by past incremental investments in demand-side programs, and current incremental spending has out-year effects. While these offset each other, the net effect is overwhelmingly to understate the per-unit costs, because there were many prior years of spending but only one, the current year's, incremental cost is included in this calculation.

These various extreme high and low flawed cases over the last 15 years differ by about a factor of 10, averaging 2.2¢ per kWh and 21.4¢ per kWh. The correct answer is somewhere in between these extremes. This is a very wide range. The conceptually "correct" per-unit cost of energy efficiency also requires some determination of a discount rate and an estimate of the economic life of demand-side programs. This is necessary to reflect the time value of money in these per-unit demand-side cost estimates.

The discount rate could convert the future expected savings to a present value. These could be reduced to a per-unit savings value by dividing them by the incremental costs that produced them. The EIA data are not reported in this manner. Instead, all prior years' programs still expected to be yielding savings and the current year's incremental savings are reported. These data are also aggregated across all energy efficiency and load management programs. Therefore, it is not possible to make specific engineering estimates of the expected amount of savings over time. Nevertheless, this time value adjustment is important when demand- and supply-side alternatives are compared to each other.

The nature of the EIA data means that a combination of assumptions and sensitivity analyses would be useful. One conceptually correct approach is to use cumulative savings in any particular year and to assume that these were caused by some reasonable number of prior years' reported incremental investments.

The EIA data begin in 1992. There were some pre-existing cumulative savings in that first year, because reported incremental kWhs saved were less than the cumulative kWhs saved in 1992. At the other end of the data, the increases reported for 2006 over the 2005 cumulative would subtract expiring savings in 2006 and add just the first year's expected incremental savings for new programs or new participants in existing programs. This would mean that, in any year, the expected future savings would not be included in the reported cumulative savings. These two data anomalies have opposite mathematical effects. One approach is to assume they cancel out one another. The following analysis does this and uses the full fifteen-year period of data to estimate the likely range of per-unit costs.

Another approach is to pick a point in the middle of the data, such as 2002, and assign a portion of the future cumulative savings to the amount spent up to 2002. This means the additional incremental savings reported after 2002 needs to be subtracted from the reported future years' cumulative savings. This will attribute just that portion of the 2003 through 2006 cumulative savings that were caused by spending in or before 2002. In this case, the assumption for economic life is that energy efficiency programs would last five years, with 2002 as the year this analysis uses as the central year.

The discount rate used should reflect the underlying nature of the cost data. If original costs are used to determine the revenue requirement equivalent, the discount rate should be about the same as a utility's supply-side investments. Three values (10 percent, 12 percent, and 15 percent) are used here to determine unit costs, when the original amounts spent for energy efficiency and load management are analyzed.

An alternative approach is to convert past spending to current 2008 dollars to adjust past data for inflation over the 15-year period. This is similar to a replacement cost approach in supply-side regulation. If this is done, somewhat lower discount rates should be used to determine the per-unit cost of demand-side activities. Three values (twelve percent, nine percent, and seven percent) are used. These effectively represent about a three percent per year estimate for inflation.

The data reporting characteristics, plus the purpose to determine the "unit" cost of the reported savings, favor using annualized cost recovery methods and not the present value of future savings. This amounts to levelizing costs and assigning a fraction of past years' investments to each year. This is comparable to how annual revenue requirements are determined for multiple years' supply-side investments.

If demand-side programs last an average of ten years, each year's annual "cost" would reflect one-tenth of the prior ten years' levelized recovery "on" (the discount rate or opportunity cost) and recovery "of" (depreciation or amortization). Sinking fund depreciation would yield the same annual cost assignment for each year of the investment's life to collect a return "on" and "of" the investment. This levelized approach is the same method used for mortgages and is sometimes called an annualized or amortization method.

At the low end, the fixed costs could simply be divided by the investment's assumed economic life. This would be the same as assuming there is no time value of money, that the appropriate discount rate to convert future savings to a present value equivalent is zero, and that there should not be any recovery of a return "on" the investment. The expected length of life would simply allocate the corresponding incremental investments to the corresponding future years. Therefore, a five-year life would assign twenty percent to the current year and each of the next four years.

Table 14-2 shows the detail of the first method, when original costs are used and the assumption is that in 2006 all the 15 prior years are contributing one-fifteenth of the costs for the units saved in 2006. As explained, this approach understates savings, because it ignores future savings. However, it also likely overstates savings, because it does not remove any 2006 savings that reflect pre-1992 investments.

TABLE 14-2
THE PER-UNIT COST OF ENERGY EFFICIENCY IN ORIGINAL DOLLARS WITH A 15-YEAR LEVELIZED COST RECOVERY

N	15		
ROR	15%	12%	10%
2006 EE MWh SAVINGS:	63,075,683		

YEAR	(A) NOMINAL EE COSTS* ($000's)	(B) EE COSTS / N	(C) AMORTIZED PAYMENTS ACCRUED - 15%	(D) AMORTIZED PAYMENTS ACCRUED - 12%	(E) AMORTIZED PAYMENTS ACCRUED - 10%
1992	$1,550,574.20	103,371.61	265,174.63	227,661.88	203,859.85
1993	$2,011,368.79	134,091.25	609,152.99	522,979.57	468,302.01
1994	$1,969,342.86	131,289.52	945,944.20	812,126.84	727,219.04
1995	$1,811,247.64	120,749.84	1,255,698.44	1,078,061.90	965,350.61
1996	$1,327,444.75	88,496.32	1,482,714.12	1,272,962.96	1,139,874.78
1997	$1,084,905.90	72,327.06	1,668,251.53	1,432,253.45	1,282,511.46
1998	$885,026.61	59,001.77	1,819,606.18	1,562,196.81	1,398,869.25
1999	$934,464.71	62,297.65	1,979,415.58	1,699,398.88	1,521,726.86
2000	$1,060,899.67	70,726.64	2,160,847.51	1,855,164.66	1,661,207.34
2001	$1,233,992.77	82,266.18	2,371,881.32	2,036,344.71	1,823,445.03
2002	$1,179,701.17	78,646.74	2,573,630.34	2,209,553.44	1,978,544.80
2003	$903,264.44	60,217.63	2,728,103.96	2,342,174.56	2,097,300.39
2004	$994,850.63	66,323.38	2,898,240.38	2,488,242.74	2,228,097.16
2005	$1,265,662.66	84,377.51	3,114,690.28	2,674,072.70	2,394,498.61
2006	$1,356,150.29	90,410.02	3,346,615.10	2,873,188.44	2,572,796.81
CUMULATIVE AMT:		1,304,593.14	3,346,615.10	2,873,188.44	2,572,796.81
COST PER kWh:		$0.0207	$0.0531	$0.0456	$0.0408

* Indirect costs are allocated according to the proportion of direct costs for EE & LM.

These data reflect all the savings for all the utilities with measurable energy efficiency savings over the past 15-year period. Table 14-2 is based on savings of about 63 billion kWhs in 2006. If there is no discount rate and the time value of money is ignored, the per-unit cost would be about 2¢ per kWh. This is shown in Column B. When more realistic discounting is applied, the per-unit costs increase to just a little less than 5¢ per kWh saved averaging Columns C, D, and E. This would be so, if either the future savings are discounted or the levelized approach shown in Table 14-2 is used.

Table 14-3 shows the corresponding results using the same 15-year period for cumulative cost recovery and levelized or sinking fund depreciation based on 2008 replacement costs.

TABLE 14-3
THE PER-UNIT COST OF ENERGY EFFICIENCY IN 2008 DOLLARS WITH A 15-YEAR LEVELIZED COST RECOVERY

N	15		
ROR	12%	9%	7%
2006 EE MWh SAVINGS:	63,075,683		

YEAR	(A) REAL EE COSTS* ($000's 2008)	(B) EE COSTS / N	(C) AMORTIZED PAYMENTS ACCRUED - 12%	(D) AMORTIZED PAYMENTS ACCRUED - 9%	(E) AMORTIZED PAYMENTS ACCRUED - 7%
1992	$2,382,778.31	158,851.89	349,849.61	295,604.81	261,616.25
1993	$3,001,045.75	200,069.72	790,475.87	667,911.20	591,114.94
1994	$2,864,981.92	190,998.79	1,211,124.67	1,023,337.65	905,674.56
1995	$2,562,368.72	170,824.58	1,587,342.50	1,341,222.25	1,187,008.87
1996	$1,824,073.22	121,604.88	1,855,160.67	1,567,514.74	1,387,282.30
1997	$1,457,356.46	97,157.01	2,069,135.92	1,748,312.75	1,547,292.21
1998	$1,170,624.16	78,041.61	2,241,011.92	1,893,539.08	1,675,820.45
1999	$1,209,307.27	80,620.48	2,418,567.54	2,043,564.39	1,808,595.89
2000	$1,328,280.89	88,552.06	2,613,591.37	2,208,349.43	1,954,433.99
2001	$1,503,100.80	100,206.72	2,834,283.01	2,394,822.43	2,119,466.38
2002	$1,413,805.30	94,253.69	3,041,863.89	2,570,217.54	2,274,694.60
2003	$1,058,390.29	70,559.35	3,197,261.24	2,701,520.26	2,390,900.16
2004	$1,135,467.42	75,697.83	3,363,975.38	2,842,385.07	2,515,568.38
2005	$1,397,219.00	93,147.93	3,569,121.00	3,015,722.50	2,668,975.52
2006	$1,450,327.39	96,688.49	3,782,064.22	3,195,648.50	2,828,213.67
CUMULATIVE AMT:		1,717,275.12	3,782,064.22	3,195,648.50	2,828,213.67
COST PER kWh:		$0.0272	$0.0600	$0.0507	$0.0448

* Indirect costs are allocated according to the proportion of direct costs for EE & LM.

The per-unit cost increases with no discounting to about 2.7¢. The corresponding discounted per-unit costs also increase, but not as much proportionally, because lower "real" discount rates of twelve percent, nine percent, and seven percent were used in these calculations. The replacement cost estimates are just above about 5¢ per kWh saved.

The second approach shrinks the time period to five years and uses 2002 as the reference year. The purpose is to take a rolling five prior years of energy efficiency investments starting in 1998 and to assign one-fifth of these levelized costs to 2002.

The savings remove all cumulative savings prior to 1998 that were reported as cumulative savings in 1997 from the cumulative amount reported in 2002. The cumulative amount reported in 2006, less the incremental savings for 2003 through 2006, would represent the savings in 2002 from investments beginning in 1998 and lasting to 2006. Under this approach, each year's savings would, in effect, drop the sixth year's prior effect and add one more future year to replace it.

Table 14-4 shows the per-unit cost of energy efficiency savings estimates for 2002, based on this five-year rolling method and original costs. If discounting is ignored, Column B shows the per-unit cost would be about 2.5¢ for 2002 based upon five prior years of investment and five current and future years of savings.

TABLE 14-4
THE PER-UNIT COST OF ENERGY EFFICIENCY IN ORIGINAL DOLLARS
WITH A 5-YEAR ROLLING LEVELIZED COST RECOVERY

N	5		
ROR	15%	12%	10%
Diff of 2002 and 1997 cumulative kWhs saved added to 2006 cumulative kWhs saved minus 2003-2006 incremental savings.	42,084,693		

YEAR	(A) NOMINAL DSM COSTS* ($000's)	(B) DSM COSTS / N	(C) AMORTIZED PAYMENTS ACCRUED - 15%	(D) AMORTIZED PAYMENTS ACCRUED - 12%	(E) AMORTIZED PAYMENTS ACCRUED - 10%
1998	$885,026.61	177,005.32	264,017.20	245,514.99	233,467.79
1999	$934,464.71	186,892.94	542,782.56	504,744.60	479,977.23
2000	$1,060,899.67	212,179.93	859,265.43	799,048.49	759,839.89
2001	$1,233,992.77	246,798.55	1,227,384.66	1,141,370.01	1,085,364.07
2002	$1,179,701.17	235,940.23	1,579,307.87	1,468,630.68	1,396,566.27
CUMULATIVE AMT:		1,058,816.99	1,579,307.87	1,468,630.68	1,396,566.27
COST PER kWh:		$0.0252	$0.0375	$0.0349	$0.0332

* Indirect costs are allocated according to the proportion of direct costs for EE & LM.

Chapter 14: The Per-Unit Costs of Energy Efficiency and Load Management

The total amount of savings assigned under this second approach to 2002 equals about 42 billion kWhs. When the more conceptually sound discount rates are included, the estimated per-unit cost increases to about 3.5¢ per kWh saved.

Table 14-5 shows the same range of estimates using 2008 dollars for the rolling levelized 2002 costs and savings. The undiscounted per-unit cost would equal about 3.15¢ per kWh saved. The more conceptually correct estimates would be about 4¢ per kWh saved using real discount rates of between seven percent and twelve percent.

TABLE 14-5

THE PER-UNIT COST OF ENERGY EFFICIENCY IN 2008 DOLLARS AND A 5-YEAR ROLLING LEVELIZED COST RECOVERY

N	5		
ROR	12%	9%	7%
Diff of 2002 and 1997 cumulative kWhs saved added to 2006 cumulative kWhs saved minus 2003-2006 incremental savings.	42,084,693		

YEAR	(A) REAL DSM COSTS* ($000's 2008)	(B) DSM COSTS / N	(C) AMORTIZED PAYMENTS ACCRUED - 12%	(D) AMORTIZED PAYMENTS ACCRUED - 9%	(E) AMORTIZED PAYMENTS ACCRUED - 7%
1998	$1,170,624.16	234,124.83	324,742.53	300,958.64	285,504.34
1999	$1,209,307.27	241,861.45	660,216.14	611,862.42	580,443.13
2000	$1,328,280.89	265,656.18	1,028,694.18	953,353.41	904,398.47
2001	$1,503,100.80	300,620.16	1,445,668.97	1,339,789.29	1,270,990.77
2002	$1,413,805.30	282,761.06	1,837,872.32	1,703,267.97	1,615,804.73
CUMULATIVE AMT:		1,325,023.68	1,837,872.32	1,703,267.97	1,615,804.73
COST PER kWh:		$0.0315	$0.0437	$0.0405	$0.0384

* Indirect costs are allocated according to the proportion of direct costs for EE & LM.

Tables 14-6 through 14-9 show the same sensitivity analyses, when all the demand-side costs are used, including load management. This may need to be done, because the EIA data do not allocate indirect costs separately to energy efficiency and load management. This application adds relatively more dollars than kWh savings, because load management does not necessarily save energy. It is more likely either to shift kWh to less costly times or to avoid purchases, if the rates customers pay are less than the wholesale prices. These utility operational decisions would likely reduce kWhs. However, these savings would be relatively small and the purpose of load management, particularly in prior years, was more directly focused on system reliability and avoiding capacity investments.

TABLE 14-6

THE PER-UNIT COST OF ALL DEMAND-SIDE SAVINGS USING ORIGINAL DOLLARS WITH A 15-YEAR LEVELIZED COST RECOVERY

N	15		
ROR	15%	12%	10%
2006 DSM MWh SAVINGS:	63,941,071		

YEAR	(A) NOMINAL DSM COSTS ($000's)	(B) DSM COSTS / N	(C) AMORTIZED PAYMENTS ACCRUED - 15%	(D) AMORTIZED PAYMENTS ACCRUED - 12%	(E) AMORTIZED PAYMENTS ACCRUED - 10%
1992	$2,401,988.00	160,132.53	410,780.91	352,670.06	315,798.43
1993	$2,881,625.00	192,108.33	903,587.92	775,762.46	694,656.56
1994	$2,836,513.00	189,100.87	1,388,680.02	1,192,231.33	1,067,583.63
1995	$2,633,925.00	175,595.00	1,839,126.11	1,578,955.36	1,413,875.70
1996	$2,124,330.00	141,622.00	2,202,422.76	1,890,858.50	1,693,169.39
1997	$1,647,930.00	109,862.00	2,484,246.89	2,132,814.57	1,909,828.97
1998	$1,440,134.00	96,008.93	2,730,534.37	2,344,261.15	2,099,168.83
1999	$1,437,842.00	95,856.13	2,976,429.87	2,555,371.21	2,288,207.35
2000	$1,578,980.00	105,265.33	3,246,462.37	2,787,203.74	2,495,801.81
2001	$1,639,424.00	109,294.93	3,526,831.83	3,027,910.93	2,711,343.08
2002	$1,656,908.00	110,460.53	3,810,191.36	3,271,185.18	2,929,183.03
2003	$1,340,686.00	89,379.07	4,039,471.52	3,468,030.39	3,105,448.08
2004	$1,564,967.00	104,331.13	4,307,107.57	3,697,805.48	3,311,200.20
2005	$1,942,860.00	129,524.00	4,639,369.76	3,983,064.42	3,566,635.34
2006	$2,075,561.00	138,370.73	4,994,326.08	4,287,807.08	3,839,517.19
CUMULATIVE AMT:		1,946,911.53	4,994,326.08	4,287,807.08	3,839,517.19
COST PER kWh:		$0.0304	$0.0781	$0.0671	$0.0600

TABLE 14-7
THE PER-UNIT COST OF ALL DEMAND-SIDE SAVINGS USING 2008 DOLLARS
WITH A 15-YEAR LEVELIZED COST RECOVERY

N	15		
ROR	12%	9%	7%
2006 DSM MWh SAVINGS:	63,941,071		

YEAR	(A) REAL DSM COSTS ($000's 2008)	(B) DSM COSTS / N	(C) AMORTIZED PAYMENTS ACCRUED - 12%	(D) AMORTIZED PAYMENTS ACCRUED - 9%	(E) AMORTIZED PAYMENTS ACCRUED - 7%
1992	$3,691,151.91	246,076.79	541,950.57	457,920.18	405,268.64
1993	$4,299,504.15	286,633.61	1,173,222.00	991,311.86	877,331.08
1994	$4,126,533.08	275,102.21	1,779,097.08	1,503,244.95	1,330,402.23
1995	$3,726,208.86	248,413.92	2,326,194.87	1,965,514.25	1,739,519.94
1996	$2,919,092.08	194,606.14	2,754,788.34	2,327,653.56	2,060,020.56
1997	$2,213,667.96	147,577.86	3,079,808.46	2,602,278.73	2,303,069.40
1998	$1,904,864.36	126,990.96	3,359,488.72	2,838,594.07	2,512,213.27
1999	$1,860,736.71	124,049.11	3,632,689.97	3,069,434.99	2,716,512.16
2000	$1,976,934.31	131,795.62	3,922,951.85	3,314,691.25	2,933,568.92
2001	$1,996,948.10	133,129.87	4,216,152.23	3,562,430.40	3,152,823.08
2002	$1,985,710.75	132,380.72	4,507,702.70	3,808,775.46	3,370,843.45
2003	$1,570,934.25	104,728.95	4,738,353.93	4,003,663.81	3,543,323.59
2004	$1,786,166.68	119,077.78	5,000,606.50	4,225,253.65	3,739,435.09
2005	$2,144,806.02	142,987.07	5,315,516.01	4,491,335.89	3,974,923.26
2006	$2,219,697.18	147,979.81	5,641,421.36	4,766,709.04	4,218,634.08
CUMULATIVE AMT:		2,561,530.43	5,641,421.36	4,766,709.04	4,218,634.08
COST PER kWh:		$0.0401	$0.0882	$0.0745	$0.0660

TABLE 14-8
THE PER-UNIT COST OF ALL DEMAND-SIDE SAVINGS USING ORIGINAL DOLLARS AND A 5-YEAR ROLLING LEVELIZED COST RECOVERY

N	5		
ROR	15%	12%	10%
Diff of 2002 and 1997 cumulative kWhs saved added to 2006 cumulative kWhs saved minus 2003-2006 incremental savings.	43,512,280		

YEAR	(A) NOMINAL DSM COSTS ($000's)	(B) DSM COSTS / N	(C) AMORTIZED PAYMENTS ACCRUED - 15%	(D) AMORTIZED PAYMENTS ACCRUED - 12%	(E) AMORTIZED PAYMENTS ACCRUED - 10%
1998	$1,440,134.00	288,026.80	429,614.37	399,507.19	379,903.72
1999	$1,437,842.00	287,568.40	858,545.00	798,378.55	759,202.82
2000	$1,578,980.00	315,796.00	1,329,579.29	1,236,402.97	1,175,733.76
2001	$1,639,424.00	327,884.80	1,818,644.97	1,691,195.14	1,608,209.69
2002	$1,656,908.00	331,381.60	2,312,926.39	2,150,837.55	2,045,297.84
CUMULATIVE AMT:		1,550,657.60	2,312,926.39	2,150,837.55	2,045,297.84
COST PER kWh:		$0.0356	$0.0532	$0.0494	$0.0470

TABLE 14-9
THE PER-UNIT COST OF ALL DEMAND-SIDE SAVINGS USING 2008 DOLLARS AND A 5-YEAR ROLLING LEVELIZED COST RECOVERY

N	5		
ROR	12%	9%	7%
Diff of 2002 and 1997 cumulative kWhs saved added to 2006 cumulative kWhs saved minus 2003-2006 incremental savings.	43,512,280		

YEAR	(A) REAL DSM COSTS ($000's 2008)	(B) DSM COSTS / N	(C) AMORTIZED PAYMENTS ACCRUED - 12%	(D) AMORTIZED PAYMENTS ACCRUED - 9%	(E) AMORTIZED PAYMENTS ACCRUED - 7%
1998	$1,904,864.36	380,972.87	528,427.91	489,726.26	464,578.69
1999	$1,860,736.71	372,147.34	1,044,614.38	968,107.63	918,395.06
2000	$1,976,934.31	395,386.86	1,593,035.20	1,476,362.53	1,400,550.94
2001	$1,996,948.10	399,389.62	2,147,008.04	1,989,762.82	1,887,588.00
2002	$1,985,710.75	397,142.15	2,697,863.52	2,500,274.08	2,371,884.37
CUMULATIVE AMT:		1,945,038.85	2,697,863.52	2,500,274.08	2,371,884.37
COST PER kWh:		$0.0447	$0.0620	$0.0575	$0.0545

Chapter 14: The Per-Unit Costs of Energy Efficiency and Load Management

Table 14-10 summarizes the sensitivity of the results using both approaches, original and 2008 dollars, the corresponding three non-zero discount rates, and the "just energy," as well as the "all demand response" saving definitions.

TABLE 14-10
SUMMARY OF SENSITIVITY ANALYSES

COSTS PER kWh SAVED
(Costs Amortized 15 Years)

	Opportunity Cost	ALL DSM (Per kWh)	Energy Efficiency (Per kWh)
Nominal Original Cost Dollars	15%	$0.0781	$0.0531
	12%	$0.0671	$0.0456
	10%	$0.0600	$0.0408
Real 2008 Dollars	12%	$0.0882	$0.0600
	9%	$0.0745	$0.0507
	7%	$0.0660	$0.0448

COSTS PER kWh SAVED
(Costs Amortized 5 Years)

	Opportunity Cost	ALL DSM (Per kWh)	Energy Efficiency (Per kWh)
Nominal Original Cost Dollars	15%	$0.0532	$0.0375
	12%	$0.0494	$0.0349
	10%	$0.0470	$0.0332
Real 2008 Dollars	12%	$0.0620	$0.0437
	9%	$0.0575	$0.0405
	7%	$0.0545	$0.0384

The range of these various sensitivity cases is approximately four cents per kWh to seven cents per kWh. These are about two to three-and-a-half times greater than the 2¢ per kWh saved that some have claimed. The actual costs would depend upon specific discount rates and regulatory decisions, particularly as these affect the portion of the total costs the utility spends to save energy.

CHAPTER 15
THE PER-UNIT COST OF LOAD MANAGEMENT

This chapter estimates the corresponding per unit cost for load management. The approach is restricted to the 15-year levelized method for two reasons. First, demand-side programs often have long lives. Second, the specifics of what utilities and EIA call load management seem to be changing from the use of tariffs and interruptible schedules (*i.e.,* potential) to smart meters and direct control (*i.e.,* actual) forms of load management. Therefore, the 2002 reference year is likely outdated, because the focus of load management is quite different today.

Table 15-1 sums the direct incremental investments for potential load management and a proportionate allocation of indirect demand-side program costs for the 15 years for which EIA published data on load management. This table uses the original dollars invested and a 15-year sinking fund or economic life. The discount rates are 10 percent, 12 percent, and 15 percent.

TABLE 15-1
THE PER-UNIT COST OF POTENTIAL LOAD MANAGEMENT USING ORIGINAL DOLLARS AND A 15-YEAR LEVELIZED APPROACH

N	15		
ROR	15%	12%	10%
2006 LM kW ACTUAL PEAK REDUCTION:	21,284,000		

YEAR	(A) NOMINAL LM COSTS* ($000's)	(B) DSM COSTS / N	(C) AMORTIZED PAYMENTS ACCRUED - 15%	(D) AMORTIZED PAYMENTS ACCRUED - 12%	(E) AMORTIZED PAYMENTS ACCRUED - 10%
1992	$749,263.80	49,950.92	128,136.89	110,010.09	98,508.54
1993	$774,523.21	51,634.88	260,593.56	223,728.87	200,338.03
1994	$783,674.14	52,244.94	394,615.21	338,791.23	303,370.63
1995	$747,424.36	49,828.29	522,437.52	448,531.24	401,637.34
1996	$675,349.25	45,023.28	637,933.75	547,688.88	490,428.05
1997	$552,310.10	36,820.67	732,388.20	628,781.39	563,042.35
1998	$541,858.39	36,123.89	825,055.22	708,339.34	634,282.52
1999	$491,725.29	32,781.69	909,148.64	780,536.53	698,931.50
2000	$507,176.33	33,811.76	995,884.44	855,002.31	765,611.89
2001	$405,431.23	27,028.75	1,065,220.09	914,529.44	818,915.46
2002	$477,206.83	31,813.79	1,146,830.60	984,594.97	881,655.64
2003	$393,945.56	26,263.04	1,214,202.00	1,042,435.73	933,449.16
2004	$562,980.37	37,532.02	1,310,481.25	1,125,094.90	1,007,466.31
2005	$670,325.34	44,688.36	1,425,118.31	1,223,514.90	1,095,596.52
2006	$710,905.71	47,393.71	1,546,695.31	1,327,893.09	1,189,061.97
CUMULATIVE AMT:		602,939.99	1,546,695.31	1,327,893.09	1,189,061.97
COST PER kWh:		$28.33	$72.67	$62.39	$55.87

* Indirect costs are allocated according to the proportion of direct costs for EE & LM.

The undiscounted estimate of $28.33 per kW saved is conceptually weak, because it does not reflect the time value of money or the various utilities' cost of capital. The discounted values shown in Table 15-1 of about $60 per kW are more reasonable estimates of the cost per potential kW saved. These would tend to be somewhat inflated, because they do not reflect benefits after 2006. This is somewhat offset by the fact that only costs before 1992 are excluded and some 2006 savings reflect savings directly tied to such prior investments.

Table 15-2 shows similar per-unit costs for potential load management using 2008 dollars. This method increases the per-unit costs to about $70 per kW. The same sorts of adjustments, discussed above, would apply.

TABLE 15-2
THE PER-UNIT COST OF POTENTIAL LOAD MANAGEMENT USING 2008 DOLLARS AND A 15-YEAR LEVELIZED APPROACH

N	15		
ROR	12%	9%	7%
2006 LM kW ACTUAL PEAK REDUCTION:	21,284,000		

YEAR	(A) REAL LM COSTS* ($000's 2008)	(B) DSM COSTS / N	(C) AMORTIZED PAYMENTS ACCRUED - 12%	(D) AMORTIZED PAYMENTS ACCRUED - 9%	(E) AMORTIZED PAYMENTS ACCRUED - 7%
1992	$1,151,398.97	76,759.93	169,053.28	142,841.27	126,417.42
1993	$1,155,620.79	77,041.39	338,726.42	286,206.29	253,298.37
1994	$1,140,081.95	76,005.46	506,118.09	427,643.59	378,473.24
1995	$1,057,379.86	70,491.99	661,367.08	558,820.95	494,567.86
1996	$928,013.38	61,867.56	797,621.94	673,949.25	596,458.74
1997	$741,919.36	49,461.29	906,553.69	765,990.94	677,917.50
1998	$716,715.76	47,781.05	1,011,784.93	854,905.90	756,609.04
1999	$636,350.38	42,423.36	1,105,216.59	933,850.81	826,476.89
2000	$635,001.26	42,333.42	1,198,450.17	1,012,628.36	896,196.62
2001	$493,847.30	32,923.15	1,270,958.93	1,073,894.51	950,418.40
2002	$571,905.46	38,127.03	1,354,928.51	1,144,844.46	1,013,210.54
2003	$461,601.43	30,773.43	1,422,702.79	1,202,110.22	1,063,891.90
2004	$642,554.63	42,836.98	1,517,045.38	1,281,824.82	1,134,440.94
2005	$740,000.73	49,333.38	1,625,695.43	1,373,628.49	1,215,689.04
2006	$760,274.16	50,684.94	1,737,322.10	1,467,947.25	1,299,163.06
CUMULATIVE AMT:		788,844.36	1,737,322.10	1,467,947.25	1,299,163.06
COST PER kWh:		$37.06	$81.63	$68.97	$61.04

* Indirect costs are allocated according to the proportion of direct costs for EE & LM.

Tables 15-3 and 15-4 show the corresponding per-unit cost estimates for actual load management. These are almost twice as high as the potential per-unit cost estimates for load management. This is consistent with the fact that actual capacity savings are more costly, because they require greater investments to accomplish actual as opposed to potential capacity savings.

TABLE 15-3
THE PER-UNIT COST OF ACTUAL LOAD CONTROL USING ORIGINAL DOLLARS AND A 15-YEAR LEVELIZED APPROACH

N	15		
ROR	15%	12%	10%
2006 LM kW ACTUAL PEAK REDUCTION:	11,291,000		

YEAR	(A) NOMINAL LM COSTS* ($000's)	(B) DSM COSTS / N	(C) AMORTIZED PAYMENTS ACCRUED - 15%	(D) AMORTIZED PAYMENTS ACCRUED - 12%	(E) AMORTIZED PAYMENTS ACCRUED - 10%
1992	$749,263.80	49,950.92	128,136.89	110,010.09	98,508.54
1993	$774,523.21	51,634.88	260,593.56	223,728.87	200,338.03
1994	$783,674.14	52,244.94	394,615.21	338,791.23	303,370.63
1995	$747,424.36	49,828.29	522,437.52	448,531.24	401,637.34
1996	$675,349.25	45,023.28	637,933.75	547,688.88	490,428.05
1997	$552,310.10	36,820.67	732,388.70	628,781.39	563,042.35
1998	$541,858.39	36,123.89	825,055.22	708,339.34	634,282.52
1999	$491,725.29	32,781.69	909,148.64	780,536.53	698,931.50
2000	$507,176.33	33,811.76	995,884.44	855,002.31	765,611.89
2001	$405,431.23	27,028.75	1,065,220.09	914,529.44	818,915.46
2002	$477,206.83	31,813.79	1,146,830.60	984,594.97	881,655.64
2003	$393,945.56	26,263.04	1,214,202.00	1,042,435.73	933,449.16
2004	$562,980.37	37,532.02	1,310,481.25	1,125,094.90	1,007,466.31
2005	$670,325.34	44,688.36	1,425,118.31	1,223,514.90	1,095,596.52
2006	$710,905.71	47,393.71	1,546,695.31	1,327,893.09	1,189,061.97
	CUMULATIVE AMT:	602,939.99	1,546,695.31	1,327,893.09	1,189,061.97
	COST PER kWh:	$53.40	$136.98	$117.61	$105.31

* Indirect costs are allocated according to the proportion of direct costs for EE & LM.

TABLE 15-4
THE PER-UNIT COST OF ACTUAL LOAD CONTROL USING 2008 DOLLARS AND A 15-YEAR LEVELIZED APPROACH

N	15		
ROR	12%	9%	7%
2006 LM kW ACTUAL PEAK REDUCTION:	11,291,000		

YEAR	(A) REAL LM COSTS* ($000's 2008)	(B) DSM COSTS / N	(C) AMORTIZED PAYMENTS ACCRUED - 12%	(D) AMORTIZED PAYMENTS ACCRUED - 9%	(E) AMORTIZED PAYMENTS ACCRUED - 7%
1992	$1,151,398.97	76,759.93	169,053.28	142,841.27	126,417.42
1993	$1,155,620.79	77,041.39	338,726.42	286,206.29	253,298.37
1994	$1,140,081.95	76,005.46	506,118.09	427,643.59	378,473.24
1995	$1,057,379.86	70,491.99	661,367.08	558,820.95	494,567.86
1996	$928,013.38	61,867.56	797,621.94	673,949.25	596,458.74
1997	$741,919.36	49,461.29	906,553.69	765,990.94	677,917.50
1998	$716,715.76	47,781.05	1,011,784.93	854,905.90	756,609.04
1999	$636,350.38	42,423.36	1,105,216.59	933,850.81	826,476.89
2000	$635,001.26	42,333.42	1,198,450.17	1,012,628.36	896,196.62
2001	$493,847.30	32,923.15	1,270,958.93	1,073,894.51	950,418.40
2002	$571,905.46	38,127.03	1,354,928.51	1,144,844.46	1,013,210.54
2003	$461,601.43	30,773.43	1,422,702.79	1,202,110.22	1,063,891.90
2004	$642,554.63	42,836.98	1,517,045.38	1,281,824.82	1,134,440.94
2005	$740,000.73	49,333.38	1,625,695.43	1,373,628.49	1,215,689.04
2006	$760,274.16	50,684.94	1,737,322.10	1,467,947.25	1,299,163.06
CUMULATIVE AMT:		788,844.36	1,737,322.10	1,467,947.25	1,299,163.06
COST PER kWh:		$69.86	$153.87	$130.01	$115.06

* Indirect costs are allocated according to the proportion of direct costs for EE & LM.

Table 15-5 summarizes the sensitivity analyses of the potential and actual per-unit costs for load management. As explained, these estimates tend to be higher than they actually would be, if future savings were included. In addition, the 15 years of economic life for load management may be on the low end of their useful life. Therefore, the money invested before 1992 is not reflected, and this decreases the per-unit cost estimate. Finally, the proportional allocation of indirect costs, particular in the potential category, is likely to be greater than the amount regulators would assign to programs that included interruptible service tariffs.

TABLE 15-5

SENSITIVITY ANALYSIS OF LOAD MANAGEMENT PER-UNIT COSTS

COSTS PER kW OF CAPACITY REDUCTION
(Costs Amortized 15 Years)

	Opportunity Cost	Actual Peak Reduction (kW)	Potential Peak Reduction (kW)
Nominal	15%	$136.98	$72.67
Original Cost	12%	$117.61	$62.39
Dollars	10%	$105.31	$55.87
Real	12%	$153.87	$81.63
2008	9%	$130.01	$68.97
Dollars	7%	$115.06	$61.04

CHAPTER 16
HOW REGULATION AFFECTS THE COSTS THAT CUSTOMERS PAY FOR DEMAND-SIDE PROGRAMS

In the early 1990s regulators permitted IOUs to earn incentives and generally to markup the direct costs of their combined demand-side programs. The relative markups permitted were not likely uniform, because there were differences in conservation and load management. For example, much of the earlier phase of load management was achieved through interruptible service offerings or tariffs. These tariffs would not typically be marked up.

In more recent years, the EIA did not require utilities to break down their indirect costs into specific categories, such as discussed above. Chart 16-1 shows, in the three most recent years, the type of desegregation that the current data reflects for IOUs with indirect costs. The most important fact shown is that the weighted average amounts IOUs spent on direct retail customer rebates (49 percent) exceeded the proportion spent directly on energy efficiency and load management (39 percent). The combined indirect costs were about 12 percent.

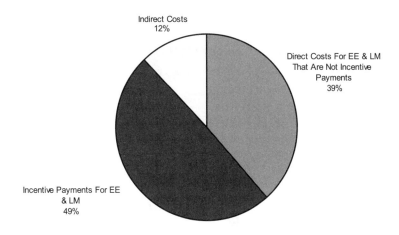

CHART 16-1
BREAKDOWN OF TOTAL COSTS
(Sum of 2004-2006, Only Includes Utilities with Positive Indirect Costs)

It is more likely that these indirect costs were attributed to direct cost of goods sold (COGS) for energy efficiency, rather than to either direct customer incentive rebates or to interruptible tariff-based load management. This would mean the direct cost markup would be more than 20 percent, and probably even more if interruptible tariffs were removed.

Charts 16-2 and 16-3 show, for example, that direct customer rebates or credits were more likely offered as part of load management (73 percent) than energy efficiency (48 percent). This result is consistent with the prior discussion. Furthermore, the EIA-reported data do not assign the indirect costs. Therefore, various estimates of retail margins or COGS markups cannot be made separately, without using assumptions to allocate indirect costs between energy efficiency and load management.

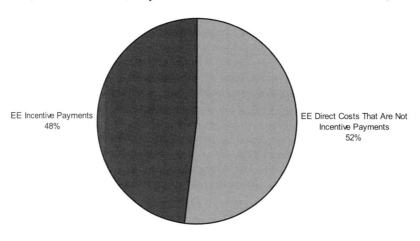

CHART 16-2
BREAKDOWN OF TOTAL EE COSTS
(Sum of 2004-2006, Only Includes Utilities with Positive Indirect Costs)

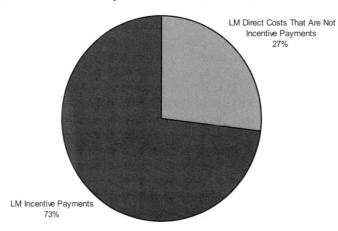

CHART 16-3
BREAKDOWN OF TOTAL LM COSTS
(Sum of 2004-2006, Only Includes Utilities with Positive Indirect Costs)

Chapter 16: How Regulation Affects the Costs that Customers Pay for Demand-Side Programs

Chart 16-4 breaks the cost data down into five categories: (1) combined indirect costs; (2) direct energy efficiency; (3) direct load management; (4) energy efficiency rebates or customer incentives; and (5) load management rebates or customer credits. When customer rebates are excluded the retail margins for energy efficiency and load management combined exceed 31 percent and equal about 14 percent, when rebates are added to the combined energy efficiency and load management COGS.

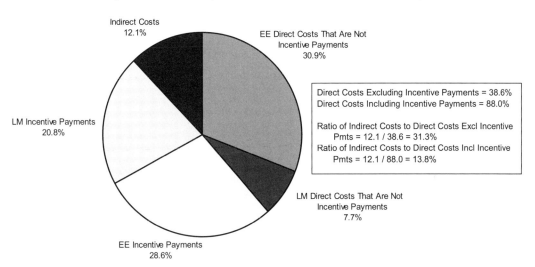

CHART 16-4
THE BREAKDOWN OF TOTAL COSTS FOR ENERGY EFFICIENCY
(Sum of 2004-2006, For Utilities with Positive Indirect Costs)

Table 16-1 shows how the various utilities' cost for demand-side management has changed over time. The various indirect costs are similar to the markups or margins that retailers add to their cost of goods sold, or COGS. Taking into account some differences in detail, these markups, shown in Columns H and I, were higher in the early 1990s, when utility-sponsored energy efficiency and load management were quite popular and often relatively novel. They both peaked about 1997, when the world price of crude oil fell in US dollar terms to about $10 per barrel. These relative markups increased once again after the 9/11 attacks until 2004, when they softened again. Current information for 2007 and 2008 is not yet available.

TABLE 16-1
DEMAND-SIDE COSTS OVER TIME

	A	B	C	D	E	F	G	H	I
	Total Cost	Indirect Cost	Administrative and General	Marketing	Monitoring	Utility Incentive Payments	Other Costs	Ratio of Indirect Costs to Total Costs	Ratio of Indirect Costs to Direct Costs
1992	1,418,939.42	335,975.77	102,790.45	33,741.27	49,645.76	79,588.03	70,210.27	0.237	0.310
1993	1,689,375.03	335,007.60	128,253.28	33,696.88	48,876.81	43,067.13	81,113.49	0.198	0.247
1994	1,633,023.63	330,311.07	125,878.54	39,109.31	51,486.50	42,011.47	71,825.24	0.202	0.254
1995	1,469,683.79	351,508.54	110,491.47	38,198.16	37,658.79	96,915.36	68,244.76	0.239	0.314
1996	1,079,712.00	247,116.65	89,500.96	29,048.44	28,627.15	82,099.43	17,840.66	0.229	0.297
1997	699,338.93	173,277.88	-	-	-	-	-	0.248	0.329
1998	569,892.02	106,236.20	-	-	-	-	-	0.186	0.229
1999	564,723.89	93,518.01						0.166	0.198
2000	644,508.13	97,534.26						0.151	0.178
2001	699,771.31	90,911.35						0.130	0.149
2002	570,038.39	97,112.85						0.170	0.205
2003	365,711.41	61,519.56						0.168	0.202
2004	378,532.57	61,944.42						0.164	0.196
2005	481,276.99	58,686.12						0.122	0.139
2006	602,989.08	56,449.40						0.094	0.103
Sum	7,290,733.87	1,599,919.63	556,914.71	173,794.07	216,295.03	343,681.41	309,234.41		

The first conclusion is quite obvious. Regulation affects the amount of money that utilities spend for energy efficiency and load management. One component is the direct cost for the specific demand-side programs that regulators either mandate or otherwise cause to occur. This joint utility and regulator choice would be modified to the extent that demand-side programs require, as most do, participating customers to contribute some time and money to increase energy efficiency and load management.

There are no magic formulas that would tell regulators how to set and perhaps vary the amount that customers pay for their own demand-reducing efforts. Undoubtedly, regulators would consider that participating customers would likely be paying lower monthly utility bills, because their expected reduction in volume would exceed any increase in electricity prices as utilities pass on their share of the costs of energy efficiency and load management to utility consumers. Neither energy efficiency nor load management is too cheap to meter. Regulators and their customers need to be aware of the fact that if utilities provide demand-side management sources, there will be cost recovery consequences.

CHAPTER 17
COST-OF-SERVICE REGULATION AND DEMAND-SIDE MANAGEMENT

Most of the nation's electricity suppliers are privately-owned businesses that need to recover their costs. This often means the need for a return "on" and "of" the amount they invest. If mandated utility programs to increase demand-side reductions in sales are to succeed, they will require sensible regulation. One regulatory goal might be to achieve some neutrality between demand- and supply-side approaches. For this to happen, regulators would need to approve some type of financial incentives for utilities.

Other factors could also cause utilities to lose interest in energy efficiency. These include lost margins or fixed cost recovery for supply-side investments that demand-side programs displace, particularly in the near and intermediate term between rate cases. This means there are obstacles, while it is often very cost effective for consumers and society broadly to replace kWhs with energy efficiency.

More than fifty years ago, the United States adopted an approach to regulation known today as "cost-of-service" (COS) regulation. A major reason why COS trumped value-of-service pricing was that COS could be more reasonably quantified. As long as electric utilities build power stations and sell the kWh produced, COS works relatively well. This makes it possible to understand what it costs to produce and deliver electricity and make informed judgments about the prices utility consumers pay.

When energy efficiency is introduced, "value" once again takes center stage. Most regulators and many customers want to pursue energy efficiency. The challenge for regulators is determining how to graft energy efficiency onto a traditional "build, own, and operate" vertically-integrated electric utility company. At first blush, the introduction of energy efficiency is akin to a hamburger joint selling tofu salad. If customers demand a healthier choice than red meat, and the price is right, the customers would switch to a salad. No restaurant would simply rebuff sales of hamburgers and give away salads. The challenge for regulators is to strike a balance that reflects the needs for shareholders and the desire to increase energy efficiency. Regulators also need to determine what is necessary to maintain a sustainable future regulatory sector.

Current COS regulation does not work very well in terms of pricing energy efficiency, due to a myriad of issues, such as (1) the relationship between the marginal cost of electricity and regulated prices; (2) the relationship between current and prospective regulated prices and the marginal cost of energy efficiency; (3) the lost revenue or lost margin consequences for fixed cost recovery and authorized utility earnings (these are sometimes combined using the generic word, "decoupling"); (4) the amount that energy efficiency program participants pay relative to any contributions from or any cost allocations to nonparticipating customers; and (5) external social benefits related to energy efficiency.

Sensible demand-side programs align the consumer benefits along with strong shareholder support and incentives to expand energy efficiency onto cost-of-service regulation, using integrated resource planning, avoided cost, regulated revenue requirements and rate riders. Progressive regulators combine these into a balanced and transparent regulatory approach designed to help shareholders and their customers leap ahead of the pack, and make energy efficiency succeed, flourish, and be sustained after public interest wanes.

The crux of regulation for either future supply-side investments or conservation efforts is the same. Utilities require an opportunity to earn a return "on" and "of" these fundamental utility choices. Traditional "hard" (*i.e.,* steel-in-the-ground) generating stations would be granted their normal rate-base cost recovery status (*i.e.,* a return "on" and "of" rate-base investments) and recovery of the unit's full effects of fuel and operating expenses. Conservation and efficiency, in effect, would be granted a new product or business venture status. This means utilities need some positive financial incentives to expand and market energy efficiency.

The utility has two solutions, new generation and conservation, for meeting their customers' energy needs. Regulation that favors this joint approach would treat shareholders similarly under either path. Regulators that favor a greener path must expect to tilt the rewards for shareholders towards energy efficiency and load management. This does not remove the past difficulties related to lost revenue and nonparticipants. However, if energy efficiency and conservation are viewed as a new business, this becomes less difficult to do.

If a near substitute costs more (*e.g.,* the regulated price of electricity exceeds the cost of energy efficiency), it should be easier to sell a lower-priced product and cost-effective substitute (*e.g.,* energy efficiency services and products). Hoping for this result and making it happen are two different matters. There are risks, uncertainty, packaging, and marketing costs. The utility has public support for energy efficiency. To make much of this happen, the utility needs to become a retailer and it needs to be able to earn a return on retail margin. In addition, the utility is also a producer of energy efficiency and not simply a delivery service. This means that regulated margins should reflect both the production as well as the retail aspects of energy efficiency and load management.

Two ideas can come together, because energy efficiency is both an input for utility production and a new consumer product or utility service. This means that energy efficiency needs to be treated in a revenue requirements manner that is similar to a new supply-side rate base addition. This approach is very significant in the states that adhere to traditional cost-of-service regulation and also rely on some type of integrated planning regulation.

Another utility pushback can arise, when some propose to make IOUs into what would essentially be energy efficiency tax collectors and bankers that would subsidize and finance nonutility or third-party providers of energy efficiency and demand response services. Many regulated utilities have shown they can accomplish a great deal. The states that have succeeded recognize this often requires new regulatory and utility business paradigms. It is almost never sufficient to simply mandate that utilities take action. Even if this works for a relatively short period, it is almost never sustained.

This does not mean that simply collecting money to fund energy efficiency that nonutility entities design and operate is fatally flawed. Several jurisdictions are, in fact, already doing this. Nevertheless, it means that utilities will remain focused on the supply side. This may not, if history is a guide, be a sustainable approach for energy efficiency.

It could also cause energy efficiency and renewable energy to take conflicting future paths. This makes no economical, environmental, or energy policy sense. A jurisdiction that encourages utilities to pursue both energy efficiency and renewable energy business plans will likely gain the best of both approaches. The jurisdictions that, through regulation, make utilities favor one, because it earns income over the other, are less likely to strike the best "green" expansion plan without the regulators becoming even more fully engaged in a "command and control" manner.

The nation benefits from energy efficiency and load management. Regulators balance three interests: (1) participating customers; (2) nonparticipating utility customers; and (3) shareholders. First, participants could, and from an economist's perspective should, pay for the renewable and energy efficiency services they consume. Second, if the marginal cost or value of electricity exceeds the regulated price of energy, this difference could be used to reduce a portion of the cost of renewable energy that participating customers pay, because the nonparticipants would benefit as well. Third, external benefits also affect the price that other nonparticipating consumers would be willing to pay.

The regulatory challenge is to establish a policy that works for all customers. Participants should pay when they benefit directly. Participating consumers should not be worse off than they would be if they did not reduce their energy use. The challenge for regulators is how sharply to draw the tariff distinction between participating and nonparticipating customers in utility-sponsored energy efficiency. Regulators are often unwilling to embrace the "subsidy" word to describe their decisions. They do not need to do so. There is, as explained above, an alternative and economically equivalent conceptual justification for establishing how much both participants and nonparticipants should pay for utility-sponsored energy efficiency.

A customer who had been paying $80 for electricity could still pay $80 and be made better off if external benefits are present. Obviously, if the customer paid less, he/she would be even better off. Regulators could assign any shortfall to other customers, if there is a shortfall in recovering the cost of energy efficiency and load management. The nonparticipating customers would pay a portion of the shortfall, based upon the difference between the marginal cost of electricity and regulated prices, as well as a value based on reducing negative externalities. These joint payments from participating and nonparticipating customers would mean utilities could invest the difference between the marginal cost of electricity and the marginal cost of energy efficiency, plus the value assigned for reducing negative externalities. Society would also reduce its environmental footprint and internalize the cost of carbon. These savings would also extend the life of the nation's resource base and/or make it possible to import less energy and improve the nation's economic security. In effect, regulators would encourage utilities and their customers to expand energy efficiency to its socially optimal level.

The third stakeholder is the shareholder. Neutrality means that shareholders should not be required to earn less than they would have earned, if they exclusively sold supply-side based energy services. This means that regulators should pay close attention to any lost income, fixed-cost recovery, or margins when energy efficiency replaces supply-side based electricity sales. Another consideration is the potential regulatory use of direct financial incentives or COGS markups or margins. The following sections of these analyses will address both of these regulatory approaches in depth. Before doing this, the next chapter reviews how COS regulation became so dominant and caused value-of-service pricing to take a back seat.

CHAPTER 18
WHY COST-OF-SERVICE REGULATION MAY BE GIVEN TOO MUCH CURRENT WEIGHT

Rate base or cost-of-service (COS) regulation ties utility earnings to the original cost of supply-side investments that have not been depreciated. This raises one regulatory question: How will conservation that displaces electricity sales make up the lost income and fixed-cost recovery such sales would have recovered for the utility and its shareholders? This is a very major concern. In addition, a strict adherence to COS regulation makes it difficult to adopt the value-of-service concept that many associate with conservation.

This chapter is based upon a previous paper that Charles J. Cicchetti and Colin M. Long wrote in 2006.[1] This discussion reviews some important matters related to why regulators use rate base to implement traditional cost-of-service regulation. There are three reasons why this digression is important.

First, energy efficiency is more akin to retailing. Profit margins, shared savings, and other incentives are not necessarily based on the amount of money a utility invests. Value-of-service ratemaking seems to be a better fit. This is partly because utilities buy, package, and resell energy efficiency products. These activities are more closely aligned with utility purchases of "qualifying facility" electricity or other wholesale purchases. Some of these supply-side buys do not increase utility earnings and are instead simply complements to the utility's built and owned generation. The utility that develops a comprehensive demand-side business simultaneously produces, delivers, and markets conservation. In addition, energy efficiency increases and will be sustained when the utility can earn income through expanded energy efficiency and/or can reasonably avoid losses under COS pricing, when it encourages its retail consumers to conserve.

Second, customers do not seem particularly knowledgeable as to the meaning of kWhs, how to interpret their utility bills, or the effect of their conservation on the utility bills they receive and pay weeks after they use electricity. This has led some to seek to reintroduce the concept of value of service to regulation, and to encourage consumers to think about lighting, space conditioning, appliance use, hot water, etc. In this fashion, customers would, some suspect, be more willing to pay for energy efficiency and to recognize that it is the size of their bills

---

[1] *See* Cicchetti, Charles J. and Colin M. Long. (2006, July). A Brief History of Rate Base: Necessary Foundation or Regulatory Misfit. *Public Utilities Fortnightly*, Vol. 144, Issue 7, pp. 42-47.

that matters, not the per-unit prices they read about in headlines, when utilities seek rate increases. There is, however, an historic dispute between regulatory principles based upon "cost" and "value" of service.

Third, some jurisdictions have restructured and effectively shifted much, if not all, the generation portion of rate base to competitive wholesale markets. These states are replacing COS regulation with new forms of regulation. This opens the door to sensible energy efficiency incentives and new regulatory regimes. The problem is that many jurisdictions that have retained traditional rate base regulation do not find that there are many wholesale competition lessons to be learned, except "don't go there." Despite this reticence, some traditional COS jurisdictions are adding performance-based regulation (PBR) and making other reforms.

These thoughts combine to make a digression into history worthwhile. If the current foundations of rate base COS pricing seem less secure or at least more malleable, this opens the intellectual path for regulators to become more innovative and experimental in their efforts to expand energy efficiency. This would open some minds to revisit value-of-service principles to expand demand-side activities.

In the United States, transportation and utility regulation were initially guided by the dictum in the U.S. Supreme Court's decision in *Smyth v. Ames*,[2] which established the requirement that regulation should seek and adhere to a "fair return" or "fair value." In *Smyth*, the court was called upon to decide the constitutionality of a Nebraska statute that established maximum rates for intrastate railroad transportation. Finding that the rates established by the Nebraska study would, in many instances, result in the railroads being forced to operate their local transportation business at a loss, the Court found the statute violated the Fourteenth Amendment, amounting to "the taking of private property for public use without just compensation."

The court further stated that in ascertaining what would constitute just compensation, the interests of both shareholders and consumers must be considered. This balancing cannot be accomplished, according to the court, without considering the "fair value of the property used for the public, or the fair value of the services rendered …" Thus, the "… basis of all calculations as to the reasonableness of the rates to be charged … must be the fair value of the property being used by it for the convenience of the public." The court concluded that the property's value could be ascertained by examining several factors, including:

- original construction cost;
- cost of permanent improvements made to the property;

---

[2] 169 U.S. 466 (1898).

- the market value of the property's bonds and stock;
- the present as compared with the original construction cost;
- probable earning capacity, under statutorily prescribed rates; and
- operating costs.

The key aspect of this decision is that a company is entitled to a fair return on the value of the property employed for the public convenience, and the public is entitled to demand that it pay no more than the reasonable worth of the services received. *Smyth* established a "law of the land," whereby "value" and its equivalent "fair value" formed the cornerstone of utility, pipeline, telecommunications, and transportation ratemaking. Thus, *Smyth* created a rather elaborate approach for establishing fair returns based on value, one whose basic concepts remain embedded in modern day cost-of-service regulation. If this decision stood today, energy efficiency would be valued directly and price would be set equal to the regulated assessment of the value achieved.

The *Federal Power Commission v. Hope Natural Gas*[3] case represented a paradigm shift of major proportion in utility regulation. In that case, the Supreme Court found that the "end result" of a myriad of factors and judgment should be used to establish regulated tariffs and authorized returns. *Hope* diluted the *Smyth* decision's conceptual standard of value and left regulators to consider a combination of many factors, most of which were initially non-uniform. The *Hope* decision left something of a conceptual void, which the various legal and regulatory findings that followed attempted to fill. Many regulators, including the present day Federal Energy Regulatory Commission (FERC) and its Federal Power Commission and Interstate Commerce Commission predecessors, began to search for a conceptual anchor to secure the judgments they made to alter, tweak, and adjust the formulas and accounting data used in ratemaking. Into this void, original cost rate base emerged as that anchor for most regulators. Rate base is an accounting tool to measure what was prudently invested, less depreciation. It is based on standards that regulators, not tax collectors or financial reporting, dictated.

The original cost-of-service approach to utility regulation was ushered in over Justice Jackson's dissent in the *Hope* case. Justice Jackson argued that it was ill-advised to apply the concept of a prudent investment to set the price of a commodity, such as natural gas. Justice Jackson asserted that the prudent investment theory had relative merit in setting rates for a utility that creates its service through its investment. Thus, Justice Jackson considered that the prudent investment theory could readily be applied to transportation services that

---

[3] 320 U.S. 591 (1944).

consumers receive from common carriers. Customers receive a service provided by the carrier's property, but do not take or receive any of the carrier's property. Thus, Justice Jackson saw merit in applying the prudent investment standard to the value of service provided by natural gas pipelines, which provide what is in essence a transportation service.

However, Justice Jackson found lacking the merits of applying original or prudent investment principles, in setting the price of the natural gas itself. He argued that attempting to set a rate base for the commodity was elusive and that is was easier and more logical to set the price of the commodity based upon its value. Justice Jackson argued that the prudent investment theory "has no rational application where there is no such relationship between investment and capacity to serve. There is no such relationship between investment and amount of gas produced." The value of the service provided is measured by what is taken from the ground, not by what is invested to get the gas from the ground. As Justice Jackson colorfully put it, "there is little more relation between the investment and the results than in a game of poker." In a statement prescient to current circumstances, Justice Jackson stated that "… we must fit our legal principle to the economy of the industry and not try to fit the industry to our books." Justice Jackson's reasoning could easily fit the concept that many have today with respect to treating energy efficiency as a new utility service with important value-of-service characteristics.

Nevertheless, the majority in *Hope* adopted a standard whereby the result was important, not the means by which the result was obtained. The majority ruled that "the fact that the method employed to reach that result may contain infirmities is not then important" and that it is not "important in this case to determine the various permissible ways in which any rate base on which the return is computed might be arrived at." In his dissent, Justice Jackson argued that regulating unique businesses required new approaches that required adopting "concepts of 'just and reasonable' rates and practices and of the 'public interest' that will take account of the peculiarities of the business."

Over the decades following the *Hope* decision, regulation mostly evolved into new, somewhat formula-driven regulation in which rate base and empirical estimates of return on equity became paramount. Some jurisdictions have adopted very prescriptive and narrow approaches. Others were willing to review a broader array of measurements and concepts. Most regulators interpreted the input as a starting point and engaged in a process to let their judgments, not the metrics, rule.

There has been aggressive tinkering in the more recent parts of the twentieth century in the United States, by which regulators made major modifications to the original cost structure. These were mostly transparently made. Sometimes, these deviations were favorable to utility companies, such as with PBR and rate freezes to recover merger-related acquisition premiums.

Indeed, some of the innovative proposals for energy efficiency are closely tied to PBR concepts. There is a growing need for regulators to think rather broadly, in order to usher in more energy efficiency.

The merit of COS was its relative certainty. Original cost rate base could, after all, be measured relatively objectively. Specific disallowances for imprudence, used and useful violations, and cost disallowances could be decided and memorialized. Therefore, rate base measurement was mainly a function of original cost accounting and depreciation policies that regulators established for rate-setting purposes.

The handy objective accounting way to measure rate base and the formula used for establishing returns are just two of the many variables that regulators used to yield revenue requirements and authorized returns. Nevertheless, regulation had a neat and tidy appearance. However, regulatory judgment still ruled and courts were mostly unwilling to address the various pieces or factors considered in isolation that yielded the end result. In addition, regulated authorized earnings were not reported to investors. Investors were interested in actual net income. Reporting actual net income was required and used to explain how well a regulated business performed.

Somewhat lost in this regulatory shift was the fact that rate base and fair (or fair market) value were no longer conceptually similar. The differences were further obscured, because some regulators continued to attach the words fair or fair value to rate base. Regardless, rate base was no longer conceptually close to fair value as that concept was used in condemnation and tax treatments. In the context of evaluation, Bonbright *et al.*[4] observed:

> In any event, it now seems generally agreed, at least by all experts, that a fair-value measure of the rate base is not the same thing as a fair-value standard in taxation, in the law of damages, or in most other legal appraisals.[5]

Bonbright *et al.* observed that this shift from fair value to rate base is one that represents a regulatory "shift from the realm of the appraisal engineer to the realm of the accountant."[6]

Further, Bonbright *et al.* observed that the "'value of the property' in any definitive sense of the term 'value,' cannot qualify as an acceptable measure of the rate base."[7] To this, add the observation: and *vice versa*.

---

[4] See Bonbright, James C., Albert L. Danielsen, and David R. Kamerschen. (1988). *Principles of Public Utility Rates*. Public Utilities Reports, Inc., Arlington, Va.

[5] Bonbright, p. 217.

[6] Bonbright, p. 212.

[7] Bonbright, p. 218.

In a more positive way of explaining what matters here, Bonbright *et al.* concluded:

> … for ratemaking purpose, the value of the corporate assets must cease to be identified with their market value, or their value as private property, to the corporation or to its investors.
>
> Instead, the relevant values must refer to the potential values of the assets as instruments for the production of service to the community of ratepayers. If these assets were not only utterly essential for the performance of the service, but also utterly irreplaceable, their value to the ratepayers would be set by whatever rates of charge for service the ratepayers would be willing to pay rather than go without – set, in short, by what the traffic would bear. But if the assets are replaceable, their potential value to the consuming public is limited by their replacement costs.[8]

These choices and reasoning are precisely how tax and other fair market appraisals are conceptually defined and evaluated. Bonbright *et al.* observed, in effect, that original cost rate base, which rejects all the above metrics and concepts, is devoid of any meaningful measure of value. When combined with other factors to establish regulated tariffs, rate base becomes an input to the process, not an end result measure of value since, in effect, there is no rate base that reflects an energy service provider's or energy marketer's true value to retail energy consumers, owners, or society more broadly speaking.

As Justice Jackson explained in his insightful *Hope Natural Gas* dissent, regulation should look to the benefits provided consumers and the economy, not the invested capital. During his days, this sage advice was primarily rejected and "rate base" or "prudent investment" formed the basis of traditional cost-of-service regulation. For today's hybrid restructured utility industries, Justice Jackson's advice is both prescient and very much applicable. This is particularly so, when regulators seek regulatory and judicial precedents to expand energy efficiency. This discussion means that a regulator would need no more reason than their value judgment to expand energy efficiency.

In order to understand why this is so, it is useful first to review why rate base regulation has ruled for so long. Four factors seem most relevant. First, in the last century, growth in utility sales meant more energy for people and the economy. Quality of life and economic progress were viewed quite reasonably to be important benefits. The engineering focus of energy utilities meant that we could simply "live better" with more electricity and often with lower prices.

---

[8] Bonbright. pp. 219-220.

Justice Jackson's logic that the benefits provided should be the regulatory focus of how utility returns should be set was implicitly adopted. This was because utility investments expanded energy supplies and use. Consumers and the economy benefited as utility companies invested capital. Utility earnings increased as utilities invested more and the amount the utilities invested increased and tracked societal benefits to the invested capital. Developing countries over the latter part of the twentieth century ubiquitously understand that each percentage increase in electricity investment/output would be matched to a one percent growth in gross domestic product (GDP), as well as national quality of life and economic benefits.

Second, vertically integrated, traditional utility companies are capital intensive. One measure of this fact is that the ratio of annual sales to total assets, or invested capital, is typically less than 1.0. Another measure is the fact that in the late 1970s and early 1980s, electricity and natural gas utility companies frequently invested more new capital each year than all other industries combined in the United States. When rate base is no longer the focus of regulated service businesses, it cannot serve as the sole regulatory linchpin.

The sheer size and importance of a utility company's prudently invested capital meant that a regulator could attach to the return associated with this rate base all sorts of additions and deletions. These were meant either to encourage good things (such as better performance, replacing older more costly or more polluting technologies, improving customer safety or service, etc.) or to discourage or penalize bad actions (such as cost overruns, bad choices, or rising consumer complaints). Energy efficiency undermines the comfort that rate base once provided, because another value concept has now been added to the mix.

Regardless, regulators often learned that relatively small refinements to their authorized rates of return and the frequency of their rate cases could reward or punish a utility, because these adjustments, albeit often relatively minor, would be leveraged through the huge rate base that most utility companies carried on their books.

Third, the most significant single rate case variable was rate base. When original cost was adopted, the dollars included were matters of record. In a word, they could simply be measured "objectively." Issues related to prudence and acquisition premiums existed, but rate base could be measured objectively when original cost became the regulatory standard. Accumulated depreciation, the cost recovery and reduction variable used to determine "net" invested capital or rate base, was also objectively measured, because prior rate periods reflected the depreciation taken as operating expenses. Rate base regulation persisted for decades, because it could be determined objectively and rate base remained the largest cost component of regulation.

Fourth, utility regulation began to change in the 1980s. Consumers and policy makers began to find costs, as well as benefits, in energy growth. Some fuels were deemed worse than others. Some utility and some regulatory/political mistakes were made. Perhaps the biggest change that led to restructuring in the utility industry was the end of the "benign cycle." Until the 1970s, the average or unit prices reflected in utility bills generally declined in both nominal and real terms. Technology and scale offset inflation and increasing labor costs. The two oil shocks, double-digit inflation, and extraordinarily high interest rates caused utility rate shock and ended the benign period. Some utilities fared worse than others.

These unhappy events ushered in a comprehensive review of energy companies and their virtual monopoly status. Deregulation in transportation and telecommunications were viewed as creating consumer choice and lower prices and greater quality for unbundled services. This drew attention to a similar policy of restructuring electricity and natural gas industries. Independent power producers (IPPs) began to emerge and to introduce the concept of selling MWhs as electricity is produced, not as power plants are built. Conservation, global warming, and other environmental concerns grew in importance. Today, they are beginning to dominate the domestic political debate. Energy efficiency and renewables are displacing traditional supply-side investments. COS and rate base need to be modified to accommodate value-of-service principles or be replaced.

The first regulatory response to all of this was often to tweak rate base regulation. Some jurisdictions mandated different types of purchases and investments. Some required utility companies and, therefore, consumers to subsidize specific customers that did something deemed beneficial on the customer side of the meter (*e.g.*, add solar energy, more insulation, etc.).

This tweaking added costs and often caused increases in utility prices. Tweaking often yielded to regulatory hammers when prudence, used and useful standards, and other disallowances emerged in the 1980s. Regardless, utility rate increases were ubiquitous. Regulated monopolies could keep their customers. Further, higher prices added to the calls for restructuring, unbundling, competition, and retail choice. Rate base and cost-of-service regulation began to be seen as the problem, not the goal. Imperfect regulation was compared to hybrid models that ushered in competition.

Recently, competition, deregulation, and restructuring have emerged in which formerly vertically integrated regulated transportation, telecommunications, and energy utilities have been unbundled, divested, and forced to compete. New forms of service-oriented companies have emerged that lack a traditional utility rate base. Regardless of what a particular jurisdiction does with generation regulation and/or markets, there is much to be gained from a new value of service approach to energy efficiency regulation.

It is into this mix and against this historical background that energy efficiency's relative lack of a meaningful rate base must be reviewed. The services provided and the value that energy efficiency adds are not typically investment related. Accordingly, a strict adherence to an original cost rate base would and does effectively mean that energy efficiency services would not be reasonably valued. This, in turn, translates into an undervalued and mostly inadequate return based on invested costs, not the value that is provided to customers or which is, or should be, captured in its inherent value. Margin or value-of-service proxies are needed to make energy efficiency comparable to traditional supply-side investments.

The principal concept for regulating these new energy efficiency services should be to reject rate base regulation. It does not apply. It is an anachronism, one that has failed. One goal, that is prompting jurisdictions to restructure, is to achieve consumer and economic benefits and to reduce waste and inefficiency. Interestingly, one of Justice Jackson's stated goals, in suggesting that services provided to customers be provided at fair value, was to encourage the conservation of scarce natural gas resources. One way in which these goals can be achieved is for regulators to adopt a new regulatory standard that is tethered to how comparable competitive retail firms assess performance and value.

A broad regulatory public policy purpose is to seek the cost savings of monopoly suppliers, while restricting earnings to comparable competitive industries. Accordingly, accounting standards used to establish rate base should be given no more weight than an unregulated business gives to its historic original plant costs. Profits depend upon incremental costs and current revenues, not historic original plant costs.

That said, regulators should consider replacing rate base with a benefit- or value-provided approach to regulating energy efficiency services. Rate base is not a complete regulatory standard. It does not even, if called original cost, become a new definition of the old ambiguous "fair value" concept. Regulators need to find the right alternatives to set "fair value," when they confront new issues and new circumstances. The place to look is comparable retail service firms and industries.

Justice Jackson was well ahead of his time. His logic, that the benefits provided should be the regulatory focus of how utility returns should be set, was implicitly adopted. This was because utility investments expanded energy supplies and use. Consumers and the economy benefited as utility companies invested capital. Utility earnings increased as utilities invested more and the amount the utilities invested increased benefits. This logic no longer makes sense.

As Justice Jackson explained in his insightful *Hope Natural Gas* dissent, regulation should look to the benefits provided consumers and the economy, not the invested capital. During his day, this sage advice was mostly rejected and "rate base" or "prudent investment" formed the basis of traditional cost-of-service regulation. However, for purposes of establishing a reasonable margin for energy service companies, Justice Jackson's advice is both prescient and very much applicable. A new way of thinking about how to establish a fair return for the value of energy efficiency is to focus on the value provided to customers, not rate base.

It is important to identify other firms and industries that are: (1) comparable financially; (2) have a similar customer service focus; and (3) have much more in common with energy service companies in terms of its business and risks than the more traditional or typical investment-in-the-ground utility model. For energy efficiency, retail margins can provide a reasonable way in which to regulate the newly emerging energy efficiency services provided, based on the value of the services provided to customers. Furthermore, utilities that produce energy efficiency and resell this product should share in the savings, earn an upside, and not be unduly punished for their traditional supply-side investments.

Two measures of return are given some primacy in retail businesses. These are: "margins on sales" and "margins on cost of goods sold." The former is based on the full retail purchase price. The second is based on the "cost of goods sold," which for an energy services company would include the energy and perhaps the delivery charges for distribution and transmission that would be passed through to the consumer.

Not all retail industries sell products that are comparable to energy efficiency. Nevertheless, those industries that are comparable can form the basis for establishing return margins for energy service companies. Initially, two factors seem particularly relevant in determining which industries and companies are comparable to energy service companies. These are: (1) skills and functions; and (2) the turn over ratio of sales to total assets. The first is a qualitative factor that involves judgment. The second is an often reported financial metric. A third factor discussed below is the essential service nature of energy.

Most traditional utilities were founded on, and developed around a corporate structure centered on engineering skills, financing skills, and the building of complex, capital-intensive vertically-integrated utility systems. Competition and restructuring have broken the traditional vertically-integrated utility monolith. Choice and market forces establish different prices and risk-related packages. One important result is that "customers" and their "needs" are the focus of these new entities.

The best approach for regulating new energy efficiency services would be to replace return on investments with an adder to sales or margin on sales approach. The margin would reflect both return/profit and elements of risk, not recovered elsewhere, with explicit cost-of-service adders. Utilities should also be paid profit margins when, like QF producers, they also supply the energy efficiency that they resell. Consumers expect to pay a markup for wholesale/

retail sales. Industries such as groceries, department stores, restaurants, sports, and entertainment are predicated on a retail markup. Energy efficiency services are retail services, which reduce the social costs of an essential service, energy. Customers have other choices. In competitive markets, if a retailer, regulated or not, is to stay in business, it must recover a profit margin. A margin approach such as the one we propose here would look to margins in other retail-oriented competitive industries to establish a reasonable sales margin for energy service companies.

SECTION V: TESTING THE THEORIES THAT INCENTIVES MATTER

The nation is increasingly seeking to solve the twin problems of climate change and energy/economic security. The nation's electric industry has been targeted to play a major role in expanding the nation's commitment to energy efficiency. A threshold policy question addressed here is: "Why is this necessary?"

Retail consumers, at best, compare retail electric prices to their marginal cost of energy efficiency. This is too narrow, because the marginal cost of electricity may be increasing faster than regulated prices. Furthermore, the external costs of energy use and production are not reflected in the retail customer's price of electricity. Both mean, at best, that there is a growing and significant gap between the two services about which we expect retail customers to make rational choices: electricity and energy efficiency.

There are also likely scale economies involved in energy efficiency information gathering and analyses, installation, and purchasing/marketing. The utility, not its individual customers, seems to be better placed to capture such cost-effective advantages.

This discussion, up until now, has shown that utility regulators are taking charge, but the nation tried this at least twice before in response to previous energy crises. This time, regulators are seeking to find out what works and how to sustain new efforts.

There are two types of regulatory policies: (1) decoupling; and (2) direct profit/income incentives, which are central to much of the current regulatory action and policy debate, as policymakers seek to encourage energy efficiency.

The results reported demonstrate to regulators that what they do, with respect to such incentives, matters. The analysis in this section shows with very high statistical confidence that the regulators who add incentives would expect greater energy savings and more utility spending for energy efficiency. Accordingly, regulators need to give serious consideration to decoupling and/or direct financial incentives to utilities as the best approaches to start, expand, and sustain utility-sponsored energy efficiency programs.

This section reviews how different jurisdictions cause energy efficiency and load management to succeed. It begins with a discussion of California, which has been viewed as one of the national leaders in terms of a "green" utility industry. California uses decoupling, strong shared savings or financial incentives, and a paradigm shift to create a "green" utility industry. Next, the EIA data are used to test specific hypotheses related to the effectiveness of two regulatory approaches: decoupling and direct financial incentives. The analyses demonstrate these factors matter. Therefore, this discussion explains in some detail what states have been doing in terms of these various incentives.

CHAPTER 19
CALIFORNIA SHOWS MANDATES AND RATE RIDERS WORK: A BRIEF HISTORY OF CALIFORNIA'S ENERGY EFFICIENCY EFFORTS

The nation's energy utility companies are currently contemplating their third run at utility-sponsored energy efficiency. As with the two prior efforts, high world oil prices matter. Current support seems more broad-based politically than simply concerns about high prices. Indeed, most Americans also have climate change, macro-economic, and national security concerns.

There is much at stake. The previous discussion points toward three inherited matters that past efforts to increase energy efficiency have failed to solve satisfactorily.

First, most retail energy consumers do not seem willing to pay to conserve even when their energy savings would often rapidly pay back the additional money spent to conserve and increase end-use efficiency.

Second, virtually no utility customer, particularly when energy prices are increasing, wants to help pay for someone else to become more efficient.

Third, some regulators, perhaps even a majority, seem to resent the notion of regulated companies making additional profits, if they help their customers become more energy efficient.

These obstacles remain despite the increase in potential support for energy efficiency, as well as the nation's increased climate change and security concerns. The first question is to determine whether energy efficiency is reliable and cost-effective. This chapter reviews how one state, California, has succeeded. It also answers the question "does it work?" affirmatively.

For about 25 years, California has shown that successful and "for profit" utility-sponsored energy efficiency can help the nation reduce its energy appetite and improve the environment. California's mostly uninterrupted commitment to energy efficiency was implemented before the current energy crisis and before global climate change became so politically important. California, in effect, has found a new source of electricity in the form of energy efficiency.

CHART 19-1
PER CAPITA ELECTRICITY SALES (NOT INCLUDING SELF-GENERATION)
(kWh/person) (2006 to 2008 are forecast data)

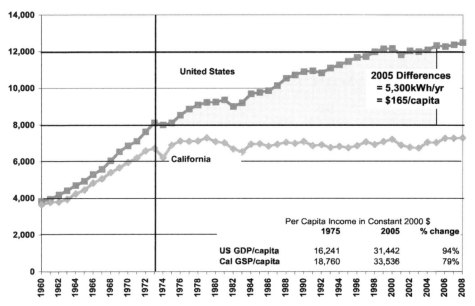

Source: Rosenfeld, Arthur H., Commissioner. (2007, May 24). "California's Success in Energy Efficiency and Climate Change: Past and Future." Presentation of the California Energy Commission (CEC). Electricite de France. Reprinted with permission.

Over the last three decades, as Chart 19-1 shows, California's per capita electricity consumption has stayed flat, while the nation's per capita electricity consumption has increased more than fifty percent. Today, California uses about half as much electricity per capita as the rest of the nation. In effect, California has found the means through energy efficiency to offset its higher utility prices with lower consumption.

There is no secret to California's success. The state has been willing to spend more money on energy efficiency than other states. Chart 19-2 shows that, on a per capita basis, California has outspent the rest of the nation.[1] This state's conservation effort has strong and widespread political support. California has learned that it spent too much for traditional energy under both cost-of-service regulation and its disastrous restructuring experience. The state's consumers would rather promote energy efficiency than return to either of its past failed approaches.

---

[1] Source for Energy Efficiency and Load Management: *http://www.eia.doe.gov/cneaf/electricity/page/eia861.html*; Source for US and CA Population Estimates 2000-2006: *http://www.census.gov/popest/states/NST-ann-est.html*; Source for US Population Estimates prior to 2000: *http://www.census.gov/popest/archives/1990s/popclockest.txt*; Source for CA Population Estimates 1990-2000: *http://www.census.gov/popest/archives/2000s/vintage_2001/CO-EST2001-12/CO-EST2001-12-06.html*

CHART 19-2
ENERGY EFFICIENCY AND LOAD MANAGEMENT PER CAPITA SPENDING

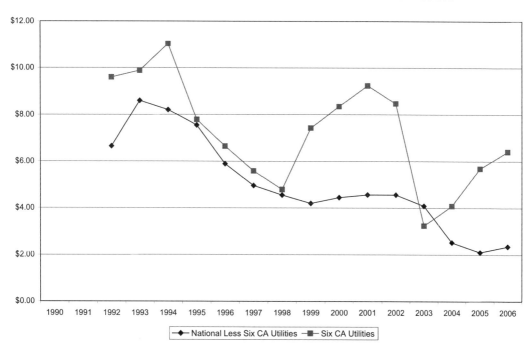

California regulators also did three straightforward things to achieve their enviable energy efficiency record. Others could benefit from California's regulatory experience that promoted energy efficiency in the state. These steps were:

- Mandate that utilities achieve specific energy efficiency objectives or targets. These have changed over time. However, after completing some expensive nuclear generation in the 1980s, California simply has preferred renewable supplies and energy efficiency over conventional generation.

- Decouple revenue and energy savings starting in 1982. Other states are recognizing the importance of protecting shareholders from the lost revenue effects of conservation. California has done so for about two decades.

- Pay for energy efficiency and reward shareholders starting in 1991. This is essential. California has paid avoided costs and premiums for wind and geothermal power stations. New renewable technologies have been particularly encouraged. Public goods charges have been added to wires charges since restructuring in 1996 under California's infamous AB 1890 legislation. These pay for conservation and energy efficiency.

Chart 19-3 shows the history of California's annual investor-owned utilities' (IOUs) spending on energy efficiency. The amounts California has spent and plans to spend for renewables, DSM, real-time/time-of-use pricing, and energy efficiency are significantly more than other states plan to spend. The highest expenditures on energy efficiency were during the state's energy crisis in 2000/2001. With the return of utility financial health, California has again returned to its reliance on energy efficiency.

CHART 19-3
CALIFORNIA IOU'S INVESTMENT IN ENERGY EFFICIENCY

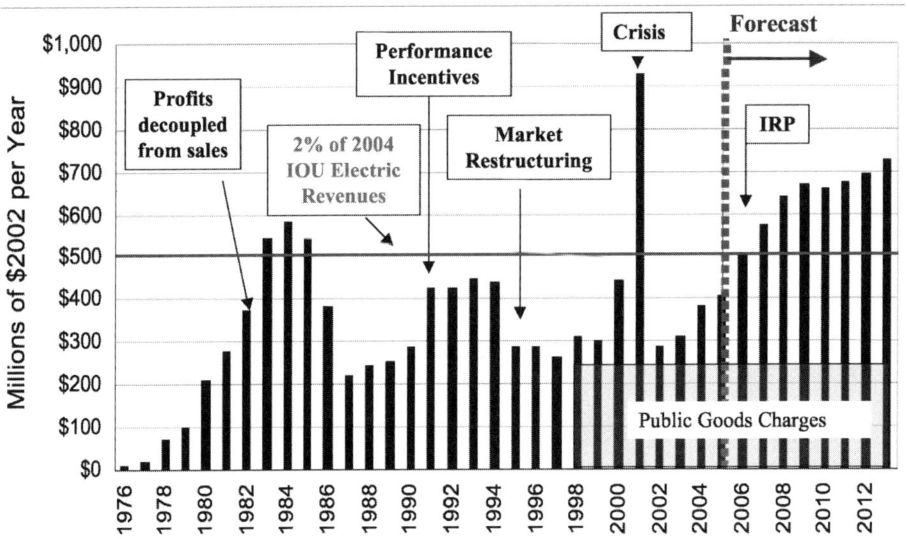

Source: Rosenfeld, Arthur H., Commissioner. (2007, May 24). "California's Success in Energy Efficiency and Climate Change: Past and Future." Presentation of the California Energy Commission (CEC). Electricite de France. Reprinted with permission.

The investment amounts shown in Chart 19-3 do not include: (1) the $80 million per year that IOUs pay for Public Interest Energy Research (about half on efficiency and half on renewables); (2) the many additional billions of dollars that IOUs spend to acquire QF supplies, such as wind, geothermal, biomass, and small hydroelectric renewable generation; (3) the additional dollars that the state's smaller municipal utilities spend on efficiency and renewables; and (4) the additional investments made in demand response, management, and metering technologies, and pricing.

In the fall of 2007, the California Public Utilities Commission (CPUC) established a revised system of "incentives and penalties" to better align utility and state interests in energy efficiency. Commissioner Dian M. Grueneich of the CPUC had estimated that

California will spend about $2.7 billion on energy efficiency in 2006-2008. She projected this would save more than $5.4 billion in direct utility costs (avoided generation, capacity, natural gas, transmission, and distribution costs) over the life of these measures. This would yield a direct utility benefit-to-cost ratio of about two to one.[2] In addition, California will soon have "smart" metering in place and plans to increase the demand-side options offered to customers, including real-time pricing and remote load control.

Chart 19-4 shows that the increasing expenditures on energy efficiency have produced significant energy savings. In 2008, California expects to "save" about 13 percent of its electric consumption. California's investor- and government-owned utilities plan to do even more in the future. These efforts include both energy efficiency and incentives for renewable energy. For example, the state plans to spend $300 million per year to achieve one million solar roofs in the state.

CHART 19-4
CALIFORNIA COMMITS TO LONG-TERM EFFICIENCY
CALIFORNIA IOU HISTORICAL AND PROJECTED ELECTRICAL EFFICIENCY SAVINGS

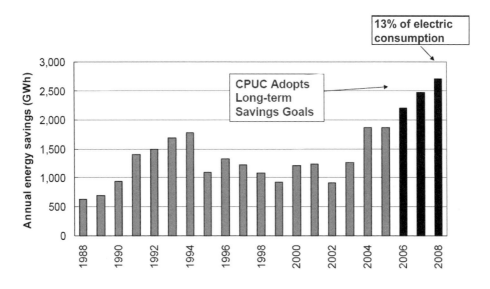

Source: Weston, Rick. (2006, January 10). "Energy Efficiency Potential: It's Always More Than You Think." the Regulatory Assistance Project. Natural Resources Council of Maine. Reprinted with permission.

---

[2] Grueneich, Dian M., Commissioner. (2007). "California's Policy Framework to Advance Demand-Side Management." California Public Utilities Commission, International Energy Agency, Paris, France (DSM Workshop).

In addition, all California utilities have accepted a 20 percent renewables portfolio standard to be achieved in the next 10 years. Chart 19-5 shows California's renewable electricity accomplishments and plans. These are in addition to the almost $800 million per year the IOUs spend on energy efficiency. Of course, California is a large state with about 35 million people, a peak load of about 64,000 MWs and sales of about 275,000 GWhs.

CHART 19-5
CALIFORNIA RENEWABLES PORTFOLIO STANDARD

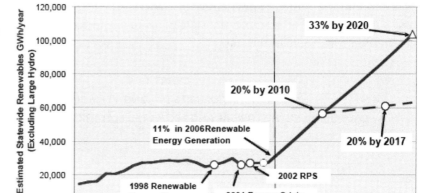

- Designed to increase diversity, reliability, public health and environmental benefits of California's energy mix.

- Current legislative goal of 20% of retail sales from renewables by 2010; increase by at least 1% per year.

- Some discussion of increasing the goal to 33% by 2020.

Source: Rosenfeld, Arthur H., Commissioner (2007, June 18). "Efficiency and Renewables in the Electricity Sector." Presentation of the California Energy Commission (CEC). Reprinted with permission.

Despite its size, California plans to continue to spend quite a lot on both what it calls energy efficiency and renewables. The state's utilities are leading the way with estimates that the state will achieve 1990 greenhouse gas emissions levels in 2020. Most impressive, the energy utility sector will achieve about 96 percent of these.[3] For example, the nearly $800 million per year that California IOUs pay just to conserve electricity use equals an all-consumer surcharge of about .39¢ per kWh each year. The additional charges for load management and renewables exceed this amount. The state also has used stiff building codes and appliance standards to achieve significant additional savings.

The California approach shows that energy efficiency, conservation, demand-side management, and renewables work. The cost effectiveness of these programs has meant California

---

[3] Grueneich, Dian M. Commissioner. (2007). "California's Policy Framework to Advance Demand-Side Management." California Public Utilities Commission, International Energy Agency, Paris, France (DSM Workshop).

can and will accommodate significant population and economic growth. Other regions should not waste their time debating the cost effectiveness of energy efficiency. California has a lot to offer to other states that are not averse to strong energy efficiency and renewable mandates, and are willing to focus on total bills (*i.e.*, the amount spent, not just unit prices).

The California approach amounts to government mandates financed through a wires charge that is comparable to using a gasoline tax to build mass transit and to pay for "greener" automobiles, trucks, etc. This "works" if and when the public is pro-environment and/or anti-new power stations. The "not so fast" caution light for others might be California's near dismissal of traditional supply alternatives and concomitant lack of concern for the resulting retail prices paid.

Most states would balk at paying the cost per kWh saved or shifted to off-peak periods in California. There would also likely be less aversion than California has to new power stations, especially when new often means cleaner, more fuel efficient, and better-for-the-climate power stations that help to retire decidedly less favorable older units. The challenge outside of California is how to expand utility-sponsored energy efficiency without paying too much and also avoiding problems that distracted regulators in the 1980s, when they last confronted these problems with a degree of urgency similar to the one they have today.

In the past, it sometimes has seemed that the least-cost case for energy efficiency has been so compelling that regulators try to make perfect choices. These attempts at perfection can result in micromanaged programs and detailed assessments that attempt to determine who pays and how much should be spent. The most important regulatory fact is none of this really matters, when a utility is confronting the need to invest in new generating stations.

Society benefits, and economic efficiency increases, when regulators and utilities consider both direct utility and external costs. The regulatory challenge is to get the best answer; to wit: decide when traditional generation must be compared to programs that promote conservation, improve demand response pricing and DSM, and expand renewable generation on the grid and customers' premises. The next section shows how past regulatory efforts help to determine what needs to be done today, if the nation is to embrace utility-sponsored energy efficiency as a core public policy.

None of this success should be surprising. California began to spend significant amounts of money on energy efficiency in the 1980s under traditional regulation. The state adopted aggressive utility-sponsored conservation, DSM, and renewable energy programs that were paid for with rate riders and sales volume adjustment mechanisms. This was widely supported because California did not want to build new utility power stations. After California restructured its electricity markets, these so-called public goods charges were continued as adders to distribution or wires charges. This meant that energy efficiency continued after restructuring. Since its restructuring debacle and subsequent energy crisis in 2000 and 2001, California has been rebuilding its institutions and changing its market design (adding long-term capacity requirements and markets), re-regulating those parts of the market particularly related to retail choice, and expanding the scope and payments for public goods charges. This means that California will retain its leadership in using energy efficiency to reduce climate change.

CHAPTER 20
MIXING MANDATES AND INCENTIVES

There are renewed efforts to expand energy efficiency in the United States. These have been mostly concentrated on electric utility companies. These businesses are either comprehensively or partially regulated. This makes it rather easy for regulators to mandate expanded energy efficiency. While spending more is likely better, there should be some considerable attention given to relative costs and benefits, as well as serious regulatory efforts to determine what works and what does not.

Regulated electric utility companies respond to incentives. The differences between firms that compete for profits and regulated businesses is how, not if, they respond to incentives to perform. As regulators currently are seeking to increase utility-sponsored energy efficiency, they are increasingly considering how to complement mandates with added inducements. The premise is that utilities would spend more for energy efficiency, when regulators add regulatory or political sweeteners. Up until now, this public policy hypothesis has not been fully tested.

Increasingly, voters and utility customers are learning that energy efficiency can improve the environment or ease climate and national security concerns. If spending more on energy efficiency is the primary goal, utility regulation is very well-suited to accomplish such a "more is better" goal. Utilities have significant cash flow and access to capital. Regulators can readily use "pipes and wires" to collect money through public benefit charges that all customers pay. With cash, regulators can easily mandate new spending requirements for energy efficiency. Some regulators might recognize that there would be economic efficiency benefits, when they improve tariff design and send sensible price signals to advance energy efficiency. These efforts would close the gap between the full marginal social costs of energy and the marginal cost of energy efficiency. Other regulators might turn the clock back and rely on value-of-service principles to recognize the value consumers place on energy efficiency. Alternatively, more informed economic and regulatory principles, based on marginal social cost, including marginal externalities, could be invoked. Regardless, regulators seem to recognize that they are invited to participate and play a leading role in the nation's energy efficiency agenda.

Put simply, spending more ratepayer money on mandated energy efficiency is the relatively easy first step. The second step is tariff design to pay for energy efficiency and to treat stakeholders fairly. This has not always been easy and has often derailed, or at least slowed, past energy efficiency efforts. That said, ironically both value of service and marginal cost principles seem to be on the ascendancy, as independent regulatory approaches that justify expanding utility-sponsored energy efficiency.

The third step is even more difficult, as past experience has proven. The essential regulatory question is how to sustain energy efficiency, if the current surge in support wanes. More seems to be needed from regulation than simply permitting electric utilities to pass on the costs of conservation to their mostly captive retail customers. About half the states do at least this much for mandated utility-sponsored energy efficiency programs. In the past, this limited regulatory approach has failed to sustain energy efficiency, when either energy prices stabilized or other public concerns replaced energy and environmental challenges.

It is also important to revisit a second question, which is "why is there any need for utility-sponsored energy efficiency?" If regulated electricity prices (P_E) exceed the marginal cost of energy efficiency (MC_{EE}), retail customers would receive a direct price signal that energy efficiency likely pays for itself. Current retail prices in the United States range from about 5¢ per kWh to more than 15¢ per kWh. The cost of many energy efficiency choices is much less, whether the customer's or the utility's cost is used. Retail customers should be educated to the fact that most of the widely available means to increase energy efficiency would save money and often have very short payback periods. For many consumers, the direct benefits of energy efficiency are easily three times the costs of energy efficiency, and this would increase when the external costs of energy use are considered.

Regulators and others increasingly recognize that many retail customers do not act and become energy efficient, despite these market and social signals. There are many reasons for this seeming customer inaction, ranging from lack of knowledge to too much conflicting information to apathy. In addition, many Americans believe the cost of electricity is affordable or not particularly expensive in relation to other consumer items. Unlike gasoline prices that consumers see when they fill up their cars, most electric utility consumers do not know what they spend for their various uses of electricity and simply pay a total monthly bill after the fact. The typical utility consumer has no real understanding of what he/she pays for specific uses, such as lighting, cooling, heating, appliances, refrigerators, audio/visual equipment, computers, etc.

In addition, many consumers look to regulated electric utilities and their regulators to achieve least-cost and high-quality electric service. Accordingly, such customers do not take the time or expend the effort "to do it themselves." When the external costs of electricity production are added, a large gap exists between what customers do on their own and what should be done because society would benefit from expanded energy efficiency.

When more investments in new generation and increasing fuel costs push marginal electricity costs (MC_E) higher, the gap expands between cost-effective efficiency and what customers seem willing to do on their own.

These factors often combine to cause regulators to consider utility-sponsored energy efficiency to close this gap between energy efficiency's potential and its reality. Mandates and cost recovery for such utility-sponsored energy efficiency are almost always in the public interest. This is particularly so when the marginal cost of electricity is increasing.

When there is more at stake, as today, regulators should not simply cause utilities to spend more on energy efficiency and seek a "just and reasonable" means to recover the costs. Instead, regulators should find out what forms of regulation cause utilities to succeed and to sustain utility-sponsored energy efficiency over the long term.

Previously, heightened utility-sponsored energy efficiency periods often focused on conceptual matters that would identify benefits and costs in the context of the revenue requirement, or cost of service, for utility companies. Regulators and stakeholders have often, as explained, performed various tests to answer basic cost effectiveness questions.

1. Would participating customers pay lower monthly bills?

2. Would the projected "savings," combined with participant payments to defer some of the cost of energy efficiency, result in "no losers"?

3. What are the net societal effects, when the marginal cost of energy is compared to the marginal cost of energy efficiency, regardless of regulated prices and who pays?

These various "tests" gave and continue to give a conceptual nod toward the question of program effectiveness in a somewhat narrow, although understandable, utility regulatory context. These previous efforts mostly missed the broader public policy mark of why consumers do not do enough on their own.

External benefits often motivated regulators to expand energy efficiency. However, these benefits were mostly not quantified. Instead, the paradigms regulators considered were based upon utility revenue requirements and regulated prices. This is also now changing. There are beginning to be some efforts to quantify external benefits, using market data from "cap-and-trade" markets for SO_2, NO_X, and CO_2, as well as to assess the economic and national security costs related to profligate energy consumption in the United States.[1]

---

[1] *See* Cicchetti, Charles J. (2008). Duke's Fifth Fuel. *Public Utilities Fortnightly*, Vol. 146, Issue 1, pp. 54-62.

Regulators are also asking questions that will ensure some degree of financial protection and incentives for utilities to increase energy efficiency over time. This all means regulation is necessary. Subsequent statistical analyses test various hypotheses, in order to determine the effectiveness of current regulatory efforts to cause utility companies to spend more, to increase energy efficiency, and investigate what can be done to sustain these efforts when, as seems to be inevitable, public support wanes.

The most basic economic concept supporting an increase in energy efficiency is found in the relationship between the marginal cost of energy efficiency (MC_{EE}) and the regulated price of electricity (P_E), which is about the same as the average total cost (ATC) of electricity. When the price of a substitute is greater than the marginal cost of an alternative product, economists expect rational consumers to substitute lower-cost products for more expensive alternatives. Electricity consumers should, therefore, prefer more energy efficiency and less electricity, other things being equal.

The first task for regulators seeking to expand energy efficiency is to help electricity consumers understand this basic fact and to help steer them through the often conflicting maze of energy efficiency products and choices. If regulators stop at this juncture, retail consumers would make better choices and energy efficiency would increase. They should not stop, because externalities also matter.

Regulation should also recognize a related matter. Electric utilities earn money when they sell electricity. This is similar to a gasoline service station that sells motor fuel. Most would recognize that gasoline service station owners would be conflicted if they were called upon to encourage their consumers to switch to either mass transit or electric cars.

In the same fashion, there are inherent challenges when seeking to make regulated electric utilities champion another competing product. The two primary regulatory solutions to any utility conflict-of-interest challenge are: (1) revenue/income decoupling, and (2) profit/income incentives. These would be coupled with a reasonable chance to recover the direct and indirect costs of demand-side programs. Put another way, regulators that add incentives are effectively transforming the utility business model to include energy efficiency, either as a new utility product or utility input, that should be able to earn a retail sales margin.

This seems both fundamental and necessary. However, there are additional considerations. The basic economics that drive utility-sponsored energy efficiency are more straightforward when the marginal cost of electricity is increasing. In such circumstances, selling more electricity would then cause regulated price increases, because the higher marginal cost of electricity would raise average total cost, which would, in turn, increase regulated electricity prices.

Regulators using an integrated resource planning proceeding would discover that electric rate increases would be needed to satisfy demand growth. This would be true under conditions of increasing marginal electricity cost, regardless of the supply technology chosen to meet traditional least-cost criteria. Before surrendering to regulated rate increases, most regulators would seek alternative approaches to avoid such rate increases.

In this context, energy efficiency becomes another choice for regulators to consider. This choice is attractive because energy efficiency typically costs less than both the marginal cost of electricity and regulated prices. In addition, the marginal cost of energy efficiency would likely have offsetting marginal external benefits. In addition, participating consumers would be willing to pay more, because they would be reducing their own monthly utility bills, when they use fewer kWhs.

Selecting energy efficiency would reduce future regulated revenue requirements relative to a traditional supply-side utility expansion that have a higher marginal cost of electricity than the utility's marginal cost for energy efficiency. Participating consumers would focus mainly on the basic economic relationship between their regulated prices, the portion they pay for energy efficiency, and the value they attach to reducing negative externalities related to supply-side options.

Regulators and other utility customers not participating in utility-sponsored energy efficiency need to consider how an increasing marginal cost of electricity would raise the regulated price of electricity. From a future total revenue requirements perspective, the most important utility cost-of-service relationship in an integrated planning analysis is the extent to which energy efficiency is less than the marginal or avoided cost of expanded electricity production.

The final, although often ignored, determination of the economically efficient amount of energy efficiency would add the marginal external costs of energy production to the marginal utility costs. This should cause regulators to urge and, as explained, incentivize utilities to expand energy efficiency.

The single challenge for regulators is that regulated prices might increase with either a traditional supply-side choice or more energy efficiency. However, the relevant regulatory factors are very different for the supply-side and energy efficiency choices. When the marginal cost of electricity is increasing and utilities respond with supply-side alternatives, revenue requirements increase, the volume of electricity sold increases, marginal cost raises average total costs, and pressure forms to increase regulated prices. These relationships are well-understood. Regulators would take notice and eventually be expected to respond, even if grudgingly, with rate increases.

When energy efficiency is used to avoid more costly supply-side choices, often with external costs, regulated prices are also likely to be affected. Indeed, they likely would also increase. This may surprise consumers who correctly think the least cost is to replace more expensive supply-side choices with cost-effective energy efficiency. The reason for this result is that regulated prices are based on average total costs. These are the ratio of total annual costs (or revenue requirements) divided by the total annual volume of electricity sold.

Energy efficiency most certainly reduces the numerator (total annual utility costs of service) when it replaces more expensive supply choices. However, energy efficiency also causes less total annual volume of electricity to be sold. The latter, other things being equal, causes average total costs to increase. The result is that regulated prices might increase.

Other factors enter the equation and regulators are beginning to take notice. When regulators and utility companies can "sell" energy efficiency to retail customers, these customers receive a "new" energy efficiency service that provides a direct benefit (*i.e.*, lower electricity bills). If these participating customers pay for the energy efficiency they consume, there is less likelihood that regulated prices would increase. Regardless, there would need to be additional revenue to cover new costs, as energy efficiency becomes a new utility product.

In addition, all customers benefit when energy efficiency reduces negative external costs of electricity production and/or when energy efficiency improves economic efficiency, because consumption is reduced and the marginal cost of electricity exceeds the regulated price of electricity. In either circumstance, regulators are correct to recognize the predominant need to expand energy efficiency. This also means they would likely consider even an increase in regulated prices due to less electricity sales and more energy savings to be more than "just and reasonable."

Nevertheless, even the acceptance or prospect of fully justified higher regulated prices for energy efficiency mandates may not be enough. Regulators that understand the current rationale should also consider various approaches to improve the success and sustainability of utility-sponsored energy efficiency. For this reason, a small number of states have gone beyond simply allowing cost recovery for mandated utility-sponsored energy efficiency. One purpose is to change the narrow utility business model and to make energy efficiency a new utility product.

More specifically, states have added two types of additional incentives to encourage energy utilities to make utility-sponsored energy efficiency programs succeed. One approach is revenue decoupling. This incentive mechanism protects utility sales revenue (or income) when energy sales decline. Accordingly, an energy utility that encourages its customers to increase conservation would be able to increase average prices to true up its revenues (or income) for lost utility sales.

This approach for the most part also adjusts other annual swings in the quantity sold related to weather/climate. Regulatory discussions related to "decoupling," as an inducement for energy efficiency among electric utilities, seems to be increasing in recent years. Over the past 16 years (1992 to 2007), eight states have utilized electricity decoupling at least for a portion of this time period (See Table 20-1) and other states are reviewing various decoupling proposals.[2]

TABLE 20-1
STATES WITH DECOUPLING MECHANISMS FOR ELECTRIC UTILITIES

State	1992	1993	1994	1995	1996	1997	1998	1999	2000	2001	2002	2003	2004	2005	2006	2007
California	1	1	1	1	1	1	0	0	0	0	0	1	1	1	1	1
New York	1	1	1	1	1	1	0	0	0	0	0	0	0	0	0	1
Washington	1	1	1	0	0	0	0	0	0	0	0	0	0	0	0	0
Maine	1	1	0	0	0	0	0	0	0	0	0	0	0	0	0	0
Minnesota	0	0	0	0	0	0	0	0	0	0	0	0	0	0	0	1
Idaho	0	0	0	0	0	0	0	0	0	0	0	0	0	0	0	1
Oregon	0	0	0	1	1	1	1	1	1	1	1	0	0	0	0	0
Rhode Island	0	0	0	0	0	0	0	0	0	0	0	0	0	0	0	1

Decoupling revenue or income from sales has also been more popular in the natural gas industry than the electric industry. This is likely because weather-related sales and natural gas commodity costs are typically more volatile for natural gas utilities than for electric utilities.

California's enviable record on utility-sponsored energy efficiency and customer premises renewable energy is at least partially the result of the state's so-called "electric revenue adjustment mechanisms" (ERAMs). These rate riders adjust the regulated prices for natural gas and electricity when either weather or conservation curtails sales.

Cost-of-service regulation relies on utility tariffs to recover a return "on" and "of" rate base. Tariffs generate more money, when sales are greater, and *vice versa*. Therefore, a "decoupling" policy or some similar consideration is deemed necessary to at least make shareholders neutral with respect to expanding conservation. The next chapter describes what various states actually have been doing with respect to "decoupling" policy.

---

[2] National Action Plan for Energy Efficiency. (2007, November). "Aligning Utility Incentives with Investment in Energy Efficiency." U.S. Environmental Protection Agency (EPA). U.S. Department of Energy and ICF International.

CHAPTER 21
HOW STATES ARE DECOUPLING[1]

The term "decoupling" can be broadly applied to a rather broad range of regulatory programs to encourage energy efficiency. At one end, the term characterizes sales or revenue neutrality in the form of a "true-up" type mechanism that adjusts for differences between predicted sales of energy (often based on a regulatory test year) and the actual annual sales. This type of adjustment might reflect a myriad of factors that would affect sales, such as weather and economic conditions, as well as energy efficiency. At best, this regulatory approach is neutral for energy efficiency. In many circumstances, the specific effects of energy efficiency are lost in the broader annual changes in sales.

At the other extreme, the decoupling term applies to lost earnings or margins on energy sales that energy efficiency displaces. The purpose of such an adjustment is to estimate the differences between sales prices and avoided (mostly fuel and variable) costs. Either actual or projected energy savings are credited with a lost margin recovery amount. This makes lost margin adjustments a form of financial inducement, based on shared savings, because the utility is able to recover the cost of energy efficiency, plus some amount that represents the margin it would have expected to recover on energy sales. The amount of sharing between shareholders and customers would depend on the program's details, particular facts, and regulatory assumptions. These specific factors can cause lost margin incentives to approach levels consistent with more direct financial incentives or fall to the more neutral incentives associated with lost sales adjustments.

There are a variety of approaches used. This chapter reviews some specific forms drawing from a non-scientific sample.

California

California has been a leader in revenue decoupling since 1982, when Pacific Gas & Electric (PG&E) first implemented a revenue decoupling mechanism. Restructuring in California's electric industry effectively ended decoupling mechanisms for the three investor-owned utilities (IOUs) in the state. However, in April of 2001, after California's 2000-2001 electric market meltdown, the California Public Utilities Code was revised to reinstate a clause

---

[1] This chapter had major assistance from Colin Long, Mark Lowry, and Larry Kaufmann.

on revenue decoupling: "The Commission shall ensure that errors in estimates of demand elasticity or sales do not result in material over or undercollections of the electrical corporations" (Section 739.10). Consequently, the three IOUs have once again instituted various decoupling mechanisms.

(1) PG&E first decoupled electric sales in 1982. In the 1982 decision establishing its electric revenue adjustment mechanism (ERAM), the California Public Utilities Commission (CPUC) reasoned that it was "especially difficult in this period to make accurate sales estimate because the state of the economy and the inability to accurately quantify the effects of conservation which we are expecting our utilities to promote even more vigorously in the future." Thus, PG&E's ERAM was designed to compare monthly authorized base rate revenues to the monthly recorded base rate revenues and adjust for any discrepancy between the two.[2] This particular ERAM was terminated effective December 31, 1997 as part of California's electric industry restructuring.

In 2002, PG&E implemented a complementary utility generation balancing account (UGBA) to adjust for generation revenue.[3] The current ERAM was established for 2004 in Decision 04-05-055, when PG&E switched from rate indexing without decoupling to a revenue decoupling mechanism. The CPUC renewed the plan in March 2007 for the 2007-2010 regulatory period.[4]

The rate adjustment mechanisms function in the following manner: The authorized base margin is equal to the previous year's authorized base margin * (1 + forecast percentage change in the Consumer Price Index [CPI]-All Consumers). The forecast percentage change is the October Global Insight CPI for the upcoming year divided by the forecast for the current year, minus one. Notwithstanding the forecasted CPI change, the CPUC decision provides the maximum and minimum authorized adjustments relative to the previous year's authorized base margin.

(2) Southern California Edison's (SCE) electric revenue adjustment mechanism (ERAM) came somewhat later and was originally approved in late 1982 in Decision 82-12-055 (50 PUR4th 317). It was intended to recover from or return to customers the effect of conservation and weather conditions on earnings. With a few minor modifications, SCE's ERAM remained in place until 1994, when the CPUC rejected a petition for its renewal.

---

[2] Decision 93887, 7 CPUC2d 349 (1981).

[3] Decision 02-04-016.

[4] Opinion Authorizing Pacific Gas & Electric's General Rate Case Revenue Requirement for 2007-2010, Public Utilities Commission of the State of California, Decision 07-03-044, March 15, 2007.

In 2002, SCE implemented performance-based regulation for its distribution revenue requirements to add an adjustment for revenue variation.[5] In 2003, SCE expanded this to include a base revenue requirement balancing account (BRRBA) to make up for earnings attrition in the years 2004 and 2005.[6] The CPUC renewed the mechanism in 2006 for the regulatory period 2006-2008.[7] The adjustment mechanism works in the following manner: Authorized base margin equals the previous year's authorized base margin * (1 + weighted average of escalation factors for labor, non-labor, and health care costs). The labor and non-labor escalation factors are published indexes that SCE cannot control.

(3) San Diego Gas & Electric (SDG&E) electric revenue adjustment mechanism (ERAM) was first authorized by the commission in late 1981.[8] The ERAM remained in place, unchanged, until it was terminated in 1997, as part of the electric industry restructuring in California. The CPUC reinstated the ERAM in 2005 in SDG&E's most recent general rate case. The ERAM is known as a post-test year ratemaking (PTYR) plan.[9] The plan ran from January 1, 2005 through December 31, 2007. The plan operates as follows: the authorized base margin equals the previous year's authorized base margin * (1 + forecast percentage change in the CPI-All Consumers). The forecast percentage change is the October Global Insight CPI for the upcoming year divided by the forecast for the current year, minus one. There was no true-up for forecast and actual CPI, either retroactively or prospectively, for the period through 2007.

Notwithstanding the forecasted CPI change, the maximum and minimum authorized adjustments relative to the previous year's authorized base margins were restricted to between 3.8 percent and 4.8 percent in 2007.

New York

New York has also had a long and varied experience with decoupling mechanisms. In 2007, New York state ordered its electric and gas utilities to eliminate disincentives for energy efficiency related to the recovery of fixed costs via volumetric rates for "pipes" and "wires"

---

[5] Decision 02-04-055.

[6] Decision 04-07-022.

[7] Opinion on Base Rate Revenue Requirement and Other Phase 1 Issues, Decision 04-07-022, July 16, 2004; Opinion on Southern California Edison Company's Test Year 2006 General Rate Increase Request, Decision 06-05-016, May 11, 2006.

[8] Decision 93892, 7 CPUC2d 584 (1981).

[9] Decision on Southern California Gas Company and San Diego Gas & Electric Company's Phase 2 Post-Test Year 2004 Ratemaking, Earnings Sharing, Incentive Proposals, and 2004 Incentive Proposals, Decision 05-03-023, March 17, 2005.

companies. This introduced revenue decoupling mechanisms that "true-up forecast and actual delivery service revenues."[10]

(1) Consolidated Edison Co. of New York, as a vertically integrated utility, had a revenue adjustment mechanism (RAM) in place from 1993 to 1997. The revenue effect of this mechanism was minimal. In 1993, a revenue deficit of less than three percent was collected from customers.[11] From 1994 to 1997, over-collections were returned to customers, with less than a one percent effect. Nevertheless, the New York Public Service Commission (NYPSC) eliminated this decoupling mechanism in 1997. Coincidentally, the company's energy saving investments also were substantially decreased.

(2) New York State Electric & Gas (NYSEG) first implemented a revenue decoupling mechanism (RDM) on August 31, 1993 in the settlement for the second stage of its electric rate filing.[12] In the follow-up settlement on August 15, 1994,[13] NYSEG updated its RDM to include a 100 percent true-up for all retail sales, except large general service, with time-of-use metering and competitive alternative industrial service.

The commission terminated NYSEG's RDM with its 1995 "Opinion and Order Concerning Electric Revenues and Rate Design."[14] The "Multi-Year Settlement Agreement Concerning Electric Rates of New York State Electric & Gas Corporation" was replaced in the third year. This reduced the anticipated third year attrition increase. The NYPSC eliminated the RDM to promote price stability in the electricity market, stating that the anticipated "third year rate increase was too high."

(3) Niagara Mohawk Power Corp. (NM) had two decoupling mechanisms in place beginning in 1989: the demand-side management (DSM) investment and revenue adjustment mechanism (DIRAM) to recover DSM-related lost revenues, and the Niagara Mohawk electric revenue adjustment mechanism (NERAM) to reconcile lost margin on electric sales.[15] In 1993, the commission altered the DIRAM to recover indirect DSM program costs – such as labor, promotion, development, and support costs – through base rates on a current basis.[16]

---

[10] Cases 03-E-0640 and 06-G-0746, 256 PUR4th 477 at 479 (2007).

[11] Case 96-G-0548; Opinion 97-1.

[12] Cases 92-E-1084 and 92-E-1085.

[13] Opinion No. 94-19; Cases 94-E-0601 and 94-M-0349, 155 PUR4th 337 (1994).

[14] Opinion 95-17, Cases 94-M-0349 et al., 165 PUR4th 309 (1995).

[15] Opinion No. 89-29, Case 89-E-041, 107 PUR4th 233 (1989).

[16] Opinion 93-3, Cases 92-E-0108 and 92-E-109, 140 PUR4th 481 (1993).

In 1995, the commission approved NM's request to discontinue the NERAM, but allowed the DIRAM to continue. NM proposed replacing the NERAM with a "statistical recoupling" proposal that would shift weather and economic risks to NM, while decoupling recovery of fixed costs from sales.[17] The commission rejected the proposal, calling the statistical modeling "overly complex and subject to gaming."

(4) Rochester Gas & Electric (RG&E) instituted an electric revenue decoupling mechanism (ERAM) in 1993.[18] On September 26, 1996, the commission eliminated RG&E's ERAM.[19] In its opinion terminating the ERAM, the commission reasoned, "the elimination of the FCA and the ERAM should enhance rate predictability and stability, and should encourage management to control costs to the extent the flow-through aspects of these mechanisms will no longer be available."

Washington

Puget Sound Power & Light (Puget) was the only utility in the Pacific Northwest during the 1990s that used a formal revenue decoupling mechanism. The commission approved a periodic rate adjustment mechanism (PRAM) in 1991,[20] and extended it for three more years in September of 1993.[21] The PRAM was applied annually and consisted of two components: (1) "Base" costs were set equal to total base costs divided by the number of customers on Puget's system, which determined the authorized revenue per customer. Puget's authorized revenue changed in response to the number of customers. Adjustments between the authorized and collected revenue were reconciled in an annual periodic rate adjustment proceeding. This mechanism was designed to remove the incentive for Puget to sell additional kWhs on a per customer basis, while recognizing that more customers would mean more sales. (2) "Resource" costs were also recovered in a manner intended to make Puget whole for certain types of expenses related to energy resource or supply-side acquisitions.

In September 1995, the commission ended the PRAM at Puget's request.[22] In deciding whether to terminate the PRAM, the commission evaluated whether the PRAM was: (1) measurable; (2) simple to administer: (3) simple to explain to customers; and (4) an improvement, on balance, over current methods of regulation. The commission stated, "On

---

[17] Cases 94-E-0098, 94-E-0099; 1995 N.Y. PUC Lexis 711.
[18] Opinion No. 93-19.
[19] Opinion No. 96-27.
[20] Docket No. UE-901184-P, 121 PUR4th 477 (1991).
[21] Eleventh Supplemental Order, Docket UE-921262, 147 PUR4th 80 (1993).
[22] Third Supplemental Order, Docket No. UE-950618, 163 PUR4th 604 (1995).

the first count … it is difficult, if not impossible to judge definitely whether the PRAM can be shown to be measurably a success or a failure. … On the second and third principles, the PRAM proceedings did become complex to administer, controversial, and difficult to explain to customers. … On the fourth and final principle, we believe … a decoupling mechanism with a resource cost adjustment mechanism was an awkward marriage."

Maine

Central Maine Power's (CMP) initial decoupling mechanism worked in conjunction with an attrition mechanism in place for Maine electric utilities.[23] The initial term for the plan was March 1, 1991 – February 28, 1994. CMP calculated the monthly ERAM-per-customer amount by allocating total (historical) test year, adjusted non-fuel revenues to each month during the particular test month of the test year, and then dividing the monthly amounts from the test year by actual customers for each month in the test year, to determine the allowed monthly revenue per customer. CMP determined the monthly accruals. CMP calculated the allowed revenue each month and multiplied these times the allowed monthly revenue per costumer for that month, times the actual customers for that month. CMP would determine the difference between the actual base revenue in the month (both billed and unbilled) and the allowed ERAM-per-customer revenue. CMP would accrue the monthly debit or credit differences in an accrued utility revenue account for future reconciliation in determining income.

The annual reconciliation took place at the end of the rate year, when CMP multiplied the annual allowed revenue per customer, times the actual average customers for the prior twelve months, to determine the total allowed revenue for the twelve-month period. CMP then accrued any differences between this total annual allowed revenue and the revenues accrued for the prior twelve months. At the end of the rate year, CMP reflected the balance in the accrued utility revenue account (including any interest accrual) in a surcharge or refund for the next year.

However, on January 12, 1992, after the first year under the ERAM, CMP filed a request to withdraw the rate increase due to the "recent sharp decline in interest rates." Pursuant to CMP's request, the commission terminated the ERAM as of November 30, 1993.[24]

---

[23] *Re: Investigation of Chapter 382 Filing of Central Maine Power Company*, Docket No. 90-085, May 7, 1991.

[24] Dockets Nos. 90-085-A et al., 141 PUR4th 412 (1993).

Maryland

Delmarva Power and Light (Delmarva) and Potomac Electric Power Company (PEPCO) have decoupling plans, called a bill stabilization adjustment (BSA). These were implemented following orders in July 2007.[25] There is a base distribution energy charge for all customers. For each rate class, and for each billing month, the companies multiply average (weather-normalized) monthly revenue per customer (at rates approved in the latest base rate proceeding), times the actual number of customers for the month. This calculation yields a "normalized monthly test year revenue." The companies then compute BSA on a monthly basis. They divide the differences between actual monthly revenue and the normalized monthly test year revenue, plus any applicable true-up amount from previous months. The companies rely on the forecasted kWh sales applicable to the service classification for the second succeeding month. If, in the current month, actual kWh sales exceed normalized levels, the BSA will lead to a downward adjustment in base distribution energy charges. The utility would apply any such credit to customers in the following billing month. If actual kWh sales are lower than normalized levels, the BSA would cause an upward adjustment in the base distribution energy charges that would be applied in the following month.

Idaho

Idaho Power Co.'s (IPC) three-year pilot decoupling program is called a fixed cost adjustment (FCA).[26] The plan was implemented for the period January 1, 2007 through December 31, 2009. The plan adjusts for revenue per customer. This reflects the difference between revenue requirements and corresponding fixed costs per customer for residential service (Schedules 1, 4, and 5) and small general service (Schedule 7) customers. The FCA works identically for each class. The utility multiplies the actual number of customers, times the fixed cost per customer rate (as calculated in IPC's allowed revenue requirement in the general rate case). This product represents the "allowed fixed recovery" amount. IPC then calculates the "actual fixed cost recovered" amount, as being equal to weather-normalized sales for each class, times the fixed cost per kWh rate, which is established in the company's general rate case. The "allowed fixed cost recovered" amount, minus the "actual fixed cost recovered" amount, is the fixed cost adjustment for each class, which can be either positive or negative. The FCA mechanism also incorporates a three percent cap on annual increases, with a carryover of unrecovered deferred costs to subsequent years. IPC files its FCA adjustment request on March 15 of each year. The

---

[25] *Re Delmarva Power and Light Co.*, Case No. 9093, Order No. 81518, 98 MD PSC 288 (2007); *Re Potomac Electric Power Co.*, Case No. 9012, Order No. 81517, 98 MD PSC 228, 258 PUR4th 463 (2007).

[26] Case No. IPC-E-04-15, Order No. 30267, 256 PUR4th 322 (2007).

rate adjustment will occur June 1. The plan is a pilot program and can be discontinued at any time during its initial three-year program at either the request of the commission or the utility.

Massachusetts

Massachusetts very recently mandated that all electric distribution utilities fully decouple their rates, in an effort to encourage energy efficiency measures.[27] State regulators establish the amount of revenue that each electric utility will be authorized to collect to meet its fixed costs and earn a reasonable return on equity. Utilities must return to ratepayers any amounts overcollected, if customers use more electricity than expected. Conversely, the utilities will collect any revenue shortfalls through a surcharge to customers.

Details were still to be worked out at the time of this writing, as utilities are not required to file their decoupling plans until their current rate plans, which will not be until the end of 2011. The department did, however, state its conclusion that a unified approach was necessary, and that rate decoupling "cannot properly be completed in a piecemeal fashion, such as through a stand-alone adjustment to existing rate designs." Performance-based regulation (PBR) plans can also be included to the extent a company can demonstrate that such PBR plans are warranted. Further, the department will consider any change in risk due to decoupling in determining a utility's authorized ROE. The department will also provide companies with the specific opportunity to receive lost base revenue from incremental efficiency programs.

Vermont

The Vermont Public Service Board approved a partial decoupling plan (a modified revenue cap) for Green Mountain Power Corp. at the end of 2006.[28] A memorandum of understanding predetermined the utility's base revenues for each of the program's three years. Changes in base revenues are capped at $1.25 million in 2008 and $1.5 million in 2009. Further, there is a sharing band in place for the utility's earnings. The utility keeps the first 75 basis points above or below the collar; the next 50 basis points above or below the collar are shared equally between ratepayers and the utility; and anything after that is borne by or kept by the ratepayers. Any variances in the costs of committed resources (up to $400,000

---

[27] *Re Investigation by the Department of Public Utilities on its own Motion into Rate Structures that will Promote Efficient Deployment of Demand Resources*, D.P.U. 07-50-A, July 16, 2008.

[28] *Re Petition of Green Mountain Power Corporation for Approval of an Alternative Regulation Plan*, Docket No. 7175, Order entered December 22, 2006.

per quarter) are the responsibility of the ratepayers. Any variances in the cost of market resources (up to $400,000 per quarter) are the responsibility of the utility. Any variances that exceed $400,000 per quarter are also the ratepayers' responsibility.

Summary

California and Connecticut are the only states in which all the investor-owned utilities (IOUs) have either decoupled revenues or must decouple them in the near future. Idaho, Maryland, Vermont and New York have at least one IOU that is currently operating under a decoupling order. There are ten states that have either an open commission docket to consider decoupling, or where an IOU has filed a decoupling proposal.[29] Those states are Colorado,[30] District of Columbia,[31] Delaware,[32] Hawaii,[33] Kansas,[34] Massachusetts,[35] Minnesota,[36] New Hampshire,[37] New Mexico,[38] and Wisconsin.[39]

---

[29] Shirley, Wayne, Jim Lazar, and Frederick Weston. (2008, June 30). "Revenue Decoupling Standards and Criteria: A Report to the Minnesota Public Utilities Commission." The Regulatory Assistance Project.

[30] *In the Matter of the Investigation of Regulatory and Rate Incentives for Gas and Electric Utilities*, Docket No. 081-113 EG.

[31] *See*, for example, PEPCO's Application for Authority to Increase Existing Retail Rates and Charges for Electric Distribution Service, Formal Case No. 1053.

[32] *In the Matter of the Investigation of the PSC into Revenue Decoupling Mechanisms for Potential Adoption and Implementation by Electric and Natural Gas Utilities*, PSC Regulation Docket No. 59 and PSC Docket No. 07-28.

[33] *In the Matter of Hawaiian Electric Company for Approval and/or Modification of Demand-Side and Load Management Programs and Recovery of Program Costs and DSM Utility Incentives*, Docket No. 05-0069.

[34] KCC Energy Efficiency, Docket 08-GIMX-441-GIV.

[35] *Re Investigation by the Department of Public Utilities on its own Motion into Rate Structures that will Promote Efficient Deployment of Demand Resources*, D.P.U. 07-50-A, July 16, 2008.

[36] Minnesota Statutes, Section 216B.2412 requires the Public Utilities Commission to establish "criteria and standards" so that the state's regulated utilities could implement decoupling. The statute also authorized the PUC to permit one or more of the utilities to participate in a pilot energy efficiency and conservation program.

[37] New Hampshire Public Utilities Commission, Docket Number DE 07-064.

[38] Public Service Company of New Mexico (PSNM) filed a request for approval of a decoupling mechanism, which was denied by the NMPRC. PSNM has appealed that decision to the state supreme court.

[39] The Governor's Global Task Force is considering a broad range of actions and the Public Service Commission of Wisconsin has ordered that decoupling proposals be considered within the context of the this task force.

CHAPTER 22
DIRECT FINANCIAL INCENTIVES FOR DEMAND RESPONSE PROGRAMS

The second regulatory approach to encourage electric utility-sponsored energy efficiency and load management is the "direct income" incentive. This regulatory mechanism is more focused on energy efficiency spending and/or performance. As with decoupling, there are a variety of mechanisms that regulators have implemented. Table 22-1 summarizes these various incentive plans that regulators have either approved or considered over the past decade and a half.

**TABLE 22-1
SOME REGULATORY APPROACHES TO ADD PROFITS AND
INCENTIVES FOR ENERGY EFFICIENCY**

Replace rate base earnings with retail sales margins for energy efficiency services, particularly in retail choice states.

Add virtual rate base that replaces a fraction of avoided supply-side costs.

Share the cost savings between customers and shareholders equal to the difference between supply-side and energy efficiency costs (program benefits).

Adjust ROE and/or net utility income based on the utility achieving targets for energy efficiency.

Unbundle supply-side energy and energy efficiency and either sell new energy services with a cost markup or permit customers to sell energy efficiency to the utility.

Add performance-based "management fees" based on a percentage of total program costs and performance.

Add incentives based on milestones (spending levels, program activities levels, and/or measured market effects).

There are currently 27 states (see Table 22-2) that have used or are authorized to use this more direct form of regulatory incentive to sustain electric utility-sponsored energy efficiency programs.[1]

TABLE 22-2
PERFORMANCE INCENTIVES (EXCLUDING DECOUPLING MECHANISMS) FOR STATES THAT HAVE CURRENT SHAREHOLDER INCENTIVE MECHANISMS FOR ENERGY EFFICIENCY/DSM OR REPORTED EARNING INCENTIVES FOR ENERGY EFFICIENCY 1992-1996 ON EIA FORM 861

State	1992	1993	1994	1995	1996	1997	1998	1999	2000	2001	2002	2003	2004	2005	2006	2007
Arizona	1	1	1	1	1	0	0	0	0	0	0	0	0	1	1	1
California	1	1	1	1	1	1	1	1	1	1	1	1	1	1	1	1
Colorado	0	1	1	1	1	1	1	1	1	0	0	0	0	0	0	1
Connecticut	1	1	1	1	1	1	1	1	1	1	1	1	1	1	1	1
Georgia	0	0	1	1	0	0	0	0	0	0	0	0	0	0	0	1
Hawaii	0	0	0	0	1	1	1	1	1	1	1	1	1	1	1	1
Illinois	1	1	1	1	1	0	0	0	0	0	0	0	0	0	0	0
Indiana	1	1	1	1	0	0	0	0	0	0	0	0	0	0	0	1
Iowa	1	1	1	1	1	0	0	0	0	0	0	0	0	0	0	0
Kansas	0	0	0	0	0	0	0	0	0	0	0	0	0	0	0	1
Kentucky	0	0	0	0	0	0	0	0	0	0	1	1	1	1	1	1
Maryland	0	0	1	1	1	0	0	0	0	0	0	0	0	0	0	0
Massachussetts	1	1	1	1	1	1	1	1	1	1	1	1	1	1	1	1
Minnesota	1	1	1	1	1	1	1	1	1	1	1	1	1	1	1	1
Montana	0	0	0	0	0	0	0	0	0	0	0	1	1	1	1	1
Nevada	0	0	0	0	0	0	0	0	0	1	1	1	1	1	1	1
New Hampshire	1	1	1	1	1	1	1	1	1	1	1	1	1	1	1	1
New Jersey	1	1	1	0	0	0	0	0	0	0	0	0	0	0	0	0
New York	1	1	1	1	1	1	0	0	0	0	0	0	0	0	0	0
Ohio	1	1	1	1	0	0	0	0	0	0	0	0	0	0	0	1
Rhode Island	1	1	1	1	1	1	1	1	1	1	1	1	1	1	1	1
South Carolina	0	0	0	0	0	0	0	0	0	0	0	0	0	0	0	1
Texas	0	0	0	0	0	0	0	0	0	0	0	0	0	0	0	0
Vermont	0	0	0	0	0	0	0	0	1	1	1	1	1	1	1	1
Virginia	0	0	0	0	0	0	0	0	0	0	0	0	0	0	0	0
Washington	1	1	1	1	1	0	0	0	0	0	0	0	0	0	0	0
Wisconsin	1	1	1	1	1	1	1	1	1	1	1	1	1	1	1	1

1 = Shareholder/utility performance incentive in place
0 = No shareholder/utility performance incentive in place

---

[1] The references for the source materials used to create this table are found in the footnotes in Appendix A.

Increasingly, regulators are considering similar incentives to complement cost recovery in their efforts to expand the use of energy efficiency. Some states couple these renewed efforts with integrated resource planning. This is particularly likely if a state finds a need to expand electric generating capacity and marginal electricity costs are increasing.

Table 22-3 is a current (2007) snapshot of the regulatory financial incentives for various utility-sponsored energy efficiency programs in the United States. There is also a column that designates the current status of restructuring in the particular state. Most states with electric utility energy efficiency programs have some explicit form of cost recovery. Nineteen of the 35 shown have performance or financial incentives. Five have broad revenue decoupling. Six have lost margin adjustments tied to energy efficiency. Recently, performance incentives and decoupling mechanisms have been increasing and these numbers will increase.[2]

In addition, the experience around the nation is that when regulators and investor-owned utilities expand energy efficiency, there is often an increase in similar programs for the governmentally and customer-owned utility companies in these same states. This is rather like the way public school systems respond with new programs and, perhaps, efforts when private or charter schools compete for traditional public school students. Regardless of the reason, different institutions offering similar services, with quite different ownership and governance structures, often become yardsticks to determine what services others will offer, as well as how their performance will be measured.

TABLE 22-3
STATE SURVEY OF ECONOMIC INCENTIVES FOR ENERGY EFFICIENCY
(Electric Utilities Only)
As of December 2007

STATE	RESTRUCTURING	COST RECOVERY	PERFORMANCE INCENTIVE	REVENUE DECOUPLING*	LOST MARGIN ADJUSTMENT*
Arizona	Suspended	Yes	Yes	No	No
California	Suspended	Yes	Yes	Yes	No
Colorado	No	Yes	Yes	Pending	No
Connecticut	Yes	Yes	Yes	No	Yes
Delaware	Yes	Yes	No	Pending	No
District of Columbia	Yes	Yes	No	Pending	No
Florida	No	Yes	No	No	No

(Continued)

---

[2] Appendix A provides additional source materials and details that were used to compile this table.

TABLE 22-3 (CONT'D)
STATE SURVEY OF ECONOMIC INCENTIVES FOR ENERGY EFFICIENCY
(Electric Utilities Only)
As of December 2007

STATE	RESTUCTURING	COST RECOVERY	PERFORMANCE INCENTIVE	REVENUE DECOUPLING*	LOST MARGIN ADJUSTMENT*
Georgia	No	Yes	Yes	No	No
Hawaii	No	No	Yes	Pending	No
Idaho	No	Yes	No	Yes	No
Illinois	Yes	Yes	No	No	No
Indiana	No	Yes	Yes	No	Yes
Iowa	No	Yes	No	No	No
Kansas	No	No	Yes	No	No
Kentucky	No	Yes	Yes	No	Yes
Maine	Yes	Yes	No	No	No
Maryland	Yes	No	No	Pending	No
Massachusetts	Yes	Yes	Yes	Pending	Yes
Minnesota	No	Yes	Yes	Yes	No
Montana	Suspended	Yes	Yes	No	No
Nevada	Partially Repealed	Yes	Yes	No	No
New Hampshire	Yes	Yes	Yes	Pending	No
New Jersey	Yes	Yes	No	Pending	No
New Mexico	Suspended	Yes	No	No	No
New York	Yes	Yes	No	Yes	No
North Carolina	No	No	No	No	No
Ohio	Yes	Yes	Yes	No	Yes
Oregon	Suspended	Yes	No	No	No
Rhode Island	Yes	Yes	Yes	Yes	No
South Carolina	No	No	Yes	No	No
Texas	Yes	Yes	No	No	No
Utah	No	Yes	No	No	No
Vermont	No	Yes	Yes	No	Yes
Washington	No	Yes	No	No	No
Wisconsin	No	Yes	Yes	Pending	No

* These are terms currently being used to distinguish between lost sales revenue generally and lost margins related specifically to energy efficiency.

Pacific Economics Group, LLC compiled the information contained in this table from data sourced from: the findings reported in the National Action Plan for Energy Efficiency report entitled "Aligning Utility Incentives with Investment in Energy Efficiency" (November 2007); "Aligning Utility Interests with Energy Efficiency Objectives: A Review of Recent Efforts at Decoupling and Perfomance Incentives" (Kushler, York, and Witte) (October 2006); and the Energy Information Administration's "Status of Electricity Restructuring by State" (April 2007). For additional information, contact Colin M. Long at cmlong@earthlink.net.

The next chapter analyzes the effectiveness of these various state regulatory incentives for the nation's two hundred largest electricity companies over the period from 1992 to 2006. This sample includes government and customer-owned utilities, as well as investor-owned utility (IOU) companies.

APPENDIX A
DETAILS OF THE CURRENT SURVEY OF
REGULATORY INCENTIVES FOR ENERGY EFFICIENCY[3]

---

[3] Colin M. Long prepared this Appendix with the assistance of Charles J. Cicchetti.

STATE SURVEY OF ENERGY EFFICIENCY INCENTIVES AND/OR DECOUPLING (DECEMBER 2007)

STATE	COST RECOVERY	PERFORMANCE INCENTIVES	REVENUE DECOUPLING	LOST MARGIN ADJUSTMENT
Arizona 2006/2007 Traditional COS Restructuring suspended in 2004. No retail choice.[13]	Electric Rate Cases;[1] System Benefits Charge[14]	Capped at 10% of APS' electric energy efficiency program DSM budget ($16 million annual budget).[17,9]	No.	No.
California 1990/2007 Traditional COS until 4/1/98 Restructuring and retail choice suspended October 2001.[13]	Electric and natural gas "system benefits" or "public goods" charge; additional funding through rates[1] Systems benefits charge and procurement payment to fund Demand-Side Management programs.[5]	Non-bypassable public goods charges are recovered immediately from ratepayers. In 1990, the Commission adopted experimental incentive sharing mechanisms for non-low income energy efficiency programs and low-income energy efficiency (LIEE programs). For non-low income energy efficiency programs, the shared savings rate was set at 30% of the effective earnings rate of avoided or deferred comparable supply-side alternatives. In 1997, the CPUC, for non-low income programs, replaced the shared-savings incentive mechanism with milestone incentives based on spending levels, program activities levels, and measuring market effects, with incentives capped at 7% of energy budgets. The milestone incentive programs were discontinued in program year 2002. Also in 1990, the CPUC began experimenting with LIEE incentive mechanisms, similar to management fee incentives. These incentives have been modified over time and were set at 2% of total LIEE program expenditures.	Yes for both electric and natural gas utilities. Designed to remove the disincentives for utilities to promote energy efficiency and conservation.[2] Utilities are ensured that they retain their expected earnings even as energy sales are reduced through EE and conservation.[3] Decoupling uses balancing accounts to collect forecasted revenue and uses an annual true-up.[5]	No.
California 2007		In its 2007 Order, the CPUC established a revised system of incentives and penalties to encourage IOUs to exceed state's energy savings goals and to align utility needs with state's environmental values. Earnings begin at 9% sharing rate, if utility meets 85% of the PUC's savings goals. If 100% is achieved, earnings increase from 9% to 12%. Each earnings rate is "shared savings" percentage. If savings falls to 65% or lower of goals, penalties accrue. Max reward/penalty is $450 million per year.[4]	Yes.	No.

(Continued)

STATE SURVEY OF ENERGY EFFICIENCY INCENTIVES AND/OR DECOUPLING (DECEMBER 2007) (CONT'D)

STATE	COST RECOVERY	PERFORMANCE INCENTIVES	REVENUE DECOUPLING	LOST MARGIN ADJUSTMENT
Colorado 2007 Traditional COS No retail choice.[13]	Electric rate cases.[1]	Yes.[14]	Proposal to implement a partial decoupling rate adjustment (PDRA) to reflect the non-weather related effect in the change in average actual use per customer from the previous year.	No.
Connecticut 2006 Restructured market with retail choice.[13]	Electric System Benefits Charge (3 mills/kWh) with an annual budget of $62 million.[10] Electric distribution companies recover lost revenues if earnings fall below their allowed ROR for six months. The DPUC has introduced a lost-revenue recovery mechanism for new conservation and load management, electric load response and distributed generation initiatives.[1]	Yes. Based on savings and other program goals, 5% incentive of total program costs for achieving 100% of program goals, capped at 8% of program costs (management fee).[1,9] These are called "Performance Management Fees" and are tied to performance goals established by the regulator, including energy savings, demand savings, and others. Incentives are earned for outcomes from 70–130% of savings goals.[5,7]	No.[14] But House Bill 7432 requires electric and natural gas utilities to decouple in their next rate cases."	Yes.[14]
Delaware 2007 Restructured market.[13]	Electric rate case.[14]	No.	Pending.	No.
District of Columbia 2007 Restructured market.[13]	Electric rate case.[14]	No.	Pending.	No.
Florida 2006/2007 Traditional COS Florida Energy 2020 Report issued 12/01 recommended wholesale competiton by 2007, but no action taken and no retail choice.[13]	Electric rate or tariff rider/surcharge.[1]	No.[1]	No.	No.
Georgia 2007 Traditional COS. No retail choice.[13]	In rate case.[14]	Yes. Utility receives 15% of the net benefits of the Power Credit Single Family Home program.[14]	No.	No.
Hawaii 2007 Traditional COS. No retail choice.[13]	No.[14]	Yes. To encourage energy efficiency, and recover revenue from lost revenue resulting from successful DSM programs, utilities are permitted to retain up to 5% of the net system benefits resulting from the DSM program. The utility must meet all four energy efficency goals to be eligible for incentives, which are capped at $4 million.[14,15,16]	Pending for electric.[14]	No.

STATE SURVEY OF ENERGY EFFICIENCY INCENTIVES AND/OR DECOUPLING (DECEMBER 2007) (CONT'D)

STATE	COST RECOVERY	PERFORMANCE INCENTIVES	REVENUE DECOUPLING	LOST MARGIN ADJUSTMENT
Idaho 2006/2007 Traditional COS. No retail choice.[13]	Electric rate or tariff rider/surcharge.[1]	No.[1]	Fixed Cost Adjustment, where the number of customers is multiplied by the fixed cost per customer (as determined in a general rate case), to determine the "allowed fixed cost" recovery. The difference between the actual fixed cost and the allowed fixed cost is the FCA.[3]	No.
Illinois 2006/2007 Restructured market with retail choice. ICC approved phase-in of utility rate hikes. Retail choice.[13]	Small-scale electric energy efficiency programs supported by assessment on electric utilities.[1] Electric rate case.[14]	No. Energy efficiency program administered by Dept of Commerce and Economic Opportunity (DCEO).[1]	No.[14]	No.
Indiana 2007 Traditional COS. No retail choice.[13]	In rate case.[14]	Southern Indiana Gas & Electric permitted to earn up to 2% added ROE on its DSM investments if performance targets are met. A 1% penalty applies if performance targets are not met.[14] PSI can defer until next rate case up to 20% shared savings associated with its DSM plan.[15] Duke Energy has proposed (12/07) an Energy Efficiency Program (Save-a-Watt) where traditional cost-of-service regulation will be used to establish the avoided costs from which utility earnings (up to 90% of the verified avoided costs) are derived.	None for electric utilities.[14]	Yes.[14]
Iowa 2006/2007 Traditional COS. No retail choice.[13]	Yes,[1] rate case and tariff rider/surcharge.[14]	No.[1]	No.	No.
Kansas 2007 Traditional COS. No retail choice.[13]	No.[14]	KCC can authorize up to a 200-basis-point premium in investments associated with renewables, conservation, or energy efficiency.[14,15]	No.	No.
Kentucky 2007 Traditional COS[13]	Yes.[14]	Yes.[14]	No.	Yes.[14]

(Continued)

STATE SURVEY OF ENERGY EFFICIENCY INCENTIVES AND/OR DECOUPLING (DECEMBER 2007) (CONT'D)

STATE	COST RECOVERY	PERFORMANCE INCENTIVES	REVENUE DECOUPLING	LOST MARGIN ADJUSTMENT
Maine 2006/2007 Restructured with retail choice.[13]	Public benefits assessment1 of 1.45 mills per kWh with an annual budget of $10.6 million. LD 1851 turns assessment into a floor.[10]	No. The Maine PUC administers the electric energy efficiency program.[1] LD 1851 created the Maine Energy Conservation Board.[10]	No.[14]	No.
Maryland 2007 Restructured with retail choice.[13]	No.	No.	Pending new rate mechanism allows utilities to increase power distribution rates to make up lost revenues if electricity demand drops due to EE and conservation.[6]	No.
Massachusetts 2006/2007 Restructured with retail choice.[13]	Electric System Benefits Charge (appx 2.5 mills/kWh)[1] to fund demand-side management programs5 with an annual budget of $120 million.[10]	Incentive set at 8.25% of total program costs for achieving 100% of target goals, capped at 9% of electric energy efficiency expenditures before taxes (5.5% after taxes).[9] Multi-factor performance targets: savings, value, and performance.[1] Performance levels bounded from 75% to 110% of design level performance.[7]	Massachusetts recently ordered a straw proposal for a base revenue adjustment mechanism that renders electric and gas companies' revenue levels immune to changes in sales between rate proceedings; DPU Docket No. 07-50.[10]	Yes.[14]
Minnesota 2006/2007 Traditional COS with no retail choice.[13]	Electric and natural gas cases based on legislative mandate.[1] Spending set at 1.5% of gross operating revenues for electric utilities and 0.5% for natural gas utilities.[5]	Incentive set at 3.3% of program costs for achieving performance goals, capped at 30% of program costs for reaching 150% of program energy savings goals (Shared Savings Financial Incentive).[1,9] Utilities also can earn a reward (recently 1.45%) for delivering programs more cost-effectively.[1] Rewards begin at 90% of performance levels.[5] Net benefits are calculated by subtracting program costs from avoided costs.	NSP proposed a decoupling mechanism to reflect non-weather-related effects of changes in actual use per customer. Opponents have suggested a generic docket on decoupling.[2] Minnesota PUC will establish criteria for pilot decoupling programs for both electric and gas utilities.[8]	No.
Montana 2006 Restructuring suspended. Limited retail choice for small and medium-sized customer.[13]	Electric System Benefits Charge. Natural gas through rate cases.[1]	Two percent additional ROE possible on capitalized demand response programs.[14]	No.[1]	No.
Nevada 2006 Restructuring statute revised in part and repealed in part July 2001. Some large customers permitted to purchase directly from IOU competitors. No other retail choice.[13]	Electric rate cases.[1] Utility required to track costs. For Commission-approved plan, utility may recover labor, overhead, materials, incentives paid to customers, marketing and evaluation costs.[7]	Additional 5% return on equity for energy efficiency investments based on program spending goals. Base ROE is 10.25%, which means utilities can earn up to 15.25% on approved DSM costs.[1,5]	No for electric.[1]	No.

STATE SURVEY OF ENERGY EFFICIENCY INCENTIVES AND/OR DECOUPLING (DECEMBER 2007) (CONT'D)

STATE	COST RECOVERY	PERFORMANCE INCENTIVES	REVENUE DECOUPLING	LOST MARGIN ADJUSTMENT
New Hampshire 2006/2007 Restructured with retail choice.[13]	Electric System Benefits Charge (3 mills/kWh)[1] with an annual budget of $16.5 million.[10]	Incentive of 8% of program costs for achieving 100% of targeted program savings. Capped at 12% of program budgets for exceeding savings and cost-effectiveness goals.[1,9] Specific target incentives are set for residential and commercial/industrial sectors.[1]	Pending PUC Docket No. DE 07-064 is investigating ratemaking alternatives including decoupling.[11]	No.
New Jersey 2006 Restructured with retail choice.[13]	Electric System Benefits Charge.[1]	No.	Pending.[14]	No.
New Mexico 2006 Restructuring suspended with repeal of Restructuring Act of 1999 (April 2003). No retail choice.[13]	Recently enacted law requires utility DSM with costs recovered through rate cases.[1]	No.[1]	No. New statute for electric and natural gas requires removing disincentives.[1] Southwestern Public Service has filed for decoupling to recover program costs, net lost revenues, and 15% of net benefits as a proxy to replace lost profit projections.[12]	No.
New York 2006/2007 Restructured with retail choice.[13]	Electric System Benefits Charge[1] to fund demand-side management programs.[5]	New York State Energy Research & Development Authority (NYSERDA) administers the electric energy efficiency programs.[1]	Yes.[14] In 2007, New York ordered its electric and gas utilities to eliminate disincentives caused by the recovery of fixed costs via volumetric rates and to propose revenue decoupling mechanisms.	No.
North Carolina 2006/2007 Traditional COS with no retail choice.[13]		Duke Energy has proposed an Energy Efficiency Program (Save-a-Watt) where traditional cost-of-service regulation will be used to establish the avoided costs from which utility earnings (up to 90% of the verified avoided costs) are derived.	No.[14]	No.
Ohio 2006/2007 Restructured with retail choice.[13]	Electric rate rider.[1]	Yes.[14]	None for electric.[14]	Yes.[14]
Oregon 2006/2007 Restructuring suspended. Large commercial and industrial customers have direct access to wholesale markets, but no retail choice for residential customers.[13]	Electric and natural gas system benefits charge.[1] Also allows capitalization of costs.[5]	No. The Energy Trust of Oregon administers programs.[1]	None for electric.[14]	No.

(Continued)

STATE SURVEY OF ENERGY EFFICIENCY INCENTIVES AND/OR DECOUPLING (DECEMBER 2007) (CONT'D)

STATE	COST RECOVERY	PERFORMANCE INCENTIVES	REVENUE DECOUPLING	LOST MARGIN ADJUSTMENT
Rhode Island 2006/2007 Restructured with retail choice.[13]	Electric System Benefits Charge[1] of 2.0 mills per kWh and an annual budget of $21.7 million.[10]	There are two components to the sharing mechanism: (1) five performance-based metrics for specific achievements; and (2) kWh savings targets by sector. For kWh savings, target incentive is 4.4% of eligible budget for achieving 100% of goals. Incentives are capped at 5.3%.[9] For energy savings by sector, the target incentive is 60% of the savings goal. Utility can earn additional incentive on savings up to 125% of target savings.[15]	Yes.[14]	No.
South Carolina 2007 Traditional COS.[13]	In rates.[14]	Yes. Duke Energy and South Carolina E&G are allowed to recover in rates DSM costs, including incentives.[14] Duke has proposed (12/07) an Energy Efficiency Program (Save-a-Watt) where traditional cost-of-service regulation will be used to establish the avoided costs from which utility earnings (up to 90% of the verified avoided costs) are derived.	No.[1]	No.
Texas 2006 Restructured with retail choice.[13]	Yes.[1]	No.[1]	No.[1]	No.
Utah 2006/2007 Traditional COS with no retail choice.[13]	Electric rate or tariff rider/surcharge.[1]	No.[1]	No.[14]	No.
Vermont 2006/2007 Traditional COS with no retail choice.[13]	Electric System Benefits Charge[1] of 2.5 mills per kWh with an annual budget of $17.5 million. S2903 and H8025 establish natural gas efficiency fund at 1.5 cents per therm.[10]	Three major types of incentives: (1) program results incentives; (2) market effects incentives; and (3) activity milestone incentives. Maximum amount of incentives is about 2% of total contract award for the period[1] or $795,000 over 3 years,[7] with 5% incentive for achieving 100% of program goals and capped at 5.5%.[9]	None, but a forecast revenue cap and costs with a balancing account and true-ups was proposed.[5]	Yes.[14]
Washington 2006/2007 Traditional COS with no retail choice.[13]	Electric rate or tariff rider/surcharge.[1] Systems benefit charge to fund demand-side management programs.[5]	No.[1]	None for electric.[14]	No.

STATE SURVEY OF ENERGY EFFICIENCY INCENTIVES AND/OR DECOUPLING (DECEMBER 2007) (CONT'D)

STATE	COST RECOVERY	PERFORMANCE INCENTIVES	REVENUE DECOUPLING	LOST MARGIN ADJUSTMENT
Wisconsin 2006/2007 Traditional COS with no retail choice.[13]	Electric System Benefits Charge and additional funding through rate cases.[1]	Department of Administration administers most programs. Utilities have the option of administering. Alliant was allowed to earn its rate of return on a "shared savings" energy efficiency program.[1] Incentive is set at 12.7% of program costs for achieving 100% of program goals.[9]	Pending for electric.[14]	No.

[1] Kushler, Martin, Dan York, and Patti Witte. (2006, October). "Aligning Utility Interests With Energy Efficiency Objectives: A Review of Recent Efforts at Decoupling and Performance Incentives." American Council for an Energy-Efficient Economy.

[2] American Gas Association. (2007, April). Update on Revenue Decoupling Mechanisms.

[3] Pacific Economics Group. (2007, August). Summary of Revenue Decoupling Mechanisms. Report. (See Appendix D).

[4] California Public Utilities Commission, *Interim Opinion on Phase 1 Issues: Shareholder Risk/Reward Incentive Mechanism for Energy Efficiency Programs*, Decision 07-09-043, R.06-04-010 (September 20, 2007).

[5] U.S. Department of Energy and U.S. Environmental Protection Agency. (2006, July). National Action Plan for Energy Efficiency, pp. 2-12.

[6] Energy Efficiency and Renewable Energy State Activities and Partnerships. (2007, July 25). "Maryland Removes Utility Disincentives for Energy Efficiency."

[7] Shirley, Wayne, Director, The Regulatory Assistance Project. (2006, November 16). "Energy Efficiency & Utility Profits: Do Your Incentives Need Alignment?" Presentation by the Third Annual Southwest Energy Efficiency Workshop, Snowbird, Utah.

[8] Strom, Sheldon. (2007, October 2). "Minnesota's Next Generation Energy Act of 2007." Presentation by the Center for Energy and Environment, presented at ACEEE's Fourth National Conference on "Energy Efficiency as a Resource," Berkeley, California.

[9] Testimony of the Division of Ratepayer Advocates of the California Public Utilities Commission on the Appropriate Shared Savings Rate for a Risk/Return Incentive Mechanism for Energy Efficiency, R.06-04-010 (May 3, 2007).

[10] Krasnow, Sam. (2007, October 2). "Efficiency Legislation in New England." Presentation at ACEEE's Fourth National Conference on "Energy Efficiency as a Resource," Berkeley, California.

[11] Murrow, Derek K. (2007, October 2). "Decoupling Activities in New England." Presentation by (Northeast Utilities) at ACEEE's Fourth National Conference on "Energy Efficiency as a Resource," Berkeley, California.

[12] Shirley, Wayne, Director, The Regulatory Assistance Project. (2007, October 2). "Energy Efficiency Incentives in the Southwest." Presentation at ACEEE's Fourth National Conference on "Energy Efficiency as a Resource," Berkeley, California.

[13] Energy Information Administration. (2007, April). "Status of Electricity Restructuring by State." Department of Energy.

[14] National Action Plan for Energy Efficiency. (2007, November). "Aligning Utility Incentives with Investment in Energy Efficiency: A Resource of the National Action Plan for Energy Efficiency." U.S. Environmental Protection Agency. U.S. Department of Energy and ICF International.

[15] Karl McDermott Testimony (Exhibit A-26) before the Michigan Public Service Commission in Case No. U-13808, June 20, 2003 (http://efile.mpsc.cis.state.mi.us/efile/docs/13808/0050.pdf)

[16] Hawaii Public Utilities Commission, *Re Hawaiian Electric Company, Inc.*, Decision and Order No 23258, Docket No 05-0069 256 PUR4th 1 (2007).

CHAPTER 23
TESTING THE STATISTICAL SIGNIFICANCE OF REGULATORY INCENTIVES FOR ENERGY EFFICIENCY

The Energy Information Administration (EIA) collects detailed annual data for all electric utilities related to the amount of energy efficiency saved and spent each year.[1] This analysis uses the data for the top 200 electric utilities, regardless of ownership or governance, for the years 1992 through 2006. Such data represent a panel or cross section of the nation's largest utilities, moving through time (15 years).

The year 2000 was used to establish the common reference point that defines the largest, in terms of revenue, 200 electric utilities for each year in the panel data. The focus of this analysis is to determine "if" and "to what extent" regulatory adjustments, that reflect concerns for a utility's bottom line, matter for the success of mandated utility-sponsored energy efficiency programs. Several utility size-specific variables are used to normalize utility spending and savings efforts. These include: (1) total annual MWhs sold; (2) summer peak (MW) load; and (3) total number of retail customers. The purpose of these variables is to adjust the results for differences in the relative size of these top 200 electric utilities.

In this analysis, it is fairly common to find that each of these "size"-related independent variables is statistically significant. In reality, they combine to reflect other operating characteristics across utilities and not just their relative size. Other utility characteristics, such as utility differences in system load factor and the amount of energy use per customer, are also reflected in the analysis. Another statistical approach to help put these various statistical tests on a more comparable or normalized basis is to convert non-zero and non-negative variables to logarithmic form. This analysis also tested this functional form. Virtually the same statistical results were found, regardless of the functional form used in the analyses.

The EIA provides two types of variables to reflect energy efficiency savings in MWhs. These are: (1) the incremental MWhs saved due to specific additional savings in a particular year; and (2) the cumulative effect that would also include continuing past savings in a particular calendar year.

---

[1] This information is reported by utilities to the EIA on Form 861. *See www.eia.doe.gov/cneaf/electricitypage/eia861.html.*

The specific analyses initially focuses on the incremental annual amount of savings, because EIA also reports spending for the two energy efficiency categories (energy efficiency and load management) on an incremental basis. When this same incremental annual spending variable is used to test hypotheses based on cumulative savings, virtually the same statistical conclusions are reached.

Two other utility characteristics potentially matter. One qualitative factor is the status of industry restructuring in various states over time.[2] Retail choice or the prospect of it could potentially affect the amount of utility-sponsored energy efficiency. Indeed, several jurisdictions followed California's lead and caused utilities that restructured to collect public benefits charges that would finance energy efficiency. The other qualitative factor is the ownership/governance status of the electric utilities. State regulation comprehensively affects investor-owned utility (IOU) companies. Therefore, state regulators potentially are more likely to mandate IOUs to expand utility-sponsored energy efficiency. However, if a state legislates utility-sponsored energy efficiency, this approach could cause municipal and customer-owned electric utilities to increase energy efficiency as well. This analysis tests the statistical significance for both restructuring and IOU status.

In this sample of the 200 largest electric utilities over fifteen years, there are some electric utilities that, at least in some years, did not have utility-sponsored energy efficiency programs. In this analysis, "not" having a program is relevant. However, a "zero" value for either the quantity of electricity saved or the amount spent could affect the statistical tests and interpretation of the results. In the following analyses, each statistical test was performed and is reported using the full sample (including utilities with "zero effort" years), as well as a partial sample analysis in which the "zero effort" observations are dropped.

These data provide a rich source of information to test various hypotheses, which are particularly relevant given the current regulatory debate "to adopt, or not, utility incentives to advance energy efficiency." There are two pairs of hypotheses tests used in this analysis. The first pair tested is the presence, or not, of "decoupling" or "income incentives." These same factors are also tested using a statistical interaction between the presence of these two effects and the utility's status, or nor, as an IOU.

The specific hypotheses tested are shown in Table 23-1. These are discussed in pairs, because the analyses reported here mostly test for both types of regulatory incentives taken together. Other analyses not reported that combine the two effects reach a similar conclusion, namely the hypotheses that regulatory incentives increase both energy efficiency spending, and the amount of energy saved cannot be reasonably rejected statistically.

---

[2] The EIA tracks and reports the current status of electric industry restructuring by state. *See* EIA, "Status of Electricity Restructuring by State" (April 2007). *www.doe.gov.cneaf/electricity/page/restructuring/restructure_elect.html.*

Chapter 23: Testing the Statistical Significance of Regulatory Incentives for Energy Efficiency

TABLE 23-1
THE INCENTIVE EFFECT HYPOTHESES

Hypothesis I
Energy Efficiency savings are greater in states with direct profit incentives.

Hypothesis II
Energy Efficiency savings are greater in states with revenue decoupling.

Hypothesis III
Expenditures on Energy Efficiency are greater in states with direct profit incentives.

Hypothesis IV
Expenditures on Energy Efficiency are greater in states with revenue decoupling.

Hypothesis V
The proportion of Energy Efficiency saved per corresponding unit of total utility output increases in states with direct profit incentives.

Hypothesis VI
The proportion of Energy Efficiency saved per corresponding unit of total utility output increases in states with revenue decoupling.

Hypothesis VII
Expenditures on Energy Efficiency relative to total utility output increase in states with direct profit incentives.

Hypothesis VIII
Expenditures on Energy Efficiency relative to total utility output increase in states with revenue decoupling.

Hypotheses I and II: Incentives Affect Energy Efficiency Savings

There are various experiments that test the most basic hypotheses to determine if incentives would increase energy efficiency. These tests use different variables to test savings. These include: (1) annual incremental energy savings for energy efficiency; (2) such savings plus energy savings related to load management (TEES); and (3) total capacity savings (TCS) for all programs.

There are two different experiments to determine the role of regulatory incentives. The first test for the statewide presence of regulatory incentives uses a proportionate allocation for multi-state utilities. The second experiment uses a statistical interaction to restrict this test to just the investor-owned utility (IOU) companies in each state with such regulatory incentives.

Table 23-2 shows the results of these hypotheses' tests for energy efficiency (MWh) savings for the full sample in the EIA data. The dependent variables in Table 23-2 are the incremental amounts of energy efficiency savings (EES) and total energy efficiency savings (TEES), including load management. The first few variables normalize these hypotheses' tests for utility size differences. The data reflect an increase in incremental savings over time (MWhs) between 1992 and 2006.

IOUs did not perform much differently than other types of electricity companies in terms of their incremental additions to energy efficiency each year. Restructured status reduces the incremental savings due to demand-side activities. The EE cost variable is jointly determined. This means, quite logically, that the more a utility spends incrementally to save energy, the greater will be the average incremental savings.

The t-statistic is shown in parentheses below each coefficient. The t-statistic reflects the probability of incorrectly accepting the stated hypothesis. For example, a t-statistic of 1.96 for a large sample would mean that there is only a one in twenty chance of incorrectly accepting the stated hypothesis. This is the typical value used in most scientific inquiries. In this analysis, the higher one-in-twenty standard is used on a minimum t-test of 1.96. In many cases, the t-tests for the regulatory incentive variables are much greater than this threshold 1.96 value. The higher the t-test value, the less likely it is that the stated hypothesis will be incorrectly accepted. For example, the chances of incorrectly accepting hypotheses with a $t = 3.0$ would be nearly one in four hundred, while the chances of incorrectly accepting hypotheses with a $t = 5.0$ would be one in more than 1.5 million, and the chances of incorrectly accepting hypotheses with a $t = 10.0$ would be about one in 2.9 trillion.

The primary hypotheses tested are the effect of "decoupling" and "direct financial incentives." Both variables are strongly statistically significant, with t-tests ranging from 3.3 to 9.7. This means that, adjusting for other meaningful factors, such as size and spending, utilities that had decoupling and/or direct financial incentives were strongly statistically more likely to save more energy. Indeed, when these policies are applied to IOUs, as shown in the two columns labeled IOU/State Incentives, the t-tests range from 5.6 to 9.7. This means the chance of incorrectly accepting the hypotheses, of no meaningful effect on energy savings from financial incentives and decoupling, approaches a billion to one.

Chapter 23: Testing the Statistical Significance of Regulatory Incentives for Energy Efficiency

TABLE 23-2
ENERGY EFFICIENCY RESPONDS TO REGULATORY INCENTIVES

	State Incentives		IOU/State Incentives	
	EES	TEES	EES	TEES
Intercept	-8075.9	-10856.5	-6056.1	-4744.2
	(3.6)	(4.6)	(3.2)	(2.1)
Total MWh	-.00094	-.00181	-.00085	-.00155
	(6.0)	(11.0)	(5.6)	(9.4)
Peak MWh	.00100	.00086	.00103	.00084
	(3.7)	(2.9)	(3.9)	(2.9)
Total Customers	.03621	.04939	.03312	.04395
	(10.8)	(14.1)	(9.9)	(12.4)
Trend	556.89	811.61	514.87	738.69
	(2.6)	(3.6)	(2.5)	(2.9)
IOU	2422.6[1]	5942.1	NS	-3339.8[1]
	(1.3)	(2.9)		(1.6)
Restructuring	-19387.8	-12875.8	-18392.3	-11100.8
	(8.0)	(5.0)	(7.8)	(4.4)
EE Cost[1]	5.78919	3.62268	5.6104	3.4606
	(50.3)	(43.7)	(48.0)	(41.5)
Decoupling	16381.7	25344.3	43252.9	60918.8
	(4.1)	(6.0)	(7.3)	(9.7)
Financial Incentives	7488.6	12688.0	16480.4	30021.1
	(3.3)	(5.3)	(5.6)	(9.3)
R^2/Number of Observations	.69	.64	.69	.65
	2961	2961	2961	2961

[1] Marginally significant.

Table 23-3 shows the statistical hypotheses tests for the total incremental annual capacity (MW) savings (TCS) related to utility-sponsored energy efficiency and load management.

TABLE 23-3
TOTAL CAPACITY SAVINGS FROM ENERGY EFFICIENCY RESPOND TO REGULATORY INCENTIVES

	TCS State Incentives	TCS IOU/State Incentives
Intercept	219.97[2] (.14)	178.49[2] (.16)
Total MWh	-.00055 (4.4)	-.00057 (4.6)
Peak MWh	.00029[1] (1.3)	.00030[1] (1.3)
Total Customers	.02837 (10.5)	.02835 (10.5)
Trend	-88.793[2] (.45)	NS
Restructuring	-5396.6 (2.8)	-5626.1 (3.0)
Change in Regulated Price	64539.6[1] (1.8)	75311.4 (2.4)
EE Cost	1.0503 (16.7)	1.03875 (16.5)
Decoupling	NS NS	NS
Financial Incentives	4808.6 (2.6)	7860.9 (3.3)
R^2/Number of Observations	.34 2961	.34 2961

[1] Marginally significant.
[2] Not significant.

In these tests, the percent explained declines and some of the utility size variables are not statistically different than zero. Again, restructuring dampens regulatory efforts to reduce capacity. This is probably due to the fact that utilities that secure electricity in wholesale power markets are less concerned with load management as a policy to achieve.

The most important statistical and policy finding is that direct financial incentives, but not decoupling, have a strong statistically significant effect on incremental capacity savings. This increases when IOU status and direct financial incentives are combined.

Decoupling does not matter statistically as a factor related to capacity savings. This could be because states seeking to save capacity through load management tend to seek to avoid new investments and increase reliability. The lost revenue potential is not particularly important in such cases. Indeed, a new variable representing the change in regulated prices relative to last year is statistically significant for the regulated IOUs, in the sample. This supports the notion that regulators embrace increased load management to reduce regulated price increases through reduced generation investments.

The next phase of these experiments eliminated the utilities that do not have energy efficiency programs in specific years. This experiment is also reported as a linear functional form. Similar results were found using the logarithmic functional form. Table 23-4 shows the same conclusions are reached, when these same statistical tests were focused on just the utility companies with active energy efficiency programs.

TABLE 23-4
RESTRICTING THE ANALYSES TO UTILITIES WITH ACTIVE ENERGY EFFICIENCY PROGRAMS DOES *NOT* CHANGE THE CONCLUSIONS

	State Incentives		IOU/State Incentives	
	EES	TEES	EES	TEES
Intercept	-19063.3	-21659.2	-12557.5	-12335.7
	(4.5)	(4.8)	(3.9)	(3.6)
Total MWh	-.00209	-.00289	-.00186	-.00258
	(6.3)	(8.5)	(5.8)	(7.9)
Peak MWh	.00371	.00351	.00359	.00327
	(5.4)	(4.7)	(5.4)	(4.5)
Total Customers	.06381	.07465	.05783	.06769
	(10.9)	(12.5)	(9.9)	(11.4)
Trend	1468.3	1784.6	1315.1	1550.1
	(3.8)	(4.3)	(3.5)	(3.8)
Restructuring	-37237.3	-26129.4	-35011.9	-24374.2
	(7.4)	(5.0)	(7.2)	(4.8)
IOU	6947.5	10103.2	NS	NS
	(1.9)	(2.6)		
EE Cost[1]	4.9529	3.0696	4.7739	2.9679
	(28.1)	(24.0)	(27.3)	(23.5)
Decoupling	30615.3	35868.5	65402.6	66316.4
	(5.1)	(5.6)	(8.0)	(7.8)
Financial Incentives	21107.0	24668.2	35269.8	42130.6
	(5.3)	(5.9)	(7.6)	(8.6)
R^2/Number of Observations	.72	.66	.73	.67
	1451	1500	1451	1500

[1] Marginally significant.

In particular, there is a positive trend in the reported annual energy savings added over time. Restructuring reduces utility-sponsored energy efficiency. Spending more, a jointly determined factor, and saving more move together. Most important, both decoupling and direct regulatory incentives combined with IOU status are strongly statistically related to increased incremental energy savings due to demand-side programs.

Table 23-5 shows these same statistical tests for annual capacity savings. There is, as previously reported, a noticeable reduction in the percent of variation explained, and several factors are not statistically significant. That said, spending more on energy efficiency is related to saving more incremental capacity (MW), and although direct financial incentives matter, their statistical significance is considerably weaker than for incremental energy savings (MWh).

TABLE 23-5
TOTAL CAPACITY SAVINGS FOR UTILITIES WITH ACTIVE PROGRAMS SHOW SIMILAR RESULTS

	State Incentives	IOU/State Incentives
Intercept	668.70[2]	270.76[2]
	(.31)	(.13)
Total MWh	-.00034[1]	-.00023[2]
	(1.3)	(.86)
Peak MW	.00024[2]	.00021[2]
	(.40)	(.34)
Total Customers	.03411	.03106
	(7.1)	(6.3)
Trend	NS	NS
Restructuring	-6618.5[1]	NS
	(1.5)	
Change in Regulated Price	57018.2[2]	71123.4[1]
	(1.0)	(1.3)
EE Cost	.75882	.71889
	(7.0)	(6.5)
Decoupling	NS	14671.4[1]
		(1.9)
Financial Incentives	5146.7[1]	7818.2[1]
	(1.5)	(1.9)
R^2/Number of Observations	.34	.34
	1423	1423

[1] Marginally significant.
[2] Not Significant

Hypotheses III and IV: Incentives Affect Utility Spending on Energy Efficiency

Table 23-6 shows the results for tests of these hypotheses that utilities would spend more incrementally each year when decoupling and direct financial incentives are authorized.

TABLE 23-6
UTILITIES SPEND MORE ON ENERGY EFFICIENCY WHEN THERE ARE REGULATORY INCENTIVES

	TCAEE	Log (TCAEE)	Log (TCEEE)
Intercept	1894	-3.5972	-5.8269
	(4.6)	(9.4)	(13.1)
Total MWh	0.0000674	.51857	.71380
	(2.3)	(17.0)	(8.7)
Peak kW	0.000184	NS	-.22482
	(3.7)		(3.4)
Total Customers	0.004047	.01553[2]	-.08314
	(6.4)	(.81)	(4.2)
Trend	-263.67	-.03542	-.01940
	(5.9)	(4.2)	(2.4)
Restructuring	750.73[1]	.42831	.72068
	(1.7)	(4.6)	(8.2)
IOU Status	-942.28	NS	NS
	(2.7)		
Change in Regulated Price	14779.1[1]	.25318[2]	-.40890[2]
	(1.8)	(.18)	(.31)
EE Savings (MWh)	.10823	.42180	.63499
	(43.6)	(28.9)	(39.0)
Decoupling	1488.6	.47527	.40590
	(2.0)	(4.3)	(4.0)
Financial Incentives	847.49	.39156	.43241
	(2.0)	(5.2)	(6.1)
R^2/Number of Observations	.67	.61	.74
	2961	1479	1362

[1] Marginally significant
[2] Not significant.

There are two dependent variables tested: (1) the total incremental cost for all energy efficiency (TCAEE) programs, and (2) the incremental cost for energy savings exclusively (TCEEE). In addition, natural logs are also used for the variables that do not have zero or negative values. This is another functional form that helps to normalize the results given the difference in the size of the 200 sample utilities.

The primary difference between the previous analyses of incremental energy efficiency savings and these analyses of incremental expenditures on energy efficiency is the reversal of the effects related to two variables: (1) trend, and (2) restructuring. While the trend in energy efficiency savings showed some increases over the 15 years, the annual expenditures corrected for inflation using the Consumers Price Index (CPI) shows a rather statistically significant decline. In addition, in contrast to savings, states that restructure appear to spend more per year on energy efficiency. This could be related to regulatory mandates that require restructured utilities to collect public benefits charges and to spend more on conservation. In addition, the "higher priced" states have been more likely to restructure. Regulators seem more likely to mandate energy efficiency when there are high prices. Finally, there could be some increase in utility-sponsored energy efficiency in order to help retain retail customers that could be lost to competitors, when restructuring leads to retail choice. Regardless of the reason, incremental energy efficiency and demand response spending appears to be higher in the states undergoing restructuring.

The natural log form shows similar results as the arithmetic form for both total spending and spending focused on energy savings (MWh). This approach permits the analyst to compare the percent change in the dependent and corresponding independent variables. The primary difference in these two functional forms is that expressed as natural logs, incremental utility spending on energy efficiency is strongly statistically related to the use of direct financial incentives and/or decoupling incentives.

There is another potentially important policy insight in the coefficients for the natural logarithm of energy savings (MWh). This coefficient for the utilities that have non-zero programs is less than one, or about .635. This means that the percent change in incremental savings exceed the percent change in the incremental amount spent for energy savings (TCEEE). In other words, a 10 percent increase in energy savings (MWh) would be associated with a 6.35 percent increase in the costs of energy savings (TCEEE). This result suggests scale economies and that learning-by-doing effects are possible as individual utilities ramp up their energy savings programs. This result, while important, should not, however, be over-interpreted, because, as discussed above, the incremental amount spent and the incremental amount saved each year are likely jointly determined variables. Nevertheless, this particular result is potentially important, because, if true, it reinforces the regulatory and policy notions that utilities should use their expertise and customer relations to expand energy efficiency.

Table 23-7 shows these same variables and functional forms for utilities with active programs. The hypotheses tests are for the effect of regulatory incentives coupled with investor-owned utility (IOU) status. The regulatory incentive hypotheses cannot be rejected statistically, and the "scale" type utility spending effects remain. This means that regulatory incentives are likely to increase incremental spending, adjusting for size and other explanatory variables.

TABLE 23-7
UTILITIES WITH ACTIVE ENERGY EFFICIENCY PROGRAMS DEMONSTRATE THAT UTILITIES SPEND MORE WHEN THERE ARE REGULATORY INCENTIVES

	TCAEE	Log (TCAEE)	Log (TCEEE)
Intercept	2293.2	-5.9346	-5.0280
	(5.9)	(12.4)	(10.8)
Total MWh	0.0000758	.74667	.63119
	(2.6)	(8.6)	(7.7)
Peak kW	0.000184	-.03626[2]	-.18444
	(3.7)	(.5)	(2.7)
Total Customers	0.00390	-.05537	-.08293
	(6.1)	(2.5)	(4.0)
Trend	-269.78	-.04957	-.02112
	(6.0)	(5.9)	(2.6)
Restructuring	794.15	.43358	.73921
	(1.8)	(4.8)	(8.4)
IOU Status	-1491.5	-.43923	-.03270[2]
	(4.1)	(5.6)	(.43)
Change in Regulated Price	16075.0	-.93520[2]	-.15023[2]
	(2.0)	(.67)	(.11)
EE Savings (MWh)	.10656	.39052	.63695
	(41.5)	(26.1)	(38.1)
IOU Decoupling	2450.8	.56357	.37441
	(2.2)	(3.9)	(2.7)
IOU Financial Incentives	1999.1	.67067	.49538
	(3.5)	(7.4)	(5.7)
R^2/Number of Observations	.67	.63	.74
	2961	1479	1362

[1] Marginally significant.
[2] Not significant.

Hypotheses V and VI: The Proportion of Energy Savings to Total MWhs Sold Increases with Regulatory Incentives

Table 23-8 shows that, when annual utility savings (TEES) are normalized for utility size measured in total MWhs, the hypotheses tested (regulatory financial incentives would increase utility savings) cannot be statistically rejected. These hypotheses were tested in a very basic form using only peak (kW) demand as a potential size variable. This variable was not statistically significant. Restructuring status also did not matter statistically. A negative trend effect was found. The most important statistical aspect of this experiment is the elimination of the potentially jointly determined expenditure variable. This likely reduces the explanatory power of this equation.

None of these nuances particularly matter, because the focus in Table 23-8 is the hypothesis that regulatory incentives cause utilities to increase the share of incremental energy efficiency relative to their total annual output (MWh). This hypothesis cannot be rejected in either the linear or logarithmic functional forms, or when the tests are performed using the effect for all utilities in states over time with incentives, or just the IOUs in these same states.

TABLE 23-8
THE PROPORTION SAVED INCREASES WITH REGULATORY INCENTIVES

	TEES/Total MWh	Log (TEES/Total MWh) with IOU/State Incentives
Intercept	0.00155	-5.7844
	(10.9)	(9.3)
Peak	NS	-.07434[1]
		(1.7)
Trend	-.0000751	-.04104
	(4.8)	(3.2)
Restructuring	.0000760[2]	-.081123
	(.43)	(5.2)
Decoupling	.00386	1.1898
	(13.6)	(4.7)
Financial Incentives	.00173	1.8163
	(10.6)	(12.5)
R^2/Number of Observations	.13	.13
	2941	1500

[1] Marginally significant.
[2] Not significant.

Hypotheses VII and VIII: Utilities Increase Annual Expenditures When There Are Regulatory Incentives To Do So

Table 23-9 reports the results for these same hypotheses tests for expenditures. In this analysis, the same variables, except for the potentially jointly determined energy savings variables, are included in the equations reported. Two interesting results are the negative trend in the annual change in inflation dollars spent and the higher change in spending, when restructuring is present. These annual change-in-spending-on-energy-efficiency results are similar to the results reported for annual utility expenditures on energy efficiency.

TABLE 23-9
THE ANNUAL CHANGE IN ENERGY EFFICIENCY EXPENDITURES INCREASES WITH REGULATORY INCENTIVES

	If EE Change in Direct Total Cost (DTCEE)	If EE Change in Direct Total Cost (DTCEE) IOU/State Incentives
Intercept	2298.8 (5.0)	2721.6 (6.2)
Total MWh	-.000413 (10.2)	-.000405 (10.1)
Peak kW	.000728 (7.0)	.000753 (7.4)
Total Customers	.01253 (18.6)	.011607 (17.0)
Trend	-389.66 (7.4)	-407.27 (7.9)
Restructuring	1651.2 (2.6)	1691.6 (2.7)
Decoupling	5279.3 (6.3)	7917.0 (6.9)
Financial Incentives	2948.6 (5.7)	5739.2 (9.3)
R^2/Number of Observations	.40 1827	.42 1827

The purpose of this experiment is to focus on the two hypotheses tests. Again, the hypothesis that annual changes in incremental expenditures would respond to regulatory incentives cannot be rejected for either all the utilities in the states with regulatory incentives or just the IOUs in these states over time. Stated more conversationally, there is very strong statistical support for concluding that regulatory financial incentives are correlated with more incremental utility energy savings, as well as the incremental spending to do so.

Conclusion

The nation is increasingly seeking to solve the twin problems of climate change and energy/economic security. The nation's electric industry is increasingly being targeted to play a major role in expanding the nation's commitment to energy efficiency.

This analysis shows that utilities are more likely to save energy and spend money to do so, when they have either direct financial incentives or decoupling, or both. There are likely scale economies involved in energy efficiency information gathering and analyses, installation, and purchasing/marketing. The utility, not its individual customers, seems to be better placed to capture such cost-effective advantages.

This is all causing regulators to take charge and, in various ways, to mandate utility-sponsored energy efficiency. The next chapter takes these various statistical analyses to a second round of testing to determine how the targets, discussed previously (one percent incrementally and 10 percent cumulatively), might respond to these twin types of incentives.

APPENDIX A
LIST OF 200 UTILITIES USED IN ANALYSIS

Rank by 2000 Revenue	Utility ID	Utility Name	Ownership
1	17609	Southern California Edison Co	Investor Owned
2	14328	Pacific Gas & Electric Co	Investor Owned
3	6452	Florida Power & Light Co	Investor Owned
4	44372	TXU Electric Delivery Co	Investor Owned
5	4110	Commonwealth Edison Co	Investor Owned
6	4226	Consolidated Edison Co-NY Inc	Investor Owned
7	8901	CenterPoint Energy Houston Electric, LLC	Investor Owned
8	7140	Georgia Power Co	Investor Owned
9	5416	Duke Energy Carolinas, LLC	Investor Owned
10	5109	Detroit Edison Co	Investor Owned
11	19876	Virginia Electric & Power Co	Investor Owned
12	15477	Public Service Electric & Gas Co	Investor Owned
13	195	Alabama Power Co	Investor Owned
14	13573	Niagara Mohawk Power Corp	Investor Owned
15	3046	Progress Energy Carolinas Inc	Investor Owned
16	4254	Consumers Energy Co	Investor Owned
17	6455	Progress Energy Florida Inc	Investor Owned
18	14354	PacifiCorp	Investor Owned
19	15497	Puerto Rico Electric Power Authority	State
20	11208	City of Los Angeles	Municipal
21	11171	Long Island Power Authority	State
22	19436	Union Electric Co	Investor Owned
23	7806	Entergy Gulf States Inc	Investor Owned
24	4176	Connecticut Light & Power Co	Investor Owned
25	14940	PECO Energy Co	Investor Owned
26	13998	Ohio Edison Co	Investor Owned
27	1167	Baltimore Gas & Electric Co	Investor Owned
28	13781	Northern States Power Co	Investor Owned
29	11241	Entergy Louisiana Inc	Investor Owned
30	14715	PPL Electric Utilities Corp	Investor Owned
31	803	Arizona Public Service Co	Investor Owned
32	16609	San Diego Gas & Electric Co	Investor Owned
33	9726	Jersey Central Power & Light Co	Investor Owned

LIST OF 200 UTILITIES USED IN ANALYSIS (CONT'D)

Rank by 2000 Revenue	Utility ID	Utility Name	Ownership
34	15270	Potomac Electric Power Co	Investor Owned
35	3755	Cleveland Electric Illuminating Co	Investor Owned
36	11804	Massachusetts Electric Co	Investor Owned
37	20847	Wisconsin Electric Power Co	Investor Owned
38	3278	AEP Texas Central Co	Investor Owned
39	16572	Salt River Project	Political Subdivision
40	15466	Public Service Co of Colorado	Investor Owned
41	15500	Puget Sound Energy Inc	Investor Owned
42	1998	Boston Edison Co	Investor Owned
43	733	Appalachian Power Co	Investor Owned
44	13511	New York State Electric & Gas Corp	Investor Owned
45	14006	Ohio Power Co	Investor Owned
46	14063	Oklahoma Gas & Electric Co	Investor Owned
47	814	Entergy Arkansas Inc	Investor Owned
48	15470	Duke Energy Indiana Inc	Investor Owned
49	17539	South Carolina Electric & Gas Co	Investor Owned
50	3542	Duke Energy Ohio Inc	Investor Owned
51	18454	Tampa Electric Co	Investor Owned
52	9208	Illinois Power Co	Investor Owned
53	4062	Columbus Southern Power Co	Investor Owned
54	13407	Nevada Power Co	Investor Owned
55	15248	Portland General Electric Co	Investor Owned
56	13756	Northern Indiana Public Service Co	Investor Owned
57	12341	MidAmerican Energy Co	Investor Owned
58	9324	Indiana Michigan Power Co	Investor Owned
59	16604	City of San Antonio	Municipal
60	4922	Dayton Power & Light Co	Investor Owned
61	10000	Kansas City Power & Light Co	Investor Owned
62	5487	Duquesne Light Co	Investor Owned
63	19547	Hawaiian Electric Co Inc	Investor Owned
64	15472	Public Service Co of NH	Investor Owned
65	15474	Public Service Co of Oklahoma	Investor Owned
66	17698	Southwestern Electric Power Co	Investor Owned
67	20387	West Penn Power Co	Investor Owned
68	5027	Delmarva Power & Light Co	Investor Owned
69	12293	City of Memphis	Municipal
70	12685	Entergy Mississippi Inc	Investor Owned

(Continued)

LIST OF 200 UTILITIES USED IN ANALYSIS (CONT'D)

Rank by 2000 Revenue	Utility ID	Utility Name	Ownership
71	18997	Toledo Edison Co	Investor Owned
72	9273	Indianapolis Power & Light Co	Investor Owned
73	16534	Sacramento Municipal Utility District	Political Subdivision
74	963	Atlantic City Electric Co	Investor Owned
75	18642	Tennessee Valley Authority	Federal
76	15263	The Potomac Edison Co	Investor Owned
77	13216	Nashville Electric Service	Municipal
78	1015	Austin Energy	Municipal
79	13214	Narragansett Electric Co	Investor Owned
80	24211	Tucson Electric Power Co	Investor Owned
81	14711	Pennsylvania Electric Co	Investor Owned
82	9617	JEA	Municipal
83	17718	Southwestern Public Service Co	Investor Owned
84	9162	Interstate Power & Light Co	Investor Owned
85	12390	Metropolitan Edison Co	Investor Owned
86	19497	United Illuminating Co	Investor Owned
87	12796	Monongahela Power Co	Investor Owned
88	10005	Kansas Gas & Electric Co	Investor Owned
89	40051	Texas-New Mexico Power Co	Investor Owned
90	17166	Sierra Pacific Power Co	Investor Owned
91	10171	Kentucky Utilities Co	Investor Owned
92	7801	Gulf Power Co	Investor Owned
93	9191	Idaho Power Co	Investor Owned
94	20856	Wisconsin Power & Light Co	Investor Owned
95	3265	Cleco Power LLC	Investor Owned
96	16183	Rochester Gas & Electric Corp	Investor Owned
97	3253	Central Illinois Public Service Co	Investor Owned
98	15473	Public Service Co of NM	Investor Owned
99	20860	Wisconsin Public Service Corp	Investor Owned
100	12686	Mississippi Power Co	Investor Owned
101	22500	Westar Energy Inc	Investor Owned
102	12698	Aquila Inc	Investor Owned
103	11249	Louisville Gas & Electric Co	Investor Owned
104	5701	El Paso Electric Co	Investor Owned
105	14127	Omaha Public Power District	Political Subdivision
106	3266	Central Maine Power Co	Investor Owned
107	17543	South Carolina Public Service Authority	State

LIST OF 200 UTILITIES USED IN ANALYSIS (CONT'D)

Rank by 2000 Revenue	Utility ID	Utility Name	Ownership
108	4089	Commonwealth Electric Co	Investor Owned
109	20169	Avista Corp	Investor Owned
110	3249	Central Hudson Gas & Electric Corp	Investor Owned
111	13478	Entergy New Orleans Inc	Investor Owned
112	1738	Bonneville Power Admin	Federal
113	12647	Minnesota Power Inc	Investor Owned
114	20455	Western Massachusetts Electric Co	Investor Owned
115	3252	Central Illinois Light Co	Investor Owned
116	14154	Orange & Rockland Utilities Inc	Investor Owned
117	16868	City of Seattle	Municipal
118	20404	AEP Texas North Co	Investor Owned
119	13780	Northern States Power Co	Investor Owned
120	12825	NorthWestern Energy LLC	Investor Owned
121	10421	Knoxville Utilities Board	Municipal
122	17470	Snohomish County PUD No 1	Political Subdivision
123	3408	City of Chattanooga	Municipal
124	14610	Orlando Utilities Comm	Municipal
125	14716	Pennsylvania Power Co	Investor Owned
126	16687	Savannah Electric & Power Co	Investor Owned
127	22053	Kentucky Power Co	Investor Owned
128	9392	Interstate Power Co	Investor Owned
129	9964	Kenergy Corp	Cooperative
130	17633	Southern Indiana Gas & Electric Co	Investor Owned
131	3292	Central Vermont Public Service Corp	Investor Owned
132	9094	City of Huntsville	Municipal
133	590	City of Anaheim	Municipal
134	14015	Ohio Valley Electric Corp	Investor Owned
135	3916	Cobb Electric Membership Corp	Cooperative
136	5860	Empire District Electric Co	Investor Owned
137	9601	Jackson Electric Member Corp	Cooperative
138	3989	City of Colorado Springs	Municipal
139	9216	Imperial Irrigation District	Political Subdivision
140	3611	Citizens Communications Co	Investor Owned
141	18429	City of Tacoma	Municipal
142	14626	Pedernales Electric Coop, Inc	Cooperative
143	19446	Union Light Heat & Power Co	Investor Owned
144	12470	Middle Tennessee E M C	Cooperative

(Continued)

LIST OF 200 UTILITIES USED IN ANALYSIS (CONT'D)

Rank by 2000 Revenue	Utility ID	Utility Name	Ownership
145	17637	Southern Maryland Electric Coop Inc	Cooperative
146	20885	Withlacoochee River Electric Coop	Cooperative
147	11843	Maui Electric Co Ltd	Investor Owned
148	8287	Hawaii Electric Light Co Inc	Investor Owned
149	16655	City of Santa Clara	Municipal
150	7601	Green Mountain Power Corp	Investor Owned
151	14232	Otter Tail Power Co	Investor Owned
152	3660	PUD No 1 of Clark County	Political Subdivision
153	18445	City of Tallahassee	Municipal
154	3757	Clay Electric Cooperative, Inc	Cooperative
155	11479	Madison Gas & Electric Co	Investor Owned
156	13640	Northern Virginia Electric Coop	Cooperative
157	14534	City of Pasadena	Municipal
158	10857	Lee County Electric Coop, Inc	Cooperative
159	16088	City of Riverside	Municipal
160	10623	City of Lakeland	Municipal
161	1179	Bangor Hydro-Electric Co	Investor Owned
162	16865	Sawnee Electric Membership Corp	Cooperative
163	15296	New York Power Authority	State
164	13337	Nebraska Public Power District	Political Subdivision
165	40228	Rappahannock Electric Coop	Cooperative
166	6958	City of Garland	Municipal
167	12745	Modesto Irrigation District	Political Subdivision
168	13716	North Georgia Electric Membership Corp	Cooperative
169	3762	City of Cleveland	Municipal
170	20065	Walton Electric Member Corp	Cooperative
171	12199	MDU Resources Group Inc	Investor Owned
172	16213	Rockland Electric Co	Investor Owned
173	10704	City of Lansing	Municipal
174	7294	City of Glendale	Municipal
175	6235	Public Works Comm-City of Fayetteville	Municipal
176	11018	Lincoln Electric System	Municipal
177	40428	Guam Power Authority	State
178	17833	City of Springfield	Municipal
179	18304	Sumter Electric Coop, Inc	Cooperative
180	6909	Gainesville Regional Utilities	Municipal
181	21632	EnergyUnited Electric Member Corp	Cooperative

LIST OF 200 UTILITIES USED IN ANALYSIS (CONT'D)

Rank by 2000 Revenue	Utility ID	Utility Name	Ownership
182	4442	PUD No 1 of Cowlitz County	Political Subdivision
183	19898	Volunteer Electric Coop	Cooperative
184	4624	Cumberland Electric Member Coop	Cooperative
185	5748	Electric Energy Inc	Investor Owned
186	6022	City of Eugene	Municipal
187	18085	South Central Power Co	Cooperative
188	9996	City of Kansas City	Municipal
189	9777	City of Johnson	Municipal
190	2507	City of Burbank Water & Power	Municipal
191	5078	Denton County Electric Coop, Inc	Cooperative
192	7090	GreyStone Power Corp	Cooperative
193	7639	Greenville Utilities Comm	Municipal
194	689	Anoka Electric Coop	Cooperative
195	19545	Black Hills Power Inc	Investor Owned
196	5202	Dixie Electric Membership Corp	Cooperative
197	17684	Southwest Louisiana E M C	Cooperative
198	17647	Southern Pine Electric Power Assn	Cooperative
199	9612	City of Jackson	Municipal
200	9336	Intermountain Rural Electric Assn	Cooperative

CHAPTER 24
THE STATISTICAL SIGNIFICANCE OF INCENTIVES FOR REGULATORY POLICY TARGETS

This statistical analysis starts with the same 200 largest electric utilities in the United States, and covers the 15-year period from 1992 through 2006. Several hypotheses are tested. The purpose is to refine the statistical analyses, discussed in the previous chapter. The primary policy emphasis is to consider the two demand-side targets previously discussed. These are:

- Target I: Achieving "one percent" savings in total annual sales and peak capacity as incremental savings targets.

- Target II: Achieving "10 percent" savings in total annual sales and peak capacity as cumulative savings targets.

The variables used in this analysis are:

1. The cumulative direct cost spent on energy efficiency divided by total retail revenue. (These were also tested using incremental spending as a percent of total sales.);

2. The cumulative indirect cost or markup for energy efficiency divided by total revenue. (These were also tested using incremental spending as a percent of total revenue.);

3. The nominal residential per unit price of electricity;

4. A binary variable indicating whether the entity is investor-owned;

5. An annual trend variable;

6. The cumulative earned (profit) incentives for a sub-sample covering the years 1992 through 1996. (These specific financial incentive data were only collected during the first five years.); and

7. These same experiments were also repeated for just the utilities that had formal energy efficiency programs in a given year. (These are called "partial," while the all-included sample is denoted "full.")

In virtually every experiment, the hypothesis tested showed that the utilities that spend more directly on energy efficiency per dollar of revenue save more MWhs in both incremental annual percentages (Target I) and cumulative percentages (Target II). This is true for both the full and partial samples. Higher retail prices and IOU status generally, but not always, suggest more cumulative efficiency savings. The trends in cumulative savings are generally negative. However, the data do not include 2007 or 2008. The incremental savings trends vary.

Table 24-1 shows the statistical hypotheses tests for Target I, achieving one percent of incremental savings relative to sales volume each year. The IOUs in the sample and utilities with higher residential prices were statistically more likely to move towards the goal of achieving Target I. Very importantly, spending more incrementally as a percentage of total revenue, a jointly determined variable, is also quite obviously important.

TABLE 24-1
TARGET I: INCREMENTAL SAVINGS AS A PERCENT OF TOTAL SALES VOLUME

Variable	Full Sample		Positive Spending	
	All Years	Pre-1997	All Years	Pre-1997
Constant	-.000997	-.001561	-.002305	-.002101
	(5.9)	(5.3)	(7.3)	(4.9)
Percent of Direct Spending (EE) to Total Revenue	.23555	.24856	.22476	.24294
	(43.2)	(38.5)	(28.4)	(30.7)
Percent of Indirect Spending (DSM) to Total Revenue	.10436	.20784	.14899	.33154
	(8.4)	(6.4)	(8.1)	(8.4)
Residential Price of Electricity	.01465	.02143	.02843	.02840
	(8.3)	(6.8)	(9.6)	(7.0)
IOU Status	.000391	.000535	.000721	.000777
	(4.1)	(3.5)	(4.0)	(3.3)
Trend	-.0000205	.0000320	.0000276	.0000661
	(1.9)	(.65)	(1.4)	(.95)
R^2	.51	.66	.48	.65
Number of Observations	2867	956	1442	613

The primary hypothesis tested here is to determine the significance of higher incremental indirect costs expressed as a percentage of annual revenue. These represent financial incentives that might affect the likelihood of achieving Target I. In all four cases, this indirect cost markup expressed as a percentage of total electricity revenue is exceedingly statistically significant. This is true for the entire sample and pre-1997, when EIA provided more detailed data, as well as the sample limited to utilities with positive incremental demand response programs over both the 15-year period and pre-1997. This means that Target I is more likely to be achieved, if regulators provide markups for more recovery than just the direct program costs of efficiency.

Table 24-2 tests the statistical significance of similar variables for Target II, achieving a 10 percent cumulative energy savings goal. The results are quite similar except for the downward trend in the target being met. This is likely due to a saturation effect as utilities level off their demand-side response programs. Over the entire 15-year program, it is particularly likely that utilities would approach Target II, if they receive indirect cost markups or financial incentives. The pre-1997 results are somewhat marginal. However, this could be too short a time period (5 years) to measure cumulative effects. Both the Target I results and the full 15-year sample for Target II support such an observation.

TABLE 24-2
TARGET II: CUMULATIVE SAVINGS AS A PERCENT OF TOTAL SALES VOLUME

Variable	Full Sample		Positive Spending	
	All Years	Pre-1997	All Years	Pre-1997
Constant	.00405	-.00256	.00221	.000520
	(3.5)	(1.5)	(1.3)	(.2)
Percent of Cumulative Direct Spending (EE) to Total Revenue	.29958	.41295	.29436	.38746
	(48.0)	(32.0)	(35.9)	(23.8)
Percent of Cumulative Indirect Spending (DSM) to Total Revenue[1]	.07839	.12838	.10764	.11323
	(5.0)	(2.1)	(5.0)	(1.6)
Residential Price of Electricity	.04237	.09560	.08812	.09526
	(3.4)	(5.2)	(5.3)	(4.0)
IOU Status	.00102	.00242	.00180	.00136
	(1.5)	(2.7)	(1.7)	(1.0)
Trend	-.000706	-.000696	-.000547	-.000587
	(9.4)	(2.3)	(4.8)	(1.4)
R^2	.58	.58	.59	.52
Number of Observations	2867	956	1692	653

[1] For pre-1997, this is the Direct Financial Incentive shareholders earned.

The most important experiments reported in this chapter were to determine if the markups (indirect cost per dollar of retail revenue added to direct program costs) affect the incremental and cumulative savings. The results of these two experiments are consistent with the hypothesis that positive financial incentives would mean a utility was more likely to achieve both Target I and Target II and save more MWhs, other things being equal, than utilities that do not have such markups or financial incentives.

SECTION VI:
THE PATH AHEAD

This last section reviews the conclusions reached and makes recommendations for increasing the use of energy efficiency and load management to address climate change, environmental matters, and national and economic security. First, an innovative set of ideas that Duke Energy has proposed and called "save-a-watt" is reviewed because it embodies much of what this primer has discussed. Next, the relationship between renewable energy with similar objectives as energy efficiency is reviewed in the context of establishing a combined policy agenda. Finally, the conclusions reached are summarized and recommendations are made.

CHAPTER 25
SAVE-A-WATT: A NEW PARADIGM[1]

About three decades ago, Amory Lovins moved the conservation debate to a new level with his famous and important use of the concept he called "Negawatts." In the last year, Jim Rogers, CEO of Duke Energy, has done the same thing with a newly coined phrase he calls "save-a-watt" (SAW). His concept marries Amory Lovins' seminal conservation ideas to a rather complete paradigm shift for the nation's electric utility industry.

Duke's proposal does this with a plan to make energy efficiency a new utility service for which it could seek additional regulated income. This save-a-watt concept inspired some of this analysis, and represents a twenty-first century idea to make energy efficiency a new core utility service. Therefore, some discussion of this very big idea is warranted in this primer.

Americans are once again debating how to redirect consumers toward energy efficiency in the form of conservation, demand-side management (DSM), and renewables. The previous analysis demonstrates that regulatory jurisdictions that adopt direct financial incentives establish a formal decoupling mechanism, and have clear goals and targets are much more likely to increase energy efficiency and load management.

The core concept underlying the Duke plan is to align shareholder and retail customer interests behind an expanded least-cost approach. Jim Rogers and Duke Energy proposed a new paradigm that eschews regulation of details in favor of a conceptually less complex approach. In effect, Duke Energy seeks to use "avoided cost" to establish the value of energy efficiency services. The goal is to make conservation a profit center and to encourage trial-and-error marketing and clear incentives to succeed on a sustained basis.

The "save-a-watt" plan starts with state regulators' determination of resource requirements. The objective is to satisfy the utility's "duty to serve" at the "least" or "best" cost. Duke would treat energy efficiency as a "fifth fuel" or utility input, and pay itself a fraction (*e.g.*, 85 or 90 percent) of the avoided cost of generation, much like utilities pay qualifying facility (QF) generators.

---

[1] This chapter is based upon a larger discussion. *See* Cicchetti, Charles J. (2008, January). Duke's Fifth Fuel. *Public Utilities Fortnightly*, pp. 54-62.

Regulators generally seek to prevent one group of customers from paying too much, so that others can pay less. Past debates concerning the differing effects of various utility-sponsored energy efficiency programs on participants and nonparticipants were a major reason why earlier programs failed to be sustained. The Duke "save-a-watt" plan partially eliminates this issue by ensuring that revenue requirements will be less, and by designing programs in which most participating customers will contribute time and/or money.

The Duke "save-a-watt" plan addresses the regulated revenue requirements aspect of energy efficiency simply and transparently. This means that regulators can focus less on details for which they are less well-suited, and instead focus on what they can do best: determine the reasonable future mix of demand-side management, traditional supplies, and renewables. Regulators are rather capable, when setting the future course based upon where the utility starts. Under the Duke "save-a-watt" plan, regulators will understand the cost/benefit trade-offs of various proposals to make the future "green," "greener" or "greenest." Importantly, regulators would also consider and address the cost of service and future revenue requirements effects for a new utility input, not just a new product.

The cornerstone of the Duke "save-a-watt" plan is that it would use this effective high-level regulatory decision-making process to establish both the utility's future plans and the future course of earnings, management, and regulated returns for utility investors. This is a most important step. When regulators previously attempted to expand conservation and DSM, their primary concern was "holding harmless" nonparticipants. Regulators relied on "due discrimination" arguments to do this. Regulators understood that regulated customers, who received utility-sponsored demand-side upgrades on their premises, would pay less for energy. The size of the incentive packages or, to be more precise, the subsidy offered to accomplish the conservation objective was often based on the savings projected in terms of the "avoided costs" of the utility's other, primarily nonparticipating customers.

Utility investors were often given little or no thought in terms of how they would earn a return on these utility-sponsored conservation and customer premises renewable energy programs. Two issues were problematic: (1) how to price conservation products, and (2) how to set a utility's earnings. The Duke plan addresses both problems.

The traditional regulatory approach based utility income entirely on the amount invested, not the value of the services provided. Customers were ratepayers and choices were virtually nonexistent. Under this regulatory paradigm, there was no opportunity for utilities to earn a margin on the services that other retailers would capture. The regulatory response to evidence that this approach was bad policy and failed to create reasonable utility incentives was typically an argument that providing such utility-sponsored conservation and demand-side services was part of the utility's franchise responsibility (*i.e.*, the regulated duty to serve).

The SAW plan sidesteps these disincentives and adds a real opportunity for shareholders to make energy efficiency a new profit center.

Under traditional cost-of-service utility regulation, when energy efficiency reduces sales between rate cases, shareholders lose income and have no opportunity to recover it. The greater the success of a utility's energy efficiency program, the greater the potential shareholder losses. Regulation sometimes attempted to address this matter using some sort of revenue decoupling mechanism, whereby lost revenues would be recovered using an energy recovery adder to regulated bills. At best, such a plan would help shareholders achieve a degree of indifference or neutrality. This often simply was not enough of an incentive to sustain support for energy efficiency.

The SAW plan fixes these traditional regulatory problems, because it recognizes that reasonable incentives are the essence of market-based societies. Earning a return "on" and "of" investments is essential. It is not enough, *per se*. Earning a margin or income on developing, packaging, and marketing energy efficiency retail customer programs is at least as important. The Duke "90 percent of QF's avoided cost" concept addresses both aspects of utility earnings, and consumers pay 90 percent of the avoided cost increase they would otherwise have been assigned in higher annual revenue requirements. Only one major regulatory question remains: how will regulators protect nonparticipants?

If the narrow question is posed, many regulators might falsely conclude that asking nonparticipants to pay more in order to help others consume less would represent an unjust form of price discrimination.[2] Broadly speaking, as explained above society would benefit when energy efficiency reduces negative externalities[3] in the form of:

- *Clean Air*: reductions in sulfur dioxide (SO_2), nitrogen oxide (NO_X), and other emissions;
- *Climate Change*: reductions in carbon dioxide (CO_2) emissions; and
- *National Security*: reductions in foreign energy and improved macroeconomic conditions.

---

[2] Nonparticipants can sometimes pay less under utility-sponsored energy efficiency. The most important determinants are: marginal costs exceed average historic costs, the relative size of the program, and participants pay at least a portion of the direct costs.

[3] We also know, based on the research of Gregory E. Aliff and Branko Terzic (2007, September), The Greening of Utility Customers, *Public Utilities Fortnightly*, pp. 32-33, that more than half of respondents would pay five percent more for energy in order to expand alternative energy.

Coal-fired generation is used throughout much of the nation. Expanding energy efficiency would most likely displace either old coal-fired generation or new more efficient generation. New generation could be efficient natural gas, cleaner coal, perhaps with some form of carbon sequestration, and potentially new nuclear power stations. The existing, mostly older coals unit displaced would be particularly relevant for determining emission benefits, when energy efficiency expands before a utility is in a must-build-new-generation position. At the other extreme, if energy efficiency just replaces new generation, the estimated air emission reduction benefits should reflect the air emission and compliance costs for a newer, more efficient generating station with new source performance built into the direct cost comparison between new generation and energy efficiency.

In most of the nation, energy efficiency would reduce coal-fired electricity generation, not oil dependence. A case can be made that any energy conservation would help to increase the nation's (as well as specific businesses' and households') energy efficiency. The macroeconomic components of the oil security premiums would apply. When oil prices increase, all domestic energy prices increase. Therefore, any premium included in crude oil would likely increase other primary energy prices, such as coal and natural gas. Further, complex construction engineering costs increase when oil markets tighten. Without energy efficiency, the cost of new generating stations would likely be higher when oil prices are high, as they are today.

The nonparticipating utility consumers, as well as regulators concerned with possible undue discrimination, would need to compare the price increase nonparticipants would pay multiplied by their annual consumption to the value of energy efficiency benefits. The latter equals the amount of energy saved multiplied by the per-unit value of the marginal benefits of energy efficiency.

Each utility region of the country would need to perform its own benefit/cost analysis. Facts and conceptual differences, such as how to define costs, likely would vary. Regardless, the benefits of energy efficiency, which has mainly been underutilized, are likely to easily exceed the corresponding costs, no matter who pays for energy efficiency.

The Duke approach substitutes light-handed regulation as an alternative to what has previously worked. That said, if political attention again shifts, the more detailed regulatory approaches likely would lapse and shrink in scope. The Duke approach was designed to withstand any such change in emphasis, because it would overhaul the current utility culture. The purpose is to shift the profit and business focus of regulated utilities, in order to sustain the current commitments regulators and utilities are making to expand energy efficiency and load management.

California's experience with energy efficiency demonstrates that utilities can rely on energy efficiency and renewable energy to keep the lights burning. Duke's save-a-watt proposal shows how to make energy efficiency a fifth fuel and to treat these services as a new utility-for-profit enterprise with lower total revenue requirements than traditional generation. Finally, this analysis quantifies the benefits of energy efficiency and uses these data to demonstrate that all customers would likely gain greater benefits than their respective costs, regardless of whether they participate or not.

Some states have wires charges or may be contemplating using such charges, as California does, to finance social programs such as energy efficiency. Others may use Duke's proposed "save-a-watt" rate mechanism based on 85 percent to 90 percent of avoided supply costs. Other states are considering other means to make energy efficiency profitable to the utility. In all regulatory jurisdictions, the retail distribution end of the network remains a regulated "natural" monopoly. This means that legislators and/or regulators can look to these regulated assets and services, as a direct means to finance energy efficiency, when, as is likely, they determine that utility, consumer, and external benefits exceed costs.

Duke's approach provides another regulatory protection, because total revenue requirements would increase no more than 90 percent of the avoided supply costs. Everything else, such as avoiding negative externalities, is simply "icing on the cake" for all but nonparticipants. Regulators that factor in external benefits would press ahead. Duke would produce new "negawatts" and seek an avoided-cost payment, based on the same QF avoided-cost principles that forged the wholesale power market.

The Duke plan has been designed to work and now regulators need to decide if they really want to improve energy efficiency, to reduce the threat of climate change, to enhance national security and improve air quality. This is an easier choice for regulators, given the impressive benefit-to-cost ratios discussed in this analysis.

California shows what a well-funded and committed regulatory approach can accomplish, if it embraces energy efficiency as a utility input/service. One size seldom fits everyone.

The Duke "save-a-watt" proposal has all the necessary ingredients to appeal to regulators. It was designed using avoided supply-side cost, in order to ensure that energy efficiency would replace more costly supply-side choices. Duke's long-term goal is sustainability. This is done through an explicit commitment to minimize the regulatory details of the plan. Many regulators, particularly given the current urgency to improve energy efficiency, would very likely seek to graft utility income, decoupling, and explicit targets onto the conservation plans they approve. With time, regulators might be convinced to remove these "add-ons," if and when they gain confidence that these programs are working as planned. Regardless, regulators should consider how to sustain whatever approach they take even if public support for energy efficiency and load management wanes.

CHAPTER 26
RENEWABLES AND ENERGY EFFICIENCY

This primer addresses the details, business challenges, and regulatory necessities of energy efficiency and load management. These policies are necessary, but not sufficient. There is also a need to expand renewable energy systems, such as wind, solar, geothermal, wave power, better batteries, distributed energy, and more new technologies that reduce the use of fossil fuels. Such advances would reduce the nation's carbon footprint and permit natural gas to shift from firing electricity generators to moving vehicles.

Carbon sequestration, nuclear energy, and hydroelectricity may also be part of the dual "green" and "national security" quest to redefine the nation's energy future. These may all seem to be no more than politicians' hollow promises. History suggests otherwise. In the 1970s, most energy experts could see a future with nothing but either continued exponential energy growth or a sharply curtailed economic expansion.

Before the 1980s, energy use and the gross domestic product (GDP) grew with nearly perfect statistical correlation. Similar statistical correlations could have been found in virtually every part of the world. Since the 1980s, GDP and energy growth have continued. However, the near perfect correlation between more energy and more GDP was sharply altered.

Chart 26-1[1] shows a dramatic reduction in energy use relative to a combined expansion in GDP. Two measures, an Energy Intensity Index and the amount of Energy per Dollar of GDP, have actually fallen. This demonstrates economic growth need not be sacrificed, if the nation becomes more energy efficient. The twin goal is to invest time, money, science, and engineering to design and build more renewable energy and more energy efficiency.

---

[1] *Source: http://www1.eere.energy.gov/ba/pba/intensityindicators/total_energy.html*

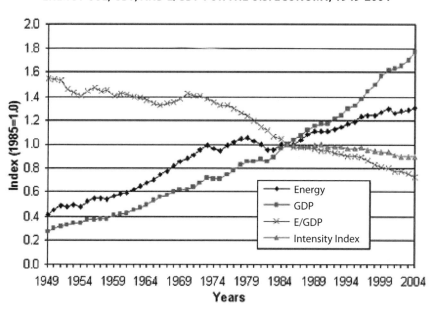

CHART 26-1
ENERGY USE, GDP, AND E/GDP FOR THE U.S. ECONOMY, 1949-2004

Once again, a second shift is not just possible, it also seems necessary. The high price of imported energy and the concomitant international dangers and instability associated with crude oil require a shift away from fossil fuel. This is not just a "green" revolution. Economic survival and national security are also co-equal drivers. Virtually every sector of the economy—transportation, industry, residential, commercial, and electric utilities—needs to become less energy-intensive. Chart 26-2[2] shows this has happened in all sectors, but the commercial sector since 1985. More is likely necessary. Energy efficiency and alternative or renewable energy make everyone's "to do" list. This conventional wisdom is not wrong.

---

[2] Source: http://www1.eere.energy.gov/ba/pba/intensityindicators/total_energy.html.

Chapter 26: Renewables and Energy Efficiency

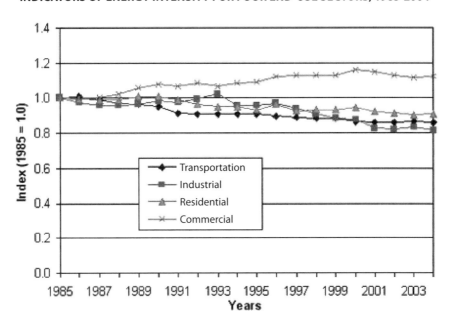

CHART 26-2
INDICATORS OF ENERGY INTENSITY FOR FOUR END-USE SECTORS, 1985-2004

Electricity is a core input or requirement to every sector of the economy, and every home, business, and farm. Accordingly, this is where the "green/security" activity must focus very significant efforts. This primer proposes a goal of "one percent" incremental annual reductions in electricity consumption. This would cut most electric utilities' growth in half from their current levels. This will not be enough to shift the nation to even a sustainable or no growth in fossil fuel future. Similarly, the "10 percent" cumulative goal for energy efficiency is also within reach economically and technically, as the top performing states and utilities have proven. Meeting this goal would be impressive, important, and probably necessary. However, it requires real state regulatory and utility leadership.

More must still be done. This is where increased supply-side use of renewables must complement energy efficiency. The regulatory challenges are different, because there are no lost revenues or margins related to utilities that "build" or "buy" renewable energy. Regulators also know how to establish direct financial incentives for supply-side expansions. Renewables would *not* be any different. One regulatory challenge is to ensure these dual efficiency and renewable approaches complement one another and do not become competitors.

Tax credits will help both green approaches. Regulators may need to consider using such credits with financially favorable conditions to increase shareholder incentives. Similarly, regulators should also consider differential authorized returns on equity and expensing the financing of construction work in progress for renewable energy investments. In restructured jurisdictions, the FERC and states should find the means to increase returns on "wires" and "pipes" to expand the purchase of renewable energy supplies.

Two policy matters are central:

- Despite the more traditional regulatory case for expanding renewables, regulators should not give energy efficiency short shrift.

- Utilities and regulators should combine their targets for energy efficiency and renewables into a combined portfolio standard.

These two political, regulatory, and utility decisions will keep utilities and customers on the parallel tracks of using less energy and beginning to end the fossil fuel addiction of the nation and the economy.

The future likely will also include a strong "halo" effect as consumers and businesses make decisions that reflect the "green" and "security" markets on their own and become less dependent on centrally organized regulated industry programs. This next major step would be accelerated as "smart metering" and real-time pricing become more commonplace across the nation. These will marry the strength of the Internet, computers, and new technology to new appliances and space conditioning products. People would, with help and information, be able to design their own "green," "secure" and money-saving energy solutions.

California is close to the "10 percent" energy efficiency goal and is also approaching a "20 percent" renewable portfolio standard. That largest state, as well as some other states, plans to increase both energy efficiency and renewables. The various goals in the leading states are becoming floors. New regulation and incentives will mean these high-end targets will be surpassed. The states that lead will save money, create jobs, and, with hope, stimulate others to emulate their successes and avoid any missteps.

Utility leaders are emerging to propose new business and financial models to increase energy efficiency and renewables. Regulators are increasingly "getting it." They need help and support. Mostly, the official case records that regulators must rely upon should neither exaggerate the potential target nor understate costs. Financial incentives and flexibility will help to sustain these current energy efficiency and renewable supply efforts and make them more effective.

CHAPTER 27
CONCLUSIONS

There are contentious debates that have emerged between various entities that all claim to favor aggressive utility-sponsored energy efficiency programs. There are three obvious problems with too much debate. First, some would let the goal of perfection trump the good. This could very well cause a stalemate, or, maybe even worse, contentious battles of words that feed on themselves, rather than chart a better course, where real actions are taken to conserve resources and save money. The better approach would also provide a strong basis for "learning by doing."

This is essential because markets and consumers are not always similar. Debates absent details about what should be done and by whom, may be interesting. Past regulatory experience with similar efforts to achieve tariff reform show these debates are, at best, distractions that impede more than they help.

Second, current interest in climate change and economic and national security has more than primed the pump for utility-sponsored energy efficiency. This is *not* the first time there has been such heightened interest in utility-sponsored energy efficiency. As the wise person has often observed: Things change. So does political opinion. Contentious debates, therefore, can do more than delay. They might actually cause competing plans to cancel out any efforts to improve energy efficiency. These delays could be the bridge to "do nothing," if the public moves on before the regulatory Cuisinart of ideas pours out a clear liquid.

Third, many past debates have given far too much attention to nonparticipating customers. Today, some larger consumers propose to "opt out" of any expanded utility-sponsored energy efficiency. Regulators must be careful not to allow the nonparticipant issue to be overstated. The economically efficient logic is rather straightforward. Nonparticipants could pay as much as the sum of the marginal value assigned to reduced externalities, plus the difference between the marginal cost of supply-side electricity and regulated prices without being worse off.

The economic logic is equally compelling and straightforward for participants, who could pay the full marginal cost of energy efficiency plus up to the difference between their own regulated prices and that marginal cost of energy efficiency, without increasing their own bills, because they would use less. In addition, participating consumers would benefit as utility customers, if their conservation savings suppress any difference between the marginal cost of supply-side electricity and current regulated prices, as well as from the value attached to less externalities.

Regulators should seek to expand energy efficiency and load management, when the marginal cost of energy, plus marginal external costs, exceeds the marginal cost of demand response programs. Regulators should also reasonably allocate the costs between nonparticipants and participants. This is not a new assignment for regulators who are familiar with the debates about value, costs, vintage, etc. Regulators are very well-suited to this cost allocation assignment. However, while they should give these matters sufficient weight, they should not be derailed by such debates.

The goals for energy efficiency seemingly increase as political leaders seek to out-promise each other. This is undoubtedly partly due to the illusion that energy efficiency pays for itself and consumers would, regardless, pay lower energy bills, if they conserve.

Energy efficiency is mostly cost-effective, but this does not mean it's too cheap to matter. Time and money must be spent. Information needs to be assembled and considered, if consumers' choices are to be made rationally.

There are some seemingly quite different contexts for considering energy efficiency. Starting with the most optimistic: Americans have sharply curtailed their energy use per dollar of gross domestic product (GDP) overall, as well as in every sector except transportation since the oil crisis of the 1970s. Building codes, appliance standards, improved technology, and better alternatives have all contributed to this undeniably beneficial achievement. Less satisfying is that part of this American gain has come from changes in America's participation in the global economy in which America has exported some of its manufacturing base. Regardless, America's gains in energy efficiency per person and per dollar of national income are real and impressive.

At the opposite end of current thinking concerning energy efficiency, there is a conventional wisdom that some strongly believe that Americans are mainly apathetic and not particularly well informed concerning their energy use, particularly their use of energy outside of the published miles per gallon that automobiles post at the point of sale. There is widespread belief that Americans lack basic knowledge, particularly with respect to domestic electricity and natural gas use. This leads many to focus attention on regulated utilities and then state regulators. The premise is that energy efficiency is too urgent a social need for the nation to wait for utility consumers to become sufficiently engaged in conservation at home.

This policy perspective has additional traction, because electricity and natural gas are primarily delivered to end-use consumers through regulated wires and pipes that nearly always have a franchise monopoly. This means that regulated utilities can often be coerced or mandated to usurp the retail customer's role in making informed and often past-due energy efficiency decisions and choices.

Some states have significantly reduced their electricity and natural gas consumption through quite just and reasonable regulation of utility-sponsored energy efficiency programs. Indeed, the states that have used financial incentives and/or that have eliminated built-in regulatory disincentives have done quite well in terms of the amount of energy and capacity saved, relative to the retail volumes sold.

The key is to design a regulatory system that can sustain utility efforts even as political and public support wanes. One state, California, has had significant economic and population growth, and flat growth in electricity per customer, while the rest of the nation has grown significantly. Some of California's enviable record is due to the state's extraordinarily high electricity prices. Some of it has also been achieved through positive spillovers for builders, commercial users, retailers and contractors that have supplemented the formal utility-sponsored programs.

Regardless of the context, states are increasingly deciding that there is a compelling role for regulated utilities to design, implement, and expand energy efficiency. Presidential candidates, the Congress and the Federal Energy Regulatory Commission are joining these choices as well.

Justice Louis Brandeis referred to the various states as the "laboratories of democracy," because history demonstrates that states can and often do address national problems and experiment before the federal government acts. This is particularly true for energy efficiency, where otherwise relatively similar states have often acted quite differently, often with equally different energy efficiency outcomes.

This analysis reviews data for the past 15 years. This includes information as to how the nation's electric utilities achieve incremental and cumulative savings in three categories: (1) MWhs of energy saved, (2) the potential MW of capacity saved, and (3) the actual MW of capacity saved. These data have also been aggregated to the state level. In addition, they have been normalized based on the MWhs sold and peak capacity requirements.

Table 27-1 shows the top 25 investor-owned utilities (IOUs) in terms of their incremental MWhs savings performance for the most recent three years. Five IOUs surpassed one percent (Target I) of incremental savings in 2006. Others did so earlier.

TABLE 27-1
TOP QUARTILE
EE INCREMENTAL MWh SAVED / TOTAL MWh SALES

Rank	Utility Name	2004	2005	2006
1	United Illuminating Co	2.86%	1.33%	1.29%
2	Narragansett Electric Co	0.71%	0.83%	1.24%
3	Massachusetts Electric Co	1.12%	0.90%	1.18%
4	Connecticut Light & Power Co	0.91%	0.95%	1.12%
5	Western Massachusetts Electric Co	1.07%	0.97%	1.09%
6	Pacific Gas & Electric Co	0.71%	1.45%	0.92%
7	Southern California Edison Co	1.13%	1.44%	0.89%
8	Interstate Power & Light Co	0.66%	0.75%	0.84%
9	Puget Sound Energy Inc	0.87%	0.76%	0.72%
10	Sierra Pacific Power Co	0.15%	0.16%	0.71%
11	Northern States Power Co MN	0.76%	0.73%	0.71%
12	Nevada Power Co	0.20%	0.34%	0.68%
13	MidAmerican Energy Co	0.52%	0.51%	0.67%
14	Public Service Co of NH	0.57%	0.70%	0.64%
15	Wisconsin Power & Light Co	0.67%	0.57%	0.63%
16	Avista Corp	0.42%	0.66%	0.53%
17	Idaho Power Co	0.06%	0.31%	0.51%
18	Hawaiian Electric Co Inc	0.00%	0.00%	0.49%
19	NorthWestern Energy LLC	0.30%	0.29%	0.39%
20	PacifiCorp	0.30%	0.30%	0.37%
21	Arizona Public Service Co	0.00%	0.00%	0.29%
22	Wisconsin Electric Power Co	0.00%	0.00%	0.20%
23	Florida Power & Light Co	0.16%	0.18%	0.19%
24	Public Service Co of Colorado	0.24%	0.38%	0.17%
25	Minnesota Power Inc	1.01%	1.51%	0.17%

Only includes the top 25 investor-owned utilities in terms of 2006 total revenue.
Source: EIA Form 861.

Table 27-2 shows the top 25 IOUs in terms of their cumulative MWh savings performance for these same three years. Six IOUs have achieved the 10 percent (Target II) goal of cumulative savings in 2006. Again, others did so earlier.

TABLE 27-2
TOP QUARTILE
EE CUMULATIVE MWh SAVED / TOTAL MWh SALES

Rank	Utility Name	2004	2005	2006
1	United Illuminating Co	11.95%	11.35%	13.00%
2	Western Massachusetts Electric Co	10.86%	11.19%	12.31%
3	Northern States Power Co MN	10.33%	10.62%	11.25%
4	Avista Corp	9.93%	10.40%	10.64%
5	Minnesota Power Inc	8.26%	9.86%	10.25%
6	Southern California Edison Co	9.76%	10.32%	10.18%
7	Massachusetts Electric Co	8.30%	8.94%	9.95%
8	Connecticut Light & Power Co	8.03%	8.52%	9.42%
9	Puget Sound Energy Inc	10.08%	9.27%	8.77%
10	Northern States Power Co WI	9.07%	8.73%	8.69%
11	Wisconsin Power & Light Co	10.24%	9.15%	8.47%
12	Pacific Gas & Electric Co	5.83%	7.61%	8.36%
13	Narragansett Electric Co	6.29%	8.51%	8.22%
14	Interstate Power & Light Co	6.60%	7.01%	7.93%
15	Potomac Electric Power Co	6.65%	6.49%	6.76%
16	Gulf Power Co	5.32%	5.43%	5.44%
17	NorthWestern Energy LLC	4.41%	4.54%	4.85%
18	PacifiCorp	3.58%	4.13%	4.23%
19	Public Service Co of NH	2.80%	2.81%	4.09%
20	Hawaiian Electric Co Inc	0.41%	0.45%	3.90%
21	Florida Power & Light Co	3.43%	3.59%	3.73%
22	MidAmerican Energy Co	2.52%	2.93%	3.39%
23	Tampa Electric Co	3.13%	3.15%	3.15%
24	Wisconsin Electric Power Co	3.08%	2.63%	2.99%
25	Public Service Co of Colorado	2.55%	2.86%	2.96%

Only includes the top 25 investor-owned utilities in terms of 2006 total revenue.
Source: EIA Form 861.

Table 27-3 shows the rankings of the various states in terms of their incremental (Target I) MWh savings relative to the volume of MWhs generated over the three most recent years. Aggregated for all utilities in a state, only Rhode Island and Connecticut achieved the one percent goal in 2006.

TABLE 27-3
EE INCREMENTAL MWh SAVED / TOTAL MWh SOLD

Rank	State	2004	2005	2006
1	Rhode Island	0.6235%	0.7153%	1.0784%
2	Connecticut	1.1790%	0.9378%	1.0549%
3	California	0.6096%	0.9321%	0.6313%
4	New Hampshire	0.5297%	0.6381%	0.6215%
5	Nevada	0.1632%	0.2531%	0.6082%
6	Iowa	0.4561%	0.4865%	0.5939%
7	Hawaii	0.0573%	0.1144%	0.4569%
8	Idaho	0.1409%	0.3219%	0.4269%
9	Massachusetts	0.4802%	0.5953%	0.4243%
10	Minnesota	0.5595%	0.6106%	0.4130%
11	Washington	0.3398%	0.3373%	0.3832%
12	Utah	0.2451%	0.2438%	0.3010%
13	Montana	0.1821%	0.1812%	0.2505%
14	Wyoming	0.1740%	0.1689%	0.2059%
15	North Dakota	0.2096%	0.2102%	0.1956%
16	Wisconsin	0.1045%	0.0900%	0.1724%
17	Oregon	0.1520%	0.1482%	0.1709%
18	Arizona	0.0011%	0.0380%	0.1686%
19	South Dakota	0.1622%	0.1562%	0.1512%
20	Florida	0.1181%	0.1249%	0.1319%
21	Colorado	0.1510%	0.2430%	0.1211%
22	Vermont	0.0701%	0.0062%	0.1174%
23	New Jersey	0.0809%	0.1453%	0.1039%
24	New York	0.0436%	0.0479%	0.0728%
25	Texas	0.0431%	0.0546%	0.0664%
26	Kentucky	0.0606%	0.0606%	0.0502%
27	Mississippi	0.0279%	0.0278%	0.0257%
28	Alabama	0.0322%	0.0269%	0.0228%
29	New Mexico	0.0342%	0.0251%	0.0214%
30	Nebraska	0.0180%	0.0296%	0.0198%
31	Alaska	0.0106%	0.0229%	0.0185%
32	South Carolina	0.0026%	0.0462%	0.0185%
33	Tennessee	0.0169%	0.0161%	0.0179%
34	Illinois	0.0123%	0.0118%	0.0160%
35	Indiana	0.0081%	0.0010%	0.0120%
36	Michigan	0.0001%	0.0001%	0.0054%
37	Maine	0.0003%	0.0027%	0.0035%
38	Arkansas	0.0022%	0.0111%	0.0022%

TABLE 27-3 (CONT'D)
EE INCREMENTAL MWh SAVED / TOTAL MWh SOLD

Rank	State	2004	2005	2006
39	North Carolina	0.0007%	0.0010%	0.0022%
40	Georgia	0.0005%	0.0009%	0.0019%
41	Louisiana	0.0014%	0.0089%	0.0017%
42	Pennsylvania	0.0015%	0.0015%	0.0014%
43	Missouri	0.0011%	0.0013%	0.0011%
44	Ohio	0.0007%	0.0005%	0.0007%
45	Oklahoma	0.0008%	0.0007%	0.0003%
46	Maryland	0.0003%	0.0003%	0.0002%
47	Virginia	0.0009%	0.0005%	0.0002%
48	Kansas	0.0055%	0.0001%	0.0001%
49	West Virginia	0.0000%	0.0000%	0.0000%
50	Delaware	0.0000%	0.0000%	0.0000%
51	District of Columbia	0.0000%	0.0000%	0.0000%

Table 27-4 shows the same state ranking for Target II, achieving an annual cumulative energy savings percentage of 10 percent. On an aggregate statewide basis, no state achieved this 10 percent cumulative goal in 2006.

TABLE 27-4
EE CUMULATIVE MWh SAVED / TOTAL MWh SOLD

Rank	State	2004	2005	2006
1	Connecticut	8.0085%	8.2991%	9.2630%
2	Rhode Island	5.5571%	7.3505%	7.1403%
3	Minnesota	6.3763%	6.7408%	7.0892%
4	California	5.4353%	6.2310%	6.3481%
5	Washington	4.9569%	5.0027%	5.5698%
6	Dist of Columbia	4.5616%	4.5171%	4.4592%
7	Iowa	3.3879%	3.7044%	4.2387%
8	Hawaii	0.8054%	1.5215%	4.1375%
9	New Hampshire	3.0163%	3.0560%	4.0183%
10	Massachusetts	3.7444%	5.6420%	3.7518%
11	Idaho	3.0185%	3.3585%	3.6140%
12	Utah	2.9097%	3.3560%	3.4306%
13	Wisconsin	3.5288%	3.1881%	3.2325%
14	Florida	2.7226%	2.7735%	2.9018%

(Continued)

TABLE 27-4 (CONT'D)
EE CUMULATIVE MWh SAVED / TOTAL MWh SOLD

Rank	State	2004	2005	2006
15	Montana	2.4756%	2.5927%	2.7540%
16	North Dakota	2.3056%	2.3987%	2.4810%
17	Wyoming	2.0551%	2.3270%	2.3461%
18	Oregon	2.1576%	2.2021%	2.2744%
19	South Dakota	1.9862%	2.0381%	2.1305%
20	Maryland	2.1532%	2.0437%	2.1047%
21	Colorado	1.4797%	1.6829%	1.7548%
22	New Jersey	1.5287%	1.8111%	1.5533%
23	Vermont	2.8094%	2.6456%	1.5228%
24	Nevada	0.2342%	0.6890%	1.2512%
25	Indiana	0.7834%	0.7639%	0.7828%
26	Texas	0.4009%	0.3877%	0.7292%
27	New York	0.3915%	0.4182%	0.5734%
28	Arizona	0.1572%	0.3651%	0.5428%
29	Kentucky	0.4064%	0.4503%	0.4148%
30	Alabama	0.5389%	0.3516%	0.3682%
31	Mississippi	0.2821%	0.3004%	0.3180%
32	Ohio	0.2193%	0.2003%	0.2148%
33	Georgia	0.2246%	0.2202%	0.2033%
34	New Mexico	0.1833%	0.1783%	0.1885%
35	Tennessee	0.1394%	0.1514%	0.1688%
36	Virginia	0.1516%	0.1473%	0.1476%
37	South Carolina	0.1330%	0.1173%	0.1240%
38	Illinois	0.0824%	0.0891%	0.1052%
39	Michigan	0.0933%	0.0814%	0.0923%
40	Alaska	0.0516%	0.0611%	0.0681%
41	Arkansas	0.0774%	0.0628%	0.0640%
42	Nebraska	0.2147%	0.0432%	0.0625%
43	Louisiana	0.0405%	0.0520%	0.0556%
44	North Carolina	0.0149%	0.0262%	0.0376%
45	Missouri	0.0284%	0.0273%	0.0320%
46	Pennsylvania	0.0104%	0.0096%	0.0098%
47	West Virginia	0.0087%	0.0085%	0.0088%
48	Oklahoma	0.1636%	0.0109%	0.0050%
49	Kansas	0.0001%	0.0006%	0.0049%
50	Maine	0.0376%	0.0420%	0.0000%
51	Delaware	0.0000%	0.0000%	0.0000%

Regulators focus on individual utilities and here both the targets are being achieved. At the state level, more needs to be done to bring along the often smaller municipal and customer-owned utilities.

This analysis also tests some important hypotheses. Some are obvious. Spending more to become more energy efficient obviously means more savings. There are three additional findings. First, the data suggest a "learning by doing" aspect of energy efficiency, where utilities that spend more save disproportionately more. Second, states with decoupling and/or direct financial incentives save and spend more on energy efficiency and load management. Third, those utilities with regulated cost markups that are added to energy efficiency, are more likely to achieve Targets I and II.

The future is uncertain. Nevertheless, the current political and consumer support for energy efficiency is very high, perhaps an all-time high. The path ahead is clear. The need to spend money and to provide incentives is now rather obvious.

The nation is poised to get it right. What it will take is for regulators literally to "overrule objections." Thoughtful leadership, that expands energy efficiency, load management and renewable energy, needs to trump debate. Some cheese in the form of utility incentives needs to be added to regulatory mandates. There is much at stake.

BIBLIOGRAPHY

Books and Publications

Aldy, Joseph E. (2007). *Assessing the Costs of Regulatory Proposals for Reducing U.S. Greenhouse Gas Emissions.* Washington, D.C.: Resources for the Future.

Aliff, Gregory E., and Branko Terzic. (2007). The Greening of Utility Customers. *Public Utilities Fortnightly*, Vol. 145, Issue 9, pp. 32-33.

Bonbright, James C., Albert L. Danielsen, and David R. Kamerschen. (1988). *Principles of Public Utility Rates*. 2d Ed. Arlington, Va.: Public Utilities Reports, Inc.

Carson, Rachel. (1962). *Silent Spring*. New York: Houghton Mifflin.

Cicchetti, Charles J. (1972). *Alaskan Oil: Alternative Routes and Markets*. Washington, D.C.: Resources for the Future.

Cicchetti, Charles J. (1989, November 7). Prepared Statement Before the Committee on Energy and Natural Resources of the United States Senate Related to the Demand-Side Provisions of the Public Utility Regulatory Policies Act of 1978 (PURPA).

Cicchetti, Charles J. (2008). Duke's Fifth Fuel. *Public Utilities Fortnightly*, Vol. 146, Issue 1, pp. 54-62.

Cicchetti, Charles J., and William Hogan. (1989, June). "Incentive Regulation: Some Conceptual and Policy Thoughts." Energy and Environmental Policy Center, Harvard University, Discussion Paper E-89-09.

Cicchetti, Charles J., and Colin M. Long. (2006). A Brief History of Rate Base: Necessary Foundation or Regulatory Misfit? *Public Utilities Fortnightly*, Vol. 144, Issue 7, pp. 42-47.

Cicchetti, Charles J., Irwin M. Stelzer, and William Hogan. (1988, July 18). Comments Before the Federal Energy Regulatory Commission in Docket No. RM88-5-000, Comments of the Energy and Environmental Policy Center, RE: Regulations Governing Bidding Programs.

Green, Kenneth P., Steven F. Hayward, and Kevin A. Hassett. (2007, June). Climate Change: Caps vs. Taxes. American Enterprise Institute for Public Policy Research. *Environmental Policy Outlook*, No. 2.

Grueneich, Dian M., Commissioner. (2007). "California's Policy Framework to Advance Demand-Side Management." California Public Utilities Commission, International Energy Agency, Paris, France (DSM Workshop).

Hardin, Garrett. (1968). The Tragedy of the Commons. *Science*, Vol. 162.

Ishii, Ken. (2005, October 26). Sempra Energy Presentation.

Keough, Miles, Director, Grants and Research, NARUC. (2006, February 21). Presentation on "Capturing the Benefits of Energy Efficiency." Little Rock, Arkansas.

Komanoff, Charles, and Dan Rosenblum (2007, April 24). "Carbon Taxes First." PowerPoint presentation for the Carbon Tax Center.

Krasnow, Sam. (2007, October 2). "Efficiency Legislation in New England." Presentation at ACEEE's Fourth National Conference on "Energy Efficiency as a Resource," Berkeley, California.

Krutilla, John V., and Charles J. Cicchetti. (1972). Evaluating Benefits of Environmental Resources With Special Application to the Hells Canyon. *Natural Resources Journal*, Vol. 12, No. 1.

Kushler, Martin. (2007, March 21). "Addressing the Crucial Issue of Utility Disincentives Regarding Energy Efficiency: Basic Concepts and Current Status." Presentation to the American Council for an Energy-Efficient Economy (ACEEE) Symposium on Market Transformation, Washington, D.C.

Kushler, Martin, Dan York, and Patti Witte. (2006, October). "Aligning Utility Interests With Energy Efficiency Objectives: A Review of Recent Efforts at Decoupling and Performance Initiatives." American Council for an Energy-Efficient Economy.

Leiby, Paul N. (2007, February 28). "Estimating The Energy Security Benefits of Reduced U.S. Oil Imports." Prepared by the Oak Ridge National Laboratory for the U.S. Department of Energy.

Leiby, Paul N., Donald W. Jones, T. Randall Curlee, and Russell Lee. (1997, November). "Oil Imports: An Assessment of Benefits and Costs." Prepared by the Oak Ridge National Laboratory for the U.S. Department of Energy.

Lovins, Amory. (1985). Saving Gigabucks with Negawatts. *Public Utilities Fortnightly*, Vol. 115, Issue 6, p. 19.

McDermott, Karl. (2003, June 20). Testimony (Exhibit A-26) before the Michigan Public Service Commission in Case No. U-13808.

Meadows, Donella H., Dennis L. Meadows, Jorgen Randers, and William W. Behrens III. (1972). *The Limits to Growth*. New York: Universe Books.

Murrow, Derek. (2007, October 2). "Decoupling Activities in New England." Presentation by (Northeast Utilities) at ACEEE's Fourth National Conference on "Energy Efficiency as a Resource," Berkeley, California.

Rosenfeld, Arthur H., Commissioner, California Energy Commission. (2007, May 24). "California's Success in Energy Efficiency and Climate Change: Past and Future." Presentation at Electricite de France.

———. (2007, June 18). "Efficiency and Renewables in the Electricity Sector." Presentation at California Clean Energy Fund Forum.

Sedano, Richard, Rick Weston, and Gordon Dunn. (2006, January 27). Presentation at an Energy Efficiency Workshop sponsored by the Kansas Corporation Commission.

Shirley, Wayne, Director, The Regulatory Assistance Project. (2006, November 16). "Energy Efficiency & Utility Profits: Do Your Incentives Need Alignment?" Presentation at the Third Annual Southwest Energy Efficiency Workshop, Snowbird, Utah.

———. (2007, October 2). "Energy Efficiency Incentives in the Southwest." Presentation at ACEEE's Fourth National Conference on "Energy Efficiency as a Resource," Berkeley, California.

Shirley, Wayne, Jim Lazar, and Frederick Weston. (2008, June 30). "Revenue Decoupling Standards and Criteria: A Report to the Minnesota Public Utilities Commission." The Regulatory Assistance Project.

Strom, Sheldon. (2007, October 2). "Minnesota's Next Generation Energy Act of 2007." Presentation by The Center for Energy and Environment at ACEEE's Fourth National Conference on "Energy Efficiency as a Resource," Berkeley, California.

Stoft, S., J. Eto, and S. Kito. (1995, January). "DSM Shareholder Incentives: Current Designs and Economic Theory." Energy and Environment Division, Lawrence Berkeley Laboratory, University of California.

Weston, Rick. (2006, January 10). "Energy Efficiency Potential: It's Always More Than You Think." The Regulatory Assistance Project for the Natural Resources Council of Maine.

Agency Publications

American Gas Association. (2007, April). "Update on Revenue Decoupling Mechanisms."

EERE State Activities and Partnerships. (2007, July). "Maryland Removes Utility Disincentives for Energy Efficiency."

Energy Information Administration. (2007). "Energy Outlook for 2007." U.S. Department of Energy.

———. (2007, April). "Status of Electricity Restructuring by State." U.S. Department of Energy.

National Academy of Sciences. (2002). *Effectiveness and Impact of Corporate Average Fuel Economy (CAFE) Standards.* Prepared by the Committee on the Effectiveness and Impact of Corporate Average Fuel Economy (CAFE) Standards, National Research Council. Washington, D.C.: National Academy Press.

Natural Resources Defense Council. (1996, September). "Risky Business: Hidden Environmental Liabilities of Power Plant Ownership." *http://www.nrdc.org/air/energy/rbr/chap3.asp.*

Pacific Economics Group. (2007, August). "Summary of Revenue Decoupling Mechanisms."

U.S. Department of Energy and U.S. Environmental Protection Agency. (2005, July). "National Action Plan for Energy Efficiency."

———. (2006, July). "National Action Plan for Energy Efficiency."

U.S. Department of Energy, U.S. Environmental Protection Agency and ICF International. (2007, November). National Action Plan for Energy Efficiency. "Aligning Utility Incentives with Investment in Energy Efficiency: A Resource of the National Action Plan for Energy Efficiency."

U.S. Department of Transportation, NHTSA. (2006, March). "Final Regulatory Impact Analysis: Corporate Average Fuel Economy and CAFE Reform for MY 2008-2011 Light Trucks." Office of Regulatory Analysis and Evaluation, National Center for Statistics and Analysis.

U.S. Environmental Protection Agency. (2007, May 7). "Cap and Trade Programs: An Update." Presentation for the Environmental Markets Association.

Government Website Sources

Source for Energy Efficiency and Load Management: *http://www.eia.doe.gov/cneaf/electricity/page/eia861.html*

Source for US and CA Population Estimates 2000-2006: *http://www.census.gov/popest/states/NST-ann-est.html*

Source for US Population Estimates prior to 2000: *http://www.census.gov/popest/archives/1990s/popclockest.txt*

Source for CA Population Estimates 1990-2000: *http://www.census.gov/popest/archives/2000s/vintage_2001/CO-EST2001-12/CO-EST2001-12-06.html*

Various State Regulatory Proceedings and Legislation

Arizona: Arizona Public Service Co., Decision 57649, 129 PUR4th 181 (1991) and Decision No. 67744, 241 PUR4th 181 (2005).

California: San Diego Gas & Electric Co., Decision 93892, 7 CPUC2d 584 (1981) and Decision 85-03-025 (1985); Pacific Gas & Electric Co., Decision 93887, 7 CPUC2d 349 (1981); Decision 04-05-055 (2004); and Decision 07-03-044 (2007); Southern California Edison Co., Decision 82-12-055, 50 PUR4th 317 (1982); Decision 85-12-076, 19 CPUC2d 453 (1985); Decision 04-07-022 (2004); and Decision 06-05-016 (2006); San Diego Gas & Electric Co., Pacific Gas & Electric Co., and Southern California Edison Co., Interim Opinion on Phase 1 Issues: Shareholder Risk/Reward

Incentive Mechanism for Energy Efficiency Programs, Decision 07-09-043, R.06-04-010 (September 20, 2007); Testimony of the Division of Ratepayer Advocates of the California Public Utilities Commission on the Appropriate Shared Savings Rate for a Risk/Return Incentive Mechanism for Energy Efficiency, R.06-04-010 (May 3, 2007).

Colorado: House Bill 1037 (2007).

Connecticut: Connecticut Light & Power Co., Docket 97-10-23 (1998); Docket No 01-01-14 (2001); and Docket No. 07-06-21, 258 PUR4th 148 (2007).

Georgia: Response of Georgia Public Service Commission to The Regulatory Assistance Project's Electric Resource Long-Range Planning Survey. (Survey dated July 2003).

Hawaii: Hawaiian Electric Co., Docket No. 94-0206 (1996); Docket No. 00-0209 (2001); Docket No. 00-0169 (2003); and Docket No. 05-0069, Decision and Order No 23258, 256 PUR4th 1 (2007).

Idaho: Idaho Power Co., Case No. IPC-E-06-32 (2006) and Case No. IPC-E-04-15, 256 PUR4th 322 (2007).

Indiana: 170 Indiana Administrative Code §§4-8-6 and 4-8-7 allow for Commission approval for lost revenues and/or other incentives for DSM programs, but Commission has not yet approved any incentives.

Kentucky: Kentucky Revised Statute 278.190.

Maine: Central Maine Power Co., Docket 91-174 (1992) and Docket 90-085-A, 141 PUR4th 412 (1993).

Maryland: Delmarva Power & Light Co., Case No. 9093, Order No. 81518, 98 MD PSC 288 (2007); Potomac Electric Power Co., Case No. 9092, Order No. 81517, 98 MD PSC 228, 258 PUR4th 463 (2007).

Massachusetts: Eastern Edison Co., D.P.U. 94-4-CC (1994); Fitchburg Gas and Electric Light Co., D.T.E. 98-48/49 (1998); *See also* D.T.E. 98-100 (2000) (generic EE and DSM proceeding); Massachusetts Electric Co., D.T.E. 03-2 (2003); D.P.U. 07-50 (2008) (generic proceeding on demand resources).

Minnesota:	Otter Tail Power Co., Docket No. E-017/M-91-457 (1993).
Montana:	Montana Code Annotated 69-8.
Nevada:	Generic proceeding reviewing resource planning regulations, Docket No. 02-5030, 233 PUR4th 281 (2004).
New Hampshire:	Order establishing guidelines for post-competition energy efficiency programs, DR 96-150, Order No. 23,574, 206 PUR4th 169 (2000); Northern Utilities, DR 97-228, Order No. 22,846 (1998).
New York:	Consolidated Edison Co. of New York, Case 96-G-0548, Opinion No. 97-1 (1997); New York State Electric & Gas Corp., Cases 92-E-1084, 92-E-1085; Case 94-M-0349, Opinion No. 94-19, 155 PUR4th 337 (1994) and Opinion No. 95-17, 165 PUR4th 309 (1995); Niagara Mohawk Power Corp., Case 89-E-041, Opinion No. 89-29, 107 PUR4th 233 (1989); Cases 92-E-0108, 92-E-0109, Opinion No. 93-3, 140 PUR4th 481 (1993); Rochester Gas & Electric Corp., Opinion No. 93-19 (1993); Cases 95-E-0673 et al., Opinion No. 96-27 (1996).
Oregon:	Portland General Electric Co., Order No. 95-322, UE 88, 160 PUR4th 201 (1995).
Rhode Island:	Narragansett Electric Co., Docket No. 1939 (1989) and Docket No. 3240 (2002).
Vermont:	Green Mountain Power Corp., Docket No. 7175 (2006).
Washington:	Puget Sound Power & Light Co., Docket UE-921262, 147 PUR4th 80 (1993) and UE-950618, 163 PUR4th 604 (1995).
Wisconsin:	Wisconsin Electric Power Co., Docket No 6630-UR-100 (1991); Wisconsin Power & Light Co., Dockets 6680-UR-102 (1992) and 6680-UR-114, 242 PUR4th 193 (2005).

INDEX

A

accumulated depreciation, 201
Alabama, 73, 76
amortization, 171
annualized savings, 169, 171
avoided costs, 15, 20, 32, 120, 131-34, 192, 211, 221, 277-81

B

benchmarking, 99-107
bidding programs, 114, 119
bill stabilization adjustment (BSA), 231
biofuels (biomass), 20, 212
Bonbright, James C., vii, 199, 200
Brandeis, Justice Louis, 289
break-even model or value, 125-28, 132-34, 143, 145-49
BSA. *See* bill stabilization adjustment
build, own, and operate, 30, 191

C

California, 11, 16, 20, 25, 29, 32, 35, 43, 46, 66, 70, 73, 107, 138, 208, 209-16, 223, 225-27, 233, 250, 281, 286, 289
cap and trade, 22, 133, 134, 137, 138, 219
capacity savings, 39, 47, 50-54, 57-61, 73, 76, 79, 82, 87, 113, 183, 251, 255, 257
Capitol Community Citizens, 11
carbon dioxide (CO_2), 131, 138, 279
carbon sequestration, 121, 280, 283
carbon taxes, 22, 137
Center for Law and Social Policy, 8

Central Maine Power Co. (CMP), 230
CFLs. *See* compact fluorescent lamps
Cicchetti, Charles, 8, 9n4,5, 11, 114, 116n4, 118n6, 119, 132n2, 195, 219n1, 277n1
clean coal, 121
climate change, 17, 19, 21, 22, 25, 28, 121, 126, 131, 133, 138, 207, 209, 216, 263, 279, 281, 287
Club of Rome, 7
coal, 21, 121, 132, 135, 141, 280
COGS. *See* cost of goods sold
Colorado, 233
compact fluorescent lamps (CFLs), 99, 153-55, 167
Connecticut, 70, 73, 76, 233, 291
Consolidated Edison Co. of New York, 228
cost allocation, 66, 111, 117, 155, 192, 288
cost of goods sold (COGS), 30, 187-89, 194, 204
cost of service, 3, 5, 12, 16, 24-25, 30-31, 112, 116, 122, 191-92, 195, 197, 200, 202, 204, 210, 219, 221, 223, 279
costs
    avoided, 15, 20, 32, 120, 131-34, 192, 211, 221, 277-81
    direct, 14, 24, 90, 93, 156, 159, 165, 187, 213, 279n2
    external, 4, 8, 12, 22, 24, 27, 28, 32, 34, 116, 125, 136, 139, 207, 215, 218, 221, 222, 288
    fixed, 8, 20, 24, 25, 32, 112, 117, 125, 167, 171, 191, 192, 194, 195, 228, 229, 231, 232

costs *(continued)*
　incremental, 169, 170, 203, 259
　indirect, 24, 31, 90, 93, 154-55, 159, 167, 175, 185, 187-89, 220, 222
　\marginal, vii, 11-16, 20, 21-22, 27, 32, 109, 114, 115, 117-20, 121-29, 134, 136, 142-43, 145, 148-49, 192-94, 207, 217-22, 279n2, 287-88
　opportunity, 151, 167-68, 171
　original, 171, 174, 195, 197-201, 203
　social, 8, 13, 23, 34, 116, 123, 124, 138, 217
　societal, 131, 134
cumulative savings, 39, 85, 100, 107, 169-70, 174, 250, 271-74, 289

D

declining block rates, 8, 11, 13, 27
decoupling, viii, 5, 16, 25, 29, 113, 116, 155, 165, 192, 207-08, 220, 222-23, 225-33, 235, 237, 250-63, 277, 279, 281, 295
Delaware, 233
Delmarva Power & Light Co., 231
demand response programs, 19, 25, 27, 104, 114n2, 151, 156, 179, 193, 212, 215, 235-39, 259, 273, 288
demand-side management (DSM), 3, 15, 27, 31, 39, 46-47, 66, 85, 90, 96, 99, 107, 111-12, 114-15, 121, 134, 151, 154, 159, 167-68, 170-71, 175, 181, 187-90, 191-94, 195, 196, 212-15, 220, 228, 252, 256, 271, 273, 277-78
direct costs, 14, 24, 90, 93, 156, 159, 165, 187, 213, 279n2
direct financial incentives, 29, 30, 32, 33, 35, 113, 194, 208, 225, 235-39, 252, 255, 257-59, 263, 277, 285, 295
distributed energy, 20, 283
District of Columbia, 70, 73, 96, 233

DSM. *See* demand-side management
Duke Energy, 275, 277

E

earnings margin, 116, 159
economic efficiency, 11-13, 22, 32, 114, 115, 116-18, 121, 215, 217, 222
economic security, 27, 121, 131, 133, 139, 194, 207, 263, 275, 287
EDF. *See* Environmental Defense Fund
EES. *See* energy efficiency savings
EIA. *See* Energy Information Administration
EIS. *See* Environmental Impact Statement
elasticity, 13, 226
electric revenue adjustment mechanism (ERAM), 223, 226-30
emission reductions, 135, 137, 138, 154, 214, 280
energy efficiency savings (EES), 33, 50, 99, 172, 174, 249, 251-52, 259
Energy Information Administration (EIA), 39, 60, 67, 151, 154, 156, 159, 167, 170, 181, 187-88, 208, 249, 273
energy intensity, 141, 283
energy not used (ENU), 21
Environmental Defense Fund (EDF), vii, 8, 10, 11, 15
Environmental Impact Statement (EIS), 9, 10
ERAM. *See* electric revenue adjustment mechanism
external benefits, 5, 116, 120, 121, 123-27, 129, 131-34, 135-43, 145, 148, 149, 193, 194, 219, 221
external costs, 4, 8, 12, 22, 24, 27, 28, 32, 34, 116, 125, 136, 139, 207, 215, 218, 221, 222, 288
externalities, 8, 11, 31, 47, 109, 120, 121-30, 131-37, 142, 143, 194, 217, 220, 221, 279, 281, 287, 288

F

fair return, 196-97, 204
fair value, 196-97, 199, 203
FCA. *See* fixed cost adjustment
Federal Energy Regulatory Commission (FERC), 8, 25, 114, 115, 197, 286, 289
Federal Power Commission (FPC), 8-9, 132, 197
Federal Power Commission v. Hope Natural Gas Co., 197, 198, 200, 203
FERC. *See* Federal Energy Regulatory Commission
fixed cost adjustment (FCA), 229, 231
FPC. *See* Federal Power Commission
Freeman, David, 11

G

geothermal power, 16, 121, 211, 212, 283
GHG. *See* greenhouse gas
Green Mountain Power Corp., 232
greenhouse gas (GHG), 21, 22, 133, 137, 154, 214
Grueneich, Dian, 29n2, 212, 213n2, 214n3

H

halo effects, 167, 286
Hawaii, 73, 233
Hell's Canyon, 8, 9, 11, 131n1, 132
Hogan, William, 114, 116n4, 118n6, 119
hydroelectric power, 8, 9, 20, 132, 212, 283

I

Idaho, 231-32, 233
Idaho Power Co. (IPC), 231

incentives
   financial, 3, 16, 29, 30, 32, 33-35, 112-14, 154-56, 191, 192, 194, 208, 225, 235-39, 252, 255, 257-59, 261, 263, 271-74, 277, 285, 286, 289, 295
   performance, 237
incremental costs, 169, 170, 203, 259
incremental savings, 43, 54, 67, 76, 107, 170, 174, 252, 259, 271-72, 289
independent power producers (IPPs), 202
indirect costs, 24, 31, 90, 93, 154-55, 159, 167, 175, 185, 187-89, 220, 222
integrated resource planning (IRP), 11, 14, 31, 129, 237
interruptible service, 15, 47, 76, 181, 185, 187, 188
Iowa, 73
IPPs. *See* independent power producers
IRP. *See* integrated resource planning

J

Jackson, Justice Robert H., 197-98, 200, 201, 203, 204
"just and reasonable" principle, 8, 32, 33, 35, 112, 129, 151, 198, 219, 222, 289

K

Kansas, 233

L

least-cost planning, 15, 111, 114, 115, 117, 119, 120, 121, 215, 218, 221, 222, 277
levelized approach, 171-75, 181
liquefied natural gas (LNG), 21, 139
LM. *See* load management
LNG. *See* liquefied natural gas
load control, 213
load curve, 135

load management (LM), 1, 11, 15, 16, 19, 39, 47, 60, 66, 76, 85, 90, 96, 99, 101, 112, 134, 154, 156, 159, 168, 175, 181, 185, 187, 188, 189, 190, 192, 194, 214, 250, 251, 252, 254, 255, 277, 280, 288, 295
Long, Colin, 195, 225n1, 238n, 240n3
lost margin, 24, 29-31, 112, 113, 123, 155, 191, 192, 225, 228, 237, 285
lost revenues, 19, 29, 112, 192, 211, 228, 255, 279, 285
lost sales, 29, 30, 168, 225
Lovins, Amory, 117, 120, 277

M

Madison Gas & Electric Co. (MG&E), 11, 14
Maine, 230
Mandatory Oil Import Quota program, 9, 10
marginal costs, vii, 11-16, 20, 21-22, 27, 32, 109, 114, 115, 117-20, 121-29, 134, 136, 142-43, 145, 148-49, 192-94, 207, 217-22, 279n2, 287-88
margins on cost of goods sold, 204
margins on sales, 204
markup, 29, 30, 33, 34, 155, 156, 159, 165, 187, 188, 189, 194, 271, 273, 274, 295
Maryland, 30, 96, 231, 233
Massachusetts, 70, 73, 232, 233
Michigan, 11
Minnesota, 66, 70, 73, 79, 233
Moskovitz, David, 16

N

NARUC. *See* National Association of Regulatory Utility Commissioners
National Association of Regulatory Utility Commissioners (NARUC), 15
National Environmental Policy Act, 9

national security, 3, 10, 17, 19, 21, 22, 25, 28, 126, 133, 139, 140, 141, 217, 219, 279, 281, 283, 284, 287
natural gas, 4, 10, 21, 115, 122, 123, 139, 141, 197, 198, 201, 202, 203, 213, 223, 280, 283, 288, 289
natural gas vehicles, 21, 283
natural monopoly, 23, 281
Natural Resources Defense Council (NRDC), 8, 10, 138n3
Nebraska, 73, 76, 79, 196
negative externalities, 8, 31, 127-29, 133, 142, 143, 194, 221, 279, 281
negawatts, 277, 281
New Hampshire, 73, 233
New Mexico, 233
New York, 11, 227-28, 233
New York State Electric & Gas Corp. (NYSE&G), 228
Niagara Mohawk Power Corp., 228
nitrogen oxide (NO_x), 131, 138, 279
No Losers Test, 117
Northern States Power Co., 43
NRDC. *See* Natural Resources Defense Council

O

oil dependence, 140, 141, 280
opportunity costs, 151, 167-68, 171
Oregon, 76
original cost, 171, 174, 195, 197-201, 203

P

Pacific Gas & Electric Co. (PG&E), 43, 225, 226
PBR. *See* performance-based regulation
performance-based regulation (PBR), 33, 112, 196, 198, 199, 227, 232
performance incentives, 237

periodic rate adjustment mechanism (PRAM), 229-30
Pickens, T. Boone, 21
Potomac Electric Power Co. (PEPCO), 231, 233n31
PRAM. *See* periodic rate adjustment mechanism
price elasticity, 12, 13
prudent investment, 197-98, 200, 203
public goods (or benefits) charge, 16, 19, 20, 25, 43, 113, 211, 215, 216, 241, 244, 250, 259
Public Utility Regulatory Policies Act (PURPA), 11, 15, 16, 20, 21, 27, 32
Puget Sound Energy, 43, 229
pure market approach, 111
PURPA. *See* Public Utility Regulatory Policies Act

Q

qualifying facilities (QFs), 20, 21, 195, 204, 212, 277, 279, 281

R

rate base, 20, 24, 27, 30, 31, 32, 112, 115, 192, 193, 195-204, 223
real-time pricing, 15, 16, 27, 32, 111, 212, 213, 286
regulatory lag, 168
renewable energy, 5, 15, 16, 19, 20, 25, 66, 193, 213, 215, 223, 281, 283, 284, 285, 286
Renewable Portfolio Standards (RPS), 20, 46, 104, 286
Resources for the Future, 8
restructuring, viii, 15, 16, 19, 20, 30, 33, 43, 66, 70, 85, 122, 196, 200, 202, 203, 204, 210, 211, 215, 216, 225, 226, 227, 237, 250, 252, 254, 256, 259, 261, 262, 286

retail choice, 23, 25, 27, 115, 123, 216, 250, 259
retail margin, 30, 35, 188, 189, 204
revenue requirements, 14, 21, 22, 24, 112, 120, 151, 159, 171, 192, 193, 199, 219, 221, 222, 278, 280
Rhode Island, 70, 73, 291
Rochester Gas & Electric Corp., 229
Rogers, Jim, 277
Rowe, John, 28
RPS. *See* Renewable Portfolio Standards

S

San Diego Gas & Electric Co. (SDG&E), 227
save-a-watt program, 277-81
savings
    cumulative, 39, 85, 100, 107, 169-70, 174, 250, 271-74, 289
    incremental, 43, 54, 67, 76, 107, 170, 174, 252, 259, 271-72, 289
    shared, 29-30, 32, 33, 195, 208, 225
sensitivity analysis, 145, 149, 170, 185
shadow price, 132
shared savings, 29-30, 32, 33, 195, 208, 225
Sierra Club, 8, 132
sinking fund, 171, 173, 181
smart metering, 181, 213, 286
Smyth v. Ames, 196, 197
social costs, 8, 13, 23, 34, 116, 123, 124, 138, 217
societal approach, 111-12, 116-17
societal costs, 131, 134
societal test, 126
solar power, 16, 20, 121, 202, 213, 283
Southern California Edison Co. (SCE), 43, 226-27
Stelzer, Irwin, 114, 116n4
subsidies, 23, 24, 25, 27, 34, 116, 117, 118, 119, 130, 194, 278

sulfur dioxide (SO$_2$), 131, 138, 279
supply-side programs (or investments), 19, 20, 21, 31, 32, 47, 112, 114, 115, 116, 119, 122, 131, 133, 143, 170, 171, 191, 192, 193, 195, 202, 203, 204, 221, 222, 229, 281, 285
system benefits charge, 241, 242, 244, 245, 246, 247

T

TAPS. *See* Trans-Alaskan Pipeline System
tax credits, 286
TCS. *See* total capacity savings
TEES. *See* total energy efficiency savings
tidal power, 121
time-of-use (TOU) pricing, vii, 11, 13, 14, 15, 16, 27, 212, 228
total capacity savings (TCS), 251, 253
total energy efficiency savings (TEES), 251, 252, 261
TOU. *See* time-of-use pricing
Trans-Alaskan Pipeline System (TAPS), 9-10
Trans-Canadian Pipeline (TCP), 9-10
true-up, 29, 31, 223, 225, 227, 228
t-statistic (*t*-test), 252

U

unbundling, 34, 111, 115, 118, 119, 122, 202
"used and useful" principle, 199, 202
Utah, 73, 76

V

value in exchange, 137
value in use, 137, 138
value of service, 191, 194, 195-96, 198, 202, 203, 218
Vermont, 232-33
virtual bid, 119, 120
volume discounts, 7, 11, 15, 27

W

Washington, 73, 229-30
wind power, 16, 20, 21, 116, 121, 211, 212, 283
wires and pipes, 25, 289
wires charges, 16, 25, 33, 66, 112, 211, 215, 216, 281
Wisconsin, 11, 12-14, 66, 233
Wisconsin Environmental Decade, 11
Wyoming, 76

periodic rate adjustment mechanism (PRAM), 229-30
Pickens, T. Boone, 21
Potomac Electric Power Co. (PEPCO), 231, 233n31
PRAM. *See* periodic rate adjustment mechanism
price elasticity, 12, 13
prudent investment, 197-98, 200, 203
public goods (or benefits) charge, 16, 19, 20, 25, 43, 113, 211, 215, 216, 241, 244, 250, 259
Public Utility Regulatory Policies Act (PURPA), 11, 15, 16, 20, 21, 27, 32
Puget Sound Energy, 43, 229
pure market approach, 111
PURPA. *See* Public Utility Regulatory Policies Act

Q

qualifying facilities (QFs), 20, 21, 195, 204, 212, 277, 279, 281

R

rate base, 20, 24, 27, 30, 31, 32, 112, 115, 192, 193, 195-204, 223
real-time pricing, 15, 16, 27, 32, 111, 212, 213, 286
regulatory lag, 168
renewable energy, 5, 15, 16, 19, 20, 25, 66, 193, 213, 215, 223, 281, 283, 284, 285, 286
Renewable Portfolio Standards (RPS), 20, 46, 104, 286
Resources for the Future, 8
restructuring, viii, 15, 16, 19, 20, 30, 33, 43, 66, 70, 85, 122, 196, 200, 202, 203, 204, 210, 211, 215, 216, 225, 226, 227, 237, 250, 252, 254, 256, 259, 261, 262, 286

retail choice, 23, 25, 27, 115, 123, 216, 250, 259
retail margin, 30, 35, 188, 189, 204
revenue requirements, 14, 21, 22, 24, 112, 120, 151, 159, 171, 192, 193, 199, 219, 221, 222, 278, 280
Rhode Island, 70, 73, 291
Rochester Gas & Electric Corp., 229
Rogers, Jim, 277
Rowe, John, 28
RPS. *See* Renewable Portfolio Standards

S

San Diego Gas & Electric Co. (SDG&E), 227
save-a-watt program, 277-81
savings
    cumulative, 39, 85, 100, 107, 169-70, 174, 250, 271-74, 289
    incremental, 43, 54, 67, 76, 107, 170, 174, 252, 259, 271-72, 289
    shared, 29-30, 32, 33, 195, 208, 225
sensitivity analysis, 145, 149, 170, 185
shadow price, 132
shared savings, 29-30, 32, 33, 195, 208, 225
Sierra Club, 8, 132
sinking fund, 171, 173, 181
smart metering, 181, 213, 286
Smyth v. Ames, 196, 197
social costs, 8, 13, 23, 34, 116, 123, 124, 138, 217
societal approach, 111-12, 116-17
societal costs, 131, 134
societal test, 126
solar power, 16, 20, 121, 202, 213, 283
Southern California Edison Co. (SCE), 43, 226-27
Stelzer, Irwin, 114, 116n4
subsidies, 23, 24, 25, 27, 34, 116, 117, 118, 119, 130, 194, 278

sulfur dioxide (SO$_2$), 131, 138, 279
supply-side programs (or investments), 19, 20, 21, 31, 32, 47, 112, 114, 115, 116, 119, 122, 131, 133, 143, 170, 171, 191, 192, 193, 195, 202, 203, 204, 221, 222, 229, 281, 285
system benefits charge, 241, 242, 244, 245, 246, 247

T

TAPS. *See* Trans-Alaskan Pipeline System
tax credits, 286
TCS. *See* total capacity savings
TEES. *See* total energy efficiency savings
tidal power, 121
time-of-use (TOU) pricing, vii, 11, 13, 14, 15, 16, 27, 212, 228
total capacity savings (TCS), 251, 253
total energy efficiency savings (TEES), 251, 252, 261
TOU. *See* time-of-use pricing
Trans-Alaskan Pipeline System (TAPS), 9-10
Trans-Canadian Pipeline (TCP), 9-10
true-up, 29, 31, 223, 225, 227, 228
t-statistic (t-test), 252

U

unbundling, 34, 111, 115, 118, 119, 122, 202
"used and useful" principle, 199, 202
Utah, 73, 76

V

value in exchange, 137
value in use, 137, 138
value of service, 191, 194, 195-96, 198, 202, 203, 218
Vermont, 232-33
virtual bid, 119, 120
volume discounts, 7, 11, 15, 27

W

Washington, 73, 229-30
wind power, 16, 20, 21, 116, 121, 211, 212, 283
wires and pipes, 25, 289
wires charges, 16, 25, 33, 66, 112, 211, 215, 216, 281
Wisconsin, 11, 12-14, 66, 233
Wisconsin Environmental Decade, 11
Wyoming, 76